FOURIER TRANSFORM RAMAN SPECTROSCOPY
Instrumentation and Chemical Applications

ELLIS HORWOOD SERIES IN ANALYTICAL CHEMISTRY

Series Editors: Dr MARY MASSON, University of Aberdeen,
and Dr JULIAN F. TYSON, Amherst, USA
Consultant Editors: Prof. J. N. MILLER, Loughborough University of Technology, and
Dr R. A. CHALMERS, University of Aberdeen

S. Alegret	Developments in Solvent Extraction
S. Allenmark	Chromatographic Enantioseparation: Methods and Applications, Second Edition
G.E. Baiulescu, P. Dumitrescu & P.Gh. Zugrăvescu	Sampling
H. Barańska, A. Łabudzińska & J. Terpiński	Laser Raman Spectrometry
G.I. Bekov & V.S. Letokhov	Laser Resonant Photoionization Spectroscopy for Trace Analysis
K. Beyermann	Organic Trace Analysis
O. Budevsky	Foundations of Chemical Analysis
J. Buffle	Complexation Reactions in Aquatic Systems: An Analytical Approach
D.T. Burns, A. Townshend & A.G. Catchpole	Inorganic Reaction Chemistry: Vol. 1: Systematic Chemical Separation
D.T. Burns, A. Townshend & A.H. Carter	Inorganic Reaction Chemistry: Vol. 2: Reactions of the Elements and their Compounds: Part A: Alkali Metals to Nitrogen, Part B: Osmium to Zirconium
J. Churáček	New Trends in the Theory & Instrumentation of Selected Analytical Methods
E. Constantin & A. Schnell	Mass Spectrometry
R. Czoch & A. Francik	Instrumental Effects in Homodyne Electron Paramagnetic Resonance Spectrometers
T.E. Edmonds	Interfacing Analytical Instrumentation with Microcomputers
Fadini, A. & Schnepel, F.-M.	Vibrational Spectroscopy
Z. Galus	Fundamentals of Electrochemical Analysis, 2nd Ed.
S. Görög	Steroid Analysis in the Pharmaceutical Industry
T. S. Harrison	Handbook of Analytical Control of Iron and Steel Production
J.P. Hart	Electroanalysis of Biologically Important Compounds
T.F. Hartley	Computerized Quality Control: Programs for the Analytical Laboratory, 2nd Ed.
Saad S.M. Hassan	Organic Analysis using Atomic Absorption Spectrometry
P. Hendra, C. Jones & G. Warnes	Fourier Transform Raman Spectroscopy: Instrumental and Chemical Applications
M.H. Ho	Analytical Methods in Forensic Chemistry
Z. Holzbecher, L. Diviš, M. Král, Š. Šůcha & F. Vláčil	Handbook of Organic Reagents in Inorganic Chemistry
A. Hulanicki	Reactions of Acids and Bases in Analytical Chemistry
David Huskins	Electrical and Magnetic Methods in On-line Process Analysis
David Huskins	Optical Methods in On-line Process Analysis
David Huskins	Quality Measuring Instruments in On-line Process Analysis
J. Inczédy	Analytical Applications of Complex Equilibria
S. Johnston	Fourier Transform Infrared: A Constantly Evolving Technology
M. Kaljurand & E. Küllik	Computerized Multiple Input Chromatography
S. Kotrlý & L. Šůcha	Handbook of Chemical Equilibria in Analytical Chemistry
J. Kragten	Atlas of Metal–Ligand Equilibria in Aqueous Solution
A.M. Krstulović	Quantitative Analysis of Catecholamines and Related Compounds
F.J. Krug & E.A.G. Zagatto	Flow-Injection Analysis in Agriculture and Environmental Science
V. Linek, V. Vacek, J. Sinkule & P. Beneš	Measurement of Oxygen by Membrane-Covered Probes
C. Liteanu, E. Hopîrtean & R. A. Chalmers	Titrimetric Analytical Chemistry
C. Liteanu & I. Rîcă	Statistical Theory and Methodology of Trace Analysis
Z. Marczenko	Separation and Spectrophotometric Determination of Elements
M. Meloun, J. Havel & E. Högfeldt	Computation of Solution Equilibria
M. Meloun, J. Militky & M. Forina	Chemometrics in Instrumental Analysis: Solved Problems for the IBM PC
O. Mikeš	Laboratory Handbook of Chromatographic and Allied Methods
J.C. Miller & J.N. Miller	Statistics for Analytical Chemistry, 2nd Ed.
J.N. Miller	Fluorescence Spectroscopy
J.N. Miller	Modern Analytical Chemistry
J. Minczewski, J. Chwastowska & R. Dybczyński	Separation and Preconcentration Methods in Inorganic Trace Analysis
D. Pérez-Bendito & M. Silva	Kinetic Methods in Analytical Chemistry
B. Ravindranath	Principles and Practice of Chromatography
V. Sediveč & J. Flek	Handbook of Analysis of Organic Solvents
O. Shpigun & Yu. A. Zolotov	Ion Chromatography in Water Analysis
R.M. Smith	Derivatization for High Pressure Liquid Chromatography
R.V. Smith	Handbook of Biopharmaceutic Analysis
K.R. Spurny	Physical and Chemical Characterization of Individual Airborne Particles
K. Štulík & V. Pacáková	Electroanalytical Measurements in Flowing Liquids
J. Tölgyessy & E.H. Klehr	Nuclear Environmental Chemical Analysis
J. Tölgyessy & M. Kyrš	Radioanalytical Chemistry, Volumes I & II
J. Urbanski, et al.	Handbook of Analysis of Synthetic Polymers and Plastics
M. Valcárcel & M.D. Luque de Castro	Flow-Injection Analysis: Principles and Applications
C. Vandecasteele	Activation Analysis with Charged Particles
F. Vydra, K. Štulík & E. Juláková	Electrochemical Stripping Analysis
N. G. West	Practical Environmental Analysis using X-ray Fluorescence Spectrometry
J. Zupan	Computer-supported Spectroscopic Databases
J. Zýka	Instrumentation in Analytical Chemistry: Volume 1

FOURIER TRANSFORM RAMAN SPECTROSCOPY
Instrumentation and Chemical Applications

PATRICK HENDRA B.Sc., Ph.D., D.Sc.
CATHERINE JONES B.Sc., Ph.D.
GAVIN WARNES B.Sc., Ph.D.
all of the Department of Chemistry, University of Southampton

with an appendix on the Fourier transformation forces by Dr Andrew Crookell late of the Department of Chemistry at Southampton and now with Perkin Elmer Ltd, Beaconsfield, UK

ELLIS HORWOOD
NEW YORK   LONDON   TORONTO   SYDNEY   TOKYO   SINGAPORE

This book is dedicated to the authors' bank managers!

First published in 1991 by
ELLIS HORWOOD LIMITED
Market Cross House, Cooper Street,
Chichester, West Sussex, PO19 1EB, England

A division of
Simon & Schuster International Group
A Paramount Communications Company

© Ellis Horwood Limited, 1991

All rights reserved. No part of this publication may be reproduced, stored in a retrieval system, or transmitted, in any form, or by any means, electronic, mechanical, photocopying, recording or otherwise, without the prior permission, in writing, of the publisher

Every effort has been made to trace all copyright holders, but if any have been inadvertently overlooked, the publishers will be pleased to make the necessary arrangements at the earliest opportunity.

Typeset in Times by Ellis Horwood Limited
Printed and bound in Great Britain
by Hartnolls, Bodmin, Cornwall

British Library Cataloguing-in-Publication Data

Hendra, Patrick
Fourier Transform Raman spectroscopy.
1. Raman spectroscopy
I. Title. II. Jones, Catherine. III. Warnes, Gavin
535.846
ISBN 0-13-327032-7

Library of Congress Cataloging-in-Publication Data

Hendra, Patrick, 1936–
Fourier transform Raman spectroscopy: instrumental and chemical applications /
Patrick Hendra, Catherine Jones, Gavin Warnes.
p. cm. — (Ellis Horwood series in analytical chemistry)
Includes bibliographical references and index.
ISBN 0-13-327032-7
1. Raman spectroscopy. 2. Fourier transform spectroscopy.
I. Jones, Catherine, 1966– . II. Warnes, Gavin, 1967– . III. Title. IV. Series.
QD96.R34H46   1990
543'.08584–dc20   90–22502
                                                          CIP

Table of contents

Preface 9

Acknowledgements 10

1. **Introduction: Historical developments** 11

2. **The vibrational behaviour of molecules** 17
   2.1   Molecular rotations and vibrations 17
           2.1.1   Spectroscopic units 21
   2.2   Infrared spectroscopy 21
   2.3   The Raman effect 24
   2.4   Infrared and Raman activity 29
   2.5   The effect of polarization and sample orientation on infrared spectra .42
   2.6   The effect of polarization and sample orientation on Raman spectra .43
   References 47
   Bibliography 48

3. **Conventional laser-Raman spectroscopy** 49
   3.1   Lasers 49
   3.2   Collection optics 51
   3.3   Scanning monochromators and spectrographs 55
   3.4   Spectral bandpass, resolution and frequency precision 60
   3.5   Detectors and detector electronics 63
   3.6   Sources of noise in spectra 66
   3.7   Sampling 72
   References 79

4. **Fourier transform vibrational spectroscopy** .80
   4.1   Introduction 80
   4.2   Principles of interferometry 80
   4.3   Data processing in FT interferometers 83
   4.4   The advantages of the interferometer in Raman spectrometry 86

Table of contents

        4.4.1   The Jacquinot advantage 87
        4.4.2   The Felgett advantage 88
        4.4.3   The Connes advantage 89
  4.5   Conclusion 89
References 89

5. The development of FT Raman spectrometry 90
  5.1   Pioneering experiments 90
  5.2   The development of analytical FT Raman spectrometers 93
  5.3   The essential elements of an FT Raman spectrometer 95
  5.4   Sampling in FT Raman spectroscopy 105
        5.4.1   Liquids 106
        5.4.2   Solids 107
        5.4.3   Gases 107
        5.4.4   Low temperatures 108
        5.4.5   High temperatures 108
        5.4.6   Electrochemical measurements 108
        5.4.7   Adsorbed species 108
        5.4.8   Microscopy 108
        5.4.9   Fibre-optic coupling 109
        5.4.10 Kinetic experiments 109
  5.5   Laser safety 110
  5.6   Purchasing an FT Raman system 112
  5.7   Survey of currently available commercial instruments 115
References 125

6. FT Raman intensities 127
  6.1   Introduction 127
  6.2   Raman intensity formulae 127
  6.3   Intensity measurements with a conventional Raman spectrometer .. 129
  6.4   Methods of quantitative experiments 131
  6.5   Measuring intensities with an FT Raman spectrometer 132
  6.6   Calibration of FT Raman spectra for instrument response 134
  6.7   Standardization of FT Raman spectra 143
  6.8   Examples of the use of FT Raman in quantitative analysis 147
  6.9   Quantitative studies on solids 152
References 154

7. The application of FT Raman spectroscopy in organic chemistry 156
  7.1   Group frequencies 156
        7.1.1   The X–H stretching region 156
        7.1.2   Saturated rings 157
        7.1.3   Unsaturated species ($\Delta v=2300-1600$ cm^{-1}) 158
        7.1.4   Aromatics 159
        7.1.5   Oxygen containing species 159
        7.1.6   Sulphur, phosphorous, silicon and halogen containing groups 160

	7.2	Specific applications	162
	7.2.1	Alkaloids	162
	7.2.2	Explosives	163
	7.2.3	Dyestuffs	168
	7.2.4	Pharmaceutical materials	169
	7.2.5	Structural studies	172
	7.2.6	Polymorphism	172
	7.3	Conclusion	173
	References		174
8.	**The analysis of polymeric materials using FT Raman spectroscopy**		**175**
	8.1	Introduction	175
	8.2	General analysis	176
	8.3	Crystallinity and tacticity studies	181
	8.4	Elastomers	187
	8.5	Paints	190
	8.6	Thermoplastics and composites	193
	8.7	Conjugated and electrically conducting polymers	197
	8.8	Miscellaneous studies	199
	References		202
9.	**Biological applications**		**204**
	9.1	Introduction	204
	9.2	Membranes	205
	9.3	Animal tissue	208
	9.4	Polypeptides	209
	9.5	Light-sensitive systems	213
	9.6	Wood and plant tissue	215
	9.7	Pharmaceuticals	217
	9.8	Foodstuffs	221
	9.9	Conclusion	225
	References		227
10.	**The application of Fourier transform Raman spectroscopy to the study of surfaces**		**228**
	10.1	Raman spectroscopy of surfaces	228
	10.2	Surface-enhanced Raman spectroscopy	228
	10.3	Near infrared FT Raman SERS spectra of electrochemical systems	230
	10.4	Near infrared FT Raman SERS spectra from metal colloids	240
	10.5	Infrared spectroscopy of catalyst systems	245
	10.6	Raman spectroscopy of catalyst surfaces	246
	References		258
11.	**The analysis and characterization of inorganic compounds**		**260**
	11.1	Introduction	260
	11.2	Small compounds	261

	11.3 Co-ordination compounds	265
	11.4 Carbonyl and related complexes	267
	References	272

12. Future developments ... 273
    12.1 Imminent developments ... 273
    12.2 Long-term developments ... 276
        12.2.1 New lasers and filters ... 276
        12.2.2 Specialist instruments ... 278
    12.3 Alternatives to FT methods for near infrared Raman spectroscopy 280
    12.4 Sources of information ... 280
    12.5 Data collections ... 281
    References ... 282

Appendix: The processing history of a spectrum — from ADC to VDU ... 283
    A.1 Introduction ... 283
    A.2 The collection of data points ... 283
    A.3 Digitization ... 285
    A.4 Signal averaging ... 286
    A.5 Apodization ... 290
    A.6 Digital filtering ... 292
    A.7 Fourier transformation ... 294
    A.8 Zero filling ... 297
    A.9 Phase correction ... 298
    A.10 Interpolation ... 298
    A.11 Ratioing ... 299
    A.12 Additional processing ... 300
        A.12.1 Non-linearity correction ... 300
        A.12.2 Fourier self-deconvolution ... 301
    A.13 Data processing for FT Raman spectra ... 301
    A.14 Glossary of terms ... 303
    References ... 305

Index ... 307

Preface

The first paper published on FT Raman spectroscopy appeared in 1964, fell on deaf ears and vanished without a trace. Nothing was written on the subject until 1986 when Dr D. B. Chase published the results of his preliminary experiments. Since then there has been a flurry of activity as conventional Raman spectroscopists have taken interferometers to bits, and instrument manufacturers have learnt what Raman spectroscopy is. At the time of writing in 1990 most multinational instrument manufacturers have announced new FT Raman spectrometers, and FT Raman spectroscopy is 'going public'. As customers flock to buy these new instruments it seems likely that a 'renaissance' in analytical Raman spectroscopy will occur.

This book has been written to inform analysts of these advances in Raman spectroscopy and enable them to capitalize on the new technology. It is not intended to be a comprehensive theoretical text, but rather a more practical guide to FT Raman spectroscopy. We provide a sufficient level of theory covering molecular vibrations and instrumentation to satisfy the requirements of the analyst. References to original works are given throughout the text.

Conventional Raman methods are described to enable those unfamiliar with the Raman technique to appreciate the advantages of the FT approach. Interferometer design and operation are explained, and the details of how the computer calculates the Fourier transform is given in an appendix (courtesy of Dr A. Crookell). Having discussed the fundamentals of FT Raman spectroscopy, we devote the remainder of the book to the diverse applications of the technique. Examples are given in the fields of polymer science, surface chemistry, biological systems, organic and inorganic analysis and quantitative methods. Readers are reminded that FT Raman is very new, and thus the applications cited here are only a taste of things to come.

Pat Hendra
Catherine Jones
Gavin Warnes
Southampton, April 1991

Acknowledgments

All members of the Applied Spectroscopy group at Southampton University contributed to the preparation of this book. Many provided assistance with the text, others by providing diagrams and drawings and some by assisting with typing and other secretarial services. Particular mention should be made of the following postgraduates, Tarik Jawhari, Colin Hodges, Pascale Le Barazer, Christopher Petty, Ganesh Sockalingum, Jonathan Agbenyega, Anne Rowlands, Christopher Dyer, Paul Evans, Peter Wallen, Ian Wesley and Elsa Corbett. More senior colleagues whose assistance is gratefully acknowledged include Dr Catherine Passingham, Dr Angeles Sanchez Blasquez, Mr Gary Ellis, Dr Bill Maddams, Dr Veronica Zichy, Mr Michael Cudby and the late Dr Harry Willis. Many secretaries have helped, of whom Mrs Jill Allen deserves special thanks. The Southampton FT Raman programme would never have 'taken off' without the assistance of Dr Ken Williams of BP, Dr Andrew Turner, Dr Michael Ford, Dr Henry Mould and Dr David Culter of Perkin Elmer Ltd, UK. It is a great pleasure to thank the US Navy, Office of Naval Research (ONR) and especially Drs Kenneth Wynn and Robert Novac for their support over many years. Their help at just the right moment made the work described in this book possible.

Catherine Jones

1

Introduction: Historical developments

Raman spectroscopy developed very rapidly after its prediction [1,2] and demonstration [3,4] in the mid 1920s and became the principal method of non-destructive analysis before the Second World War.

Very early research between 1880 and 1910 indicated that molecules possess a both complex and highly specific vibrational fingerprint [5]. Primitive infrared absorption spectra were recorded and the frequencies of several vibrations were measured. It was observed that particular chemical fragments within any molecule vibrated in specific, relatively narrow frequency ranges. These became known as 'group frequencies' (see Section 2.4). These vibrations lie between 10^{14} and 10^{12} Hz or, in spectroscopic frequency units, 4000 to 50 cm^{-1} (see Section 2.1.1). It became clear that only those vibrations which produced an oscillating dipole moment gave rise to infrared absorption.

In 1923 Smekal was one of the first of a number of scientists to predict that molecules could scatter light inelastically [1]. He also suggested that molecular polarizability often changes as particular vibrations occur. This led him to propose that the shift in frequency between the incident and scattered light would be characteristic of molecular vibrations. Raman [3], and almost concurrently Mandelstamm [4], demonstrated the effect predicted by Smekal and hence Raman spectroscopy was born. Their observation led to a frenzy of activity [6–10]. Raman research groups developed in India, France, Germany and Russia, and very rapidly the literature was flooded with Raman spectra. The reason for the immense interest and output was purely experimental. The simple apparatus required to record Raman spectra at that time already existed in most laboratories. The Raman light was excited by using a powerful mercury vapour discharge lamp, analysed with a conventional spectrograph and recorded on photographic plates. An early Raman spectrum recorded in this way is shown in Fig. 1.1. By 1939 the conventional method of studying the vibrational characteristics of compounds was Raman rather than infrared spectroscopy and a vast range of liquids, a few gases, one or two polymers and crystals had been analysed. However, following the Second World War, high sensitivity infrared detectors became available and, coupled with advances in electronics, this made the development of automatic infrared spectrometers possible. Thus, infrared spectra could be recorded routinely. Raman spectroscopy, by

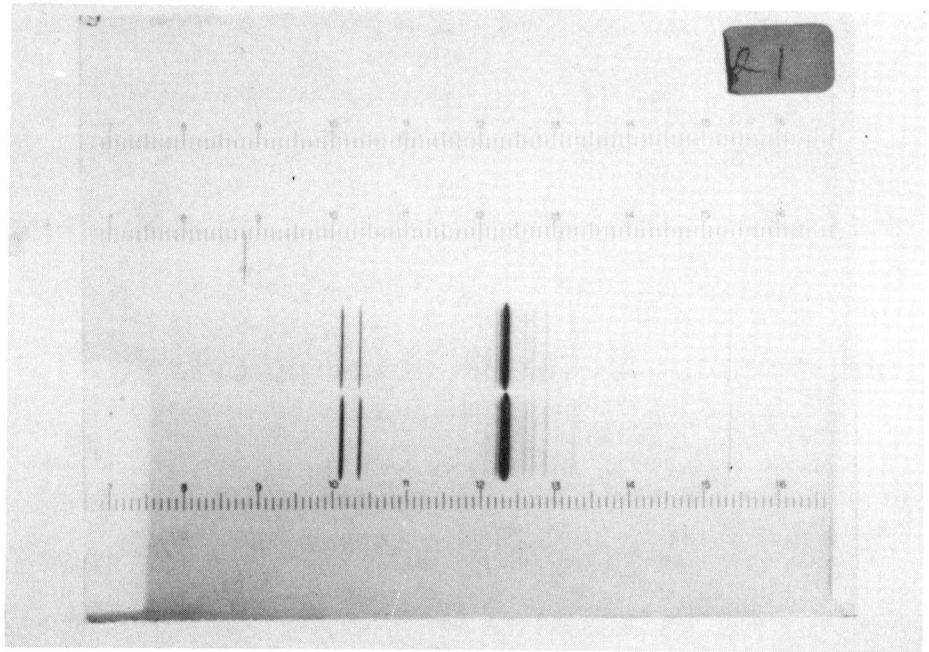

Fig. 1.1 — A vibrational Raman spectrum of carbon tetrachloride recorded on a photographic plate.

comparison, required skill and the use of darkroom facilities to develop the spectrographic plates produced.

The Raman experiment is in theory extremely attractive to the applied spectroscopist. Glass is transparent to both visible laser and Raman scattered light. Thus it is relatively easy to construct cells to study samples under a variety of different conditions. For instance, experiments can be carried out through a microscope, at high and low temperatures or pressures, on solids, liquids or gases. By comparison, infrared spectroscopy requires the use of specially designed cells incorporating windows that transmit infrared radiation. Unfortunately, materials that fit this criterion can be either highly toxic, water-soluble, brittle or very expensive. Many of the chemical reactions which are of interest to the spectroscopist occur in aqueous solutions. This presents considerable problems in the infrared since water is almost totally opaque to infrared radiation. However, water has an extremely weak visible Raman spectrum, and is transparent to visible light, making it an ideal solvent for Raman studies. To record an infrared absorption spectrum it is also necessary to prepare the sample in the form of a uniformly thin layer. As well as being time consuming, this may also perturb the structure of the sample. Raman spectroscopy has the advantage that no special sample preparation is required.

In the early 1960s commercial scanning/recording Raman spectrometers such as the Hilger & Watts E612 and the Cary 81 double-monochromator were manufactured (see Chapter 3). An example of an early Raman spectrometer is shown in

Fig. 1.2. However, these machines were still expensive and cumbersome when compared with commercial infrared spectrometers. Every well equipped analytical laboratory contained an automatic recording infrared instrument and experimenters were committed to infrared spectroscopy. As a result, less than one hundred of these innovative Raman instruments were sold.

In the mid 1960s visible lasers were developed and proved to be ideal sources for Raman experiments [11]. A renaissance in interest was widely forecast [12]. The renaissance happened, but not in the analytical laboratory [13]. Cary and Perkin Elmer both offered analytical-grade spectrometers, but neither was a commercial success. Laser Raman spectroscopy was found to be expensive in terms of both equipment and output. Samples which are pure, transparent and colourless were indeed simple to study and hence relatively economical to investigate. However, any sample which is coloured or fluorescent causes real problems and in many cases hours of work can be wasted. Fig. 1.3 shows a conventional Raman spectrum of corn oil. Although it is not usually considered to be a fluorescent material, when irradiated with a visible laser it shows a sufficiently intense level of fluorescence to mask the Raman signal. Impure materials and a vast array of pure inorganic and organic materials are fluorescent enough to prevent Raman study. As a result, even well established laboratories report very low success rates (often less than 20% of the samples submitted). The average cost per spectrum is, therefore, prohibitively high. Sample purification can sometimes improve spectral quality. Hence to an academic laboratory interested in fundamental aspects of the spectroscopic behaviour of materials, studies can be restricted to carefully purified compounds. However, for intrinsically fluorescent or coloured materials purification is futile, and for the industrial analyst it is rarely a viable option. During the 1970s and 80s other spectroscopic methods such as NMR and mass spectrometry joined infrared absorption as standard techniques for both qualitative and quantitative analysis. Thus it became clear that although visible Raman spectroscopy is valuable in specialized areas, it is not attractive as a routine method.

For many years it has been known that significant advantages are to be gained by using interferometers rather than dispersive spectrometers in the infrared [14]. If the infrared spectrum is recorded with an interferometer and the interferogram Fourier transformed into a spectrogram with a computer, the spectroscopic quality is vastly improved (see Chapter 4). During the past 15 years, analytical-grade Fourier transform infrared spectrometers (FTIRs) have completely superseded their dispersive predecessors. Every significant manufacturer now offers a range of both research and analytical-grade interferometers of this type, and the performance available today would be almost unbelievable to spectroscopists working only 15 years ago. On the other hand, although visible Raman methods have improved in many ways, their versatility is little advanced over the situation that prevailed in the early 1970s. A modern laser Raman instrument is shown in Fig. 1.4. Instrument sensitivity and computer manipulation of the data recorded have considerably improved, but the basic problems of fluorescence and colour still predominate.

In 1964 Gebbie, Chantry and Hilsum 15] demonstrated that it was possible to record near infrared excited Raman spectra by using a Michelson interferometer. However, owing to the experimental limitations of the day this initial paper failed to

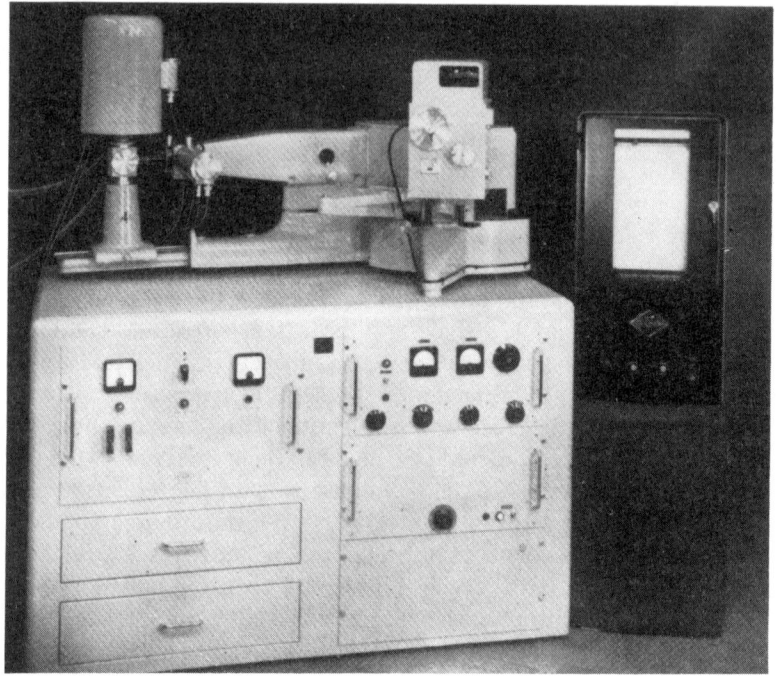

Fig. 1.2 — A Hilger & Watts E612 Raman spectrograph with scanning attachment (photograph courtesy of the Science Museum, London and Hilger Analytical Ltd).

attract interest. In fact it was not until 1986 that Chase and Hirschfeld published the second paper on the subject [16]. Laser, interferometer, detector and computer technology had all improved enormously in the intervening 22 years, making it possible to produce good quality FT Raman spectra (see Chapter 5). Several papers then appeared presenting experimental results and discussing the value of near infrared excited Fourier transform Raman methods [17–21]. These papers showed FT Raman spectra of numerous samples which were either fluorescent or absorb the excitation light when irradiated with a visible laser source. The use of interferometry and Fourier transforms enabled data to be collected rapidly and Raman light to be processed more efficiently. These developments lifted the experimental restrictions which had beleaguered Raman spectroscopy for so long, and FT Raman was set to join FTIR as a truly complimentary analytical technique.

In this volume we describe the basic elements of a FT Raman spectrometer, using a prototype instrument developed at Southampton University as a case study. The aim of our research programme at Southampton has been to develop the technique as a routine analytical tool rather than a specialist academic discipline. As a result, we have been inundated with industrial visitors who have studied a wide range of samples. It has become clear that FT Raman spectroscopy is extremely versatile and can solve a diverse range of analytical problems. Many multinational instrument

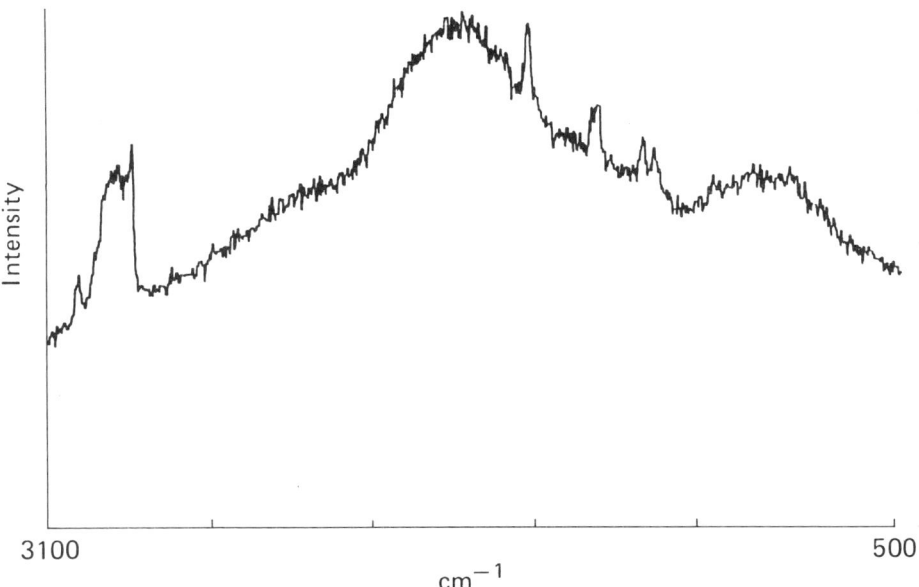

Fig. 1.3 — The Raman spectrum of corn oil recorded on a conventional scanning Raman spectrometer using an argon ion laser (514.5 nm) as the excitation source.

Fig. 1.4 — The Dilor XY Raman spectrometer with a microscope attachment, courtesy of Dilor SA.

manufacturers now offer FT Raman products. Since FT Raman and infrared spectrometers share the same basic interferometer, most commercial instruments offer both spectroscopies on a combined bench. Alternatively, FT Raman accessories have been produced for existing FTIR machines. Inevitably, this will lead a massive expansion of the applications of the Raman method. Raman spectroscopy will benefit from the innovations which have occurred in other spectroscopies such as the production of large collections or libraries of Raman spectra in a standard format, and will be able to capitalize on the development of chemometrics.

REFERENCES

[1] A. Smekal, *Naturwiss.*, 1923, **11**, 873.
[2] H. A. Kramers and W. Heisenberg, *Z. Phys.*, 1925, **31**, 681.
[3] C. V. Raman and K. S. Krishnan, *Nature*, 1928, **121**, 501.
[4] G. Landsberg and L. Mandelstam, *Naturwiss.*, 1928, **16**, 557.
[5] R. N. Jones, *European Spectroscopy News*, 1987, **70**, 10.
[6] C. V. Raman, 1928, **121**, 619.
[7] C. V. Raman, 1928, **121**, 721.
[8] G. Landsberg and L. Mandelstam, *Naturwiss.*, 1928, **16**, 772.
[9] Y. Rocard, *Compt. Rend.*, 1928, **186**, 1107.
[10] J. Cabannes, *Compt. Rend.*, 1928, **186**, 1201.
[11] H. Kogelnik and S. P. S. Porto, *J. Opt. Soc. Am.*, 1963, **53**, 1446.
[12] T. R. Gilson and P. J. Hendra, *Laser Raman Spectroscopy*, Wiley, London, 1970.
[13] H. Baranska, A. Labudzinska and J. Terpinski, *Laser Raman Spectrometry — Analytical Applications*, Ellis Horwood, Chichester, 1987.
[14] P. R. Griffiths, *Chemical Infrared Fourier Transform Spectroscopy*, Wiley, New York, 1974.
[15] G. W. Chantry, H. A. Gebbie and C. Hilsum, *Nature*, 1964, **203**, 1052.
[16] D. B. Chase and T. Hirschfeld, *Appl. Spectrosc.*, 1986, **40**, 133.
[17] D. B. Chase, *J. Am. Chem. Soc.*, 1986, **108**, 7485.
[18] V. M. Hallmark, C. G. Zimba, J. D. Swalen, J. and F. Rabolt, *Spectroscopy*, 1987, **2**, 40.
[19] D. B. Chase, *Anal. Chem.*, 1987, **59**, 881A.
[20] C. G. Zimba, V.M. Hallmark, J. D. Swalen, and J. F. Rabolt, *Appl. Spectrosc.*, 1987, **41**, 721.
[21] S. F. Parker, K. P. J. Williams, P. J. Hendra and A. J. Turner, *Appl. Spectrosc.*, 1988, **42**, 796.

2
The vibrational behaviour of molecules

This chapter discusses molecular vibrations and how they give rise to infrared and Raman spectra. It is not intended to be rigorously theoretical, but rather to be informative and easy to assimilate.

2.1 MOLECULAR ROTATIONS AND VIBRATIONS

When molecules rotate freely in the gas phase, their angular velocities about three orthogonal rotational axes can only have precisely permitted values; there is no possibility that they can rotate at intermediate speeds†. Their rotational energy is said to be quantized. Molecules can only change their rotational velocity by the injection (or removal) of a packet or quantum of energy. This energy must be exactly equal to the kinetic energy difference between two allowed states. In molecular spectroscopy we observe transitions between quantized states by studying the interaction between the sample and electromagnetic radiation.

In addition to rotation in the gas phase, molecules vibrate at all times. Molecules exhibit incredibly complex vibrational patterns, but these motions are linear combinations of much simpler fundamental modes of vibration. A mode of vibration in a molecule is a periodic contortion in which the centre of mass of the molecule or its orientation do not change as a result of the vibration and all of the atoms pass through their equilibrium position coincidentally.

The position of a molecule in three dimensional space can be described by using an x, y and z co-ordinate for each atom. Thus, for a molecule comprised of n atoms $3n$ co-ordinates are required to describe its shape, position and orientation. The motion of each atom can be described by a change Δx, Δy and Δz in these co-ordinates. There are $3n$ fundamental distinct molecular motions which are called degrees of freedom, and any other motion is some linear combination of these. Molecular motions consist of translations, rotations and vibrations, but predominantly mixtures of all three.

---

† If the velocities of motor vehicles were quantized into levels 10 mph apart, speeds of 50 and 60 mph would be possible, but any intermediate speed would be forbidden. Similarly, braking from 60 mph to rest would result in a series of six infinitely violent jerks as the kinetic energy was lost!

In consideration of molecular vibrations, movements consisting solely of translations or pure rotations must be ignored. Three of the degrees of freedom represent the translation of the whole molecule along the x, y or z axis. Translation in any other direction can be resolved into displacements along these three cartesian axes. A non-linear molecule also has three pure rotations about these axes. A linear molecule, however, only has two distinct rotations since no rotation occurs about the molecular axis. After discounting pure rotations and translations we are left with motions which must involve vibration. Thus a non-linear molecule possesses $3n - 6$ fundamental modes of vibration, whilst a linear molecule has $3n - 5$.

Many characteristics of molecular vibrations can be understood in terms of classical simple harmonic motion. A diatomic molecule can be viewed naïvely as point masses connected by a spring as shown in Fig. 2.1. To displace the masses from

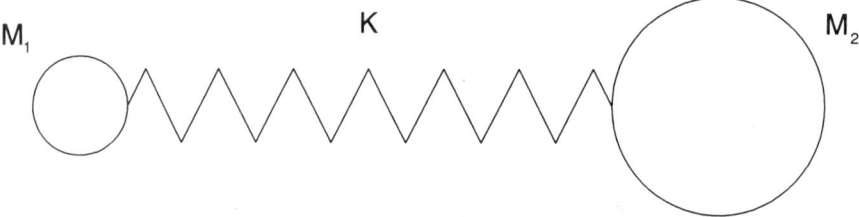

Fig. 2.1 — A molecule can be viewed as point masses connected by weightless springs.

their equilibrium position requires the application of a force, F. For simple harmonic systems, $F = kx$, where x is the displacement (in metres) and k is the force constant (N/m). This is a statement of Hooke's law. This simple mechanical system will vibrate at a frequency given by;

$$v_{vib} = \frac{1}{2\pi} \sqrt{\frac{k}{\mu}} \qquad (2.1)$$

where μ is the reduced mass

$$\mu = \frac{m_1 m_2}{m_1 + m_2} \text{ kg} \qquad (2.2)$$

This equation actually allows us to predict vibrational frequencies for diatomic molecules very well. The potential energy function for a harmonic oscillator is a parabola (see Fig. 2.2). However, although simple harmonic motion is a reasonable approximation to molecular behaviour, the potential function derived from it would never allow chemical bonds to break. The simple harmonic oscillator is only a model; real molecules are anharmonic vibrators. A more realistic model for the potential energy function is given by the Morse function:

Sec. 2.1] Molecular rotations and vibrations 19

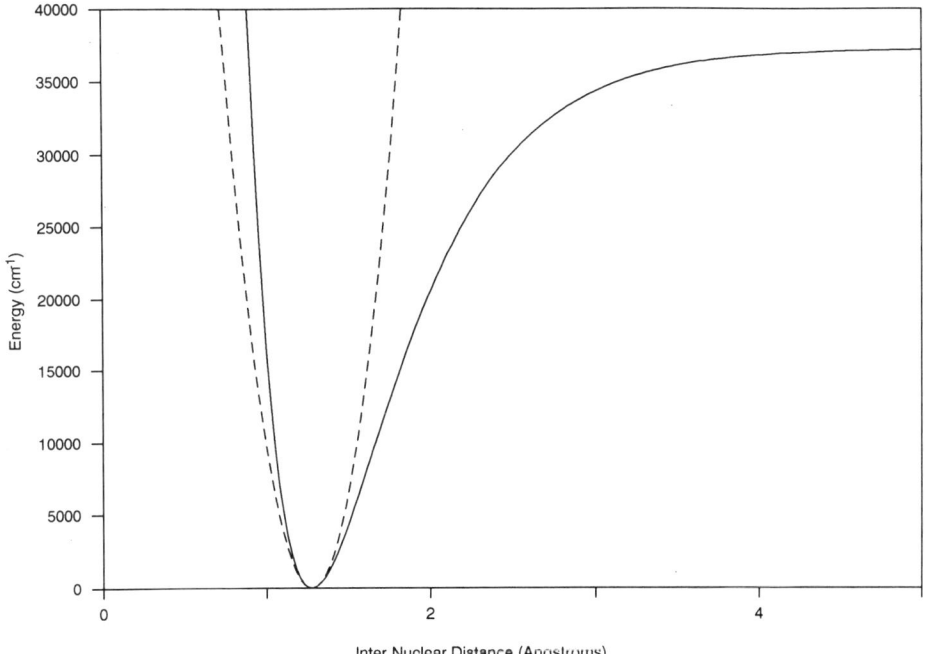

Fig. 2.2 — A Morse curve compared with a parabolic potential function for $^1H^{35}Cl$ drawn to scale.

$$E = D_{eq}[1 - \exp\{a(r_{eq} - r)\}]^2 \qquad (2.3)$$

where
- D_{eq} = dissociation energy
- r_{eq} = equilibrium bond length
- r = internuclear distance
- a = constant for a particular molecule

This is shown diagrammatically in Fig. 2.2. Compression or expansion of a chemical bond away from its equilibrium value is an endothermic process. The curvature of the bottom of the potential energy well is indicative of the stiffness of the bond; the steeper the well, the more rigid the vibrator. The molecular stiffness is described quantitatively, as referred to previously, by the force constant k.

Although the potential energy curve is a continuous function, the *total* vibrational energy of a molecule is quantized, i.e. it can possess only certain discrete values. These values are described by the vibrational quantum number, V, which can take values $V = 0, +1, +2, +3, +4...$. The expression for the quantized vibrational energy is different, depending on whether we assume the molecule is a harmonic or an anharmonic oscillator. For a simple harmonic oscillator the expression is:

$$E_{vib} = h\left(V + \frac{1}{2}\right)\nu_{vib} \tag{2.4}$$

Where
h = Planck's constant (J sec)
ν_{vib} = Vibrational frequency (Hz)
V_{vib} = Vibrational quantum number

The anharmonic oscillator model gives the Eq. (2.5). The energy values obtained compare well with experimental observations.

$$E_{vib} = h\left[\left(V + \frac{1}{2}\right)\nu_{vib} - \left(V + \frac{1}{2}\right)^2 X_e\nu_{vib}\right] \text{ J} \tag{2.5}$$

Where
X_e = Anharmonicity constant
ν_{vib} = Vibrational frequency (Hz)
V = Vibrational quantum number

The lowest vibrational energy state is represented by $V = 0$. It can be seen from Eqs. (2.4) and (2.5) that E_{vib} is never zero. Thus, even at absolute zero, molecules still vibrate. Another flaw in the simple harmonic model is that it predicts that the total vibrational energy levels will be equally spaced and experiment proves this not to be true. In reality the energy levels converge as V increases, reaching an asymptotal value at the dissociation limit of the oscillator, and this is correctly predicted by the anharmonic model. Figure 2.3 shows the vibrational energy level 'ladders' predicted by the two models.

If we plot a graph of total vibrational energy vs. interatomic separation, the allowed values for a molecule consist of a series of horizontal lines which converge at high energy. For a given quantized vibrational energy level there is a range of allowed values of internuclear distance. As the vibrational quantum number increases the amplitude of vibration increases. This is illustrated in Fig. 2.4.

The total vibrational energy is composed of kinetic and potential energy, whereas the Morse curve represents only potential energy. The points of intersection between the vibrational energy levels and the Morse curve represent the classical turning point of an oscillator where all the energy is potential and none is kinetic. At intermediate values of internuclear distance, some of the energy is potential and some kinetic. The injection (or removal) of a quantum of energy, in the form of a photon, is expressed by a change in the vibrational quantum number. These changes are observed in vibrational spectroscopy.

So far we have considered only diatomic molecules. When we move on to consider a triatomic molecule such as H_2O, the vibrational behaviour is more complex, with multiple bond stretches as well as bending vibrations. In H_2O there

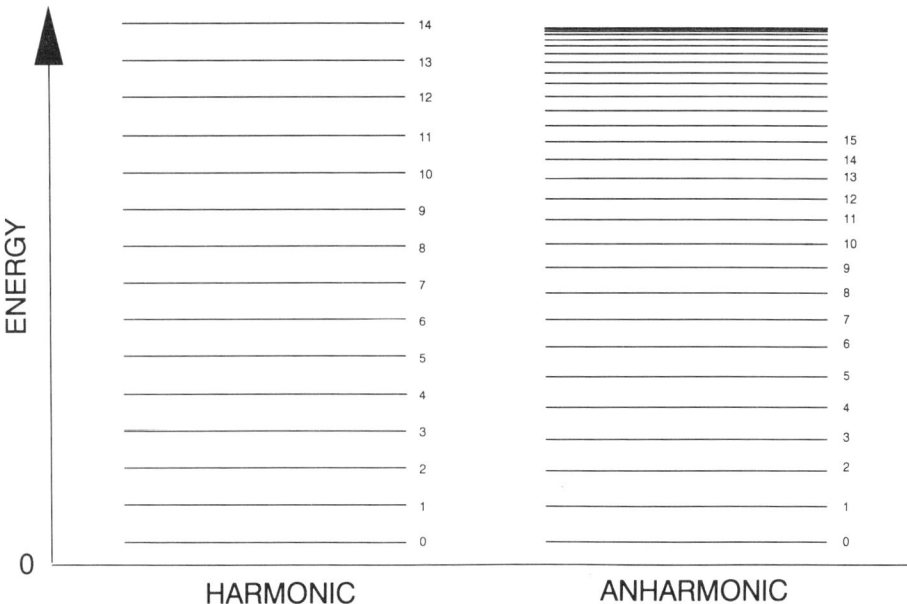

Fig. 2.3 — Energy level ladders for harmonic and anharmonic models. The energy levels are evenly spaced in the harmonic model, but converge when anharmonicity is considered.

are three atoms and hence three $(3n - 6)$ fundamental modes. The number of stretching vibrations is equal to the number of linkages in the molecule. The three modes of water are shown in Fig. 2.5.

2.1.1 Spectroscopic units
When expressed in hertz (Hz), the frequencies of molecular vibrations have very high values, typically 10^{12}–10^{14} Hz. These values proved to be inconvenient and spectroscopists have, much to the disgust of most mathematicians and physicists, settled on the use of the wavenumber (cm^{-1}), a non-SI unit. The wavenumber is defined as number of wavelengths per centimetre. The frequencies of fundamental vibrational modes lie between 4000 and 50 cm^{-1}. Strictly, the letter v should be used to denote frequencies in Hz, and $\bar{\omega}$ used for values quoted in wavenumbers. However, in the literature, v is consistently used for frequencies given in wavenumbers and we will adopt this practice in the following chapters unless specifically stated. To avoid confusion, units will be given at all times.

2.2 INFRARED SPECTROSCOPY
So far, we have avoided the spectroscopic consequences of molecular rotations and/or vibrations. If a rotation or vibration gives rise to a change in dipole as the motion occurs, then electromagnetic radiation can interact with it. The interaction is a resonance process and the energy of the incoming photon must match that of a

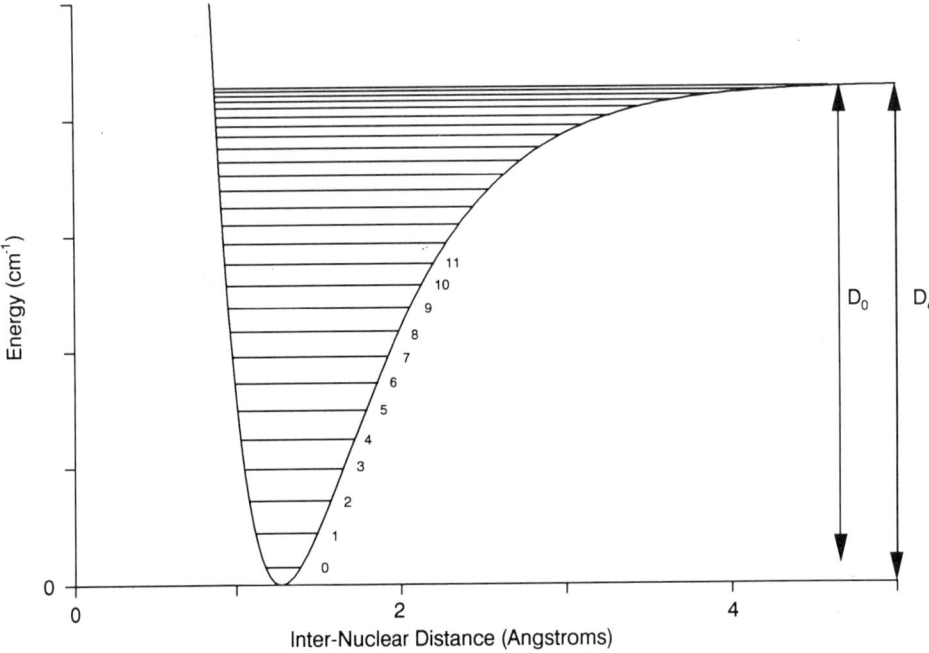

Fig. 2.4 — The energy level diagram for hydrogen chloride showing the allowed values of internuclear distance for each value of the vibrational quantum number.

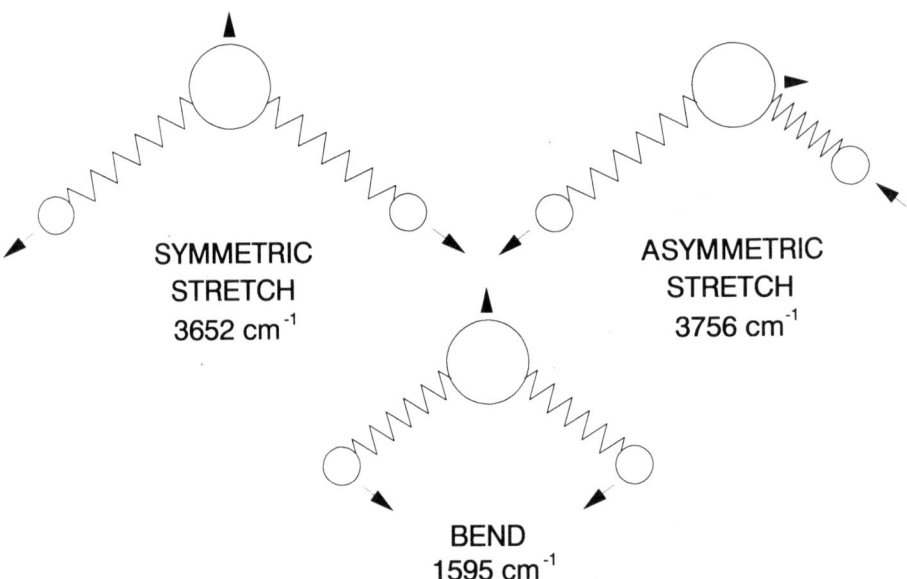

Fig. 2.5 — The fundamental vibrational modes of water. There are two stretching and one bending vibration.

vibrational or rotational energy transition. Resonance with infrared radiation results in absorption. Equally, emissions can occur if the vibrational or rotational energy falls. This is summarized in Fig. 2.6.

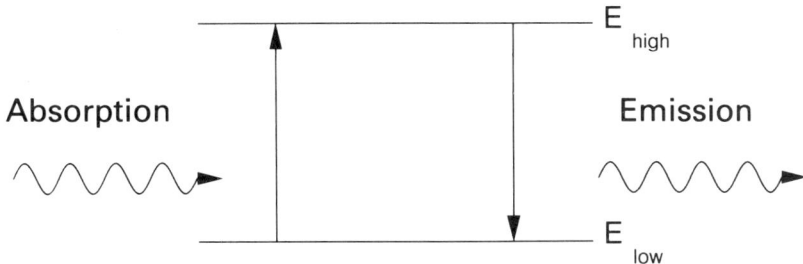

Fig. 2.6 — A simple energy level diagram for absorption and emission spectroscopy.

For our purposes it is useful to concentrate on vibrations because only gases give rise to pure rotational spectra. Although it is possible to record emission spectra, most analytical spectroscopists confine their attention exclusively to vibrational absorption spectra, such as the one shown in Fig. 2.7. Molecular vibrations give rise

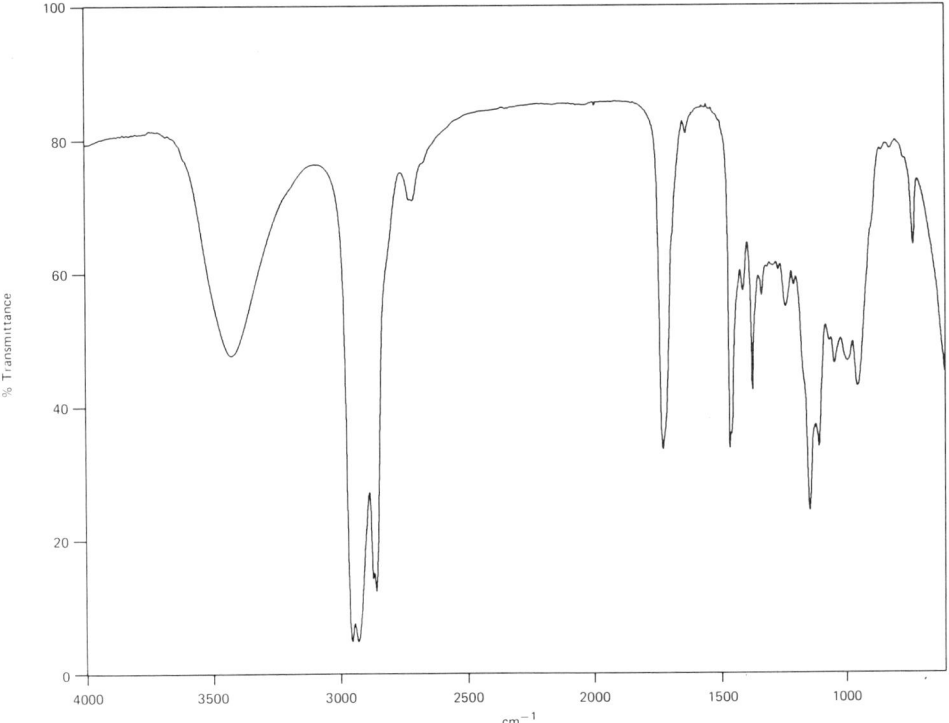

Fig. 2.7 — The infrared absorption spectrum of hexanal.

to absorptions and emissions between 100 and 4000cm-1 (ie. the infrared region on the electromagnetic spectrum). Many molecular vibrations involve no change in dipole and hence do not give rise to infrared absorptions. To observe these motions we must use Raman spectroscopy.

2.3 THE RAMAN EFFECT

All molecular species possess a polarizability†, α. Application of an electric field gradient of magnitude, E (V/m), can *induce* a dipole of magnitude, P (V/m), that is,

$$P = \alpha E \qquad (2.6)$$

This results from a slight distortion of the electron cloud. If the field is supplied by electromagnetic radiation of frequency, v_0 (Hz) then,

$$P = \alpha E_0 \cos(2\pi v_0 t) \qquad (2.7)$$

where t is time and E_0 is maximum electric field strength (V/m).

Let us now assume that this irradiated polarizable molecule is vibrating at frequency v_{vib}, that is, its distortion from its equilibrium position, q‡, will be given by,

$$q = q_0 \cos(2\pi v_{vib} t) \qquad (2.8)$$

where q_0 is the maximum distortion and v_{vib} is the vibrational frequency (Hz).

If the distortion is likely to cause an alteration in the polarizability and it is assumed that the variation will be linear and the amplitude of displacement small, the polarizability of the periodically distorting molecule will become

$$\alpha = \alpha_0 + \left(\frac{\delta \alpha}{\delta q}\right)_0 q \qquad (2.9)$$

where α_0 is the polarizability of the molecule in the equilibrium position, and $(\delta\alpha/\delta q)_0$ is the rate of change of that polarizability with distortion around the equilibrium structure. Hence the variation in polarizability as the molecule vibrates will be of the form,

$$\alpha = \alpha_0 + \left(\frac{\delta \alpha}{\delta q}\right)_0 q_0 \cos(2\pi v_{vib} t) \qquad (2.10)$$

---

† Undergraduates who find the concept difficult to retain can usually remember the idea that polarizability is a measure of the ease with which electrons can be induced to "slosh" up and down a molecule under the influence of an applied field.

‡ q can be defined in terms of the distortion from an equilibrium position as shown in, Eq. (2.8), but is more usually thought of as the progress through the vibration.

The Raman effect

The polarizability varies with time as shown diagrammatically in Fig. 2.8.
On substituting α from Eq. (2.10) into Eq. (2.7) we obtain;

$$P = \alpha_0 E_0 \cos(2\pi\nu_0 t) + \left(\frac{\delta\alpha}{\delta q}\right)_0 q_0 \cos(2\pi\nu_{vib} t)\cos(2\pi\nu_0 t) \qquad (2.11)$$

However, since

$$\cos A \cos B = \frac{1}{2}[(\cos(A+B) + \cos(A-B)] \qquad (2.12)$$

we can rearrange Eq. (2.11) to give,

$$P = \alpha_0 E_0 \cos(2\pi\nu_0 t) + \frac{1}{2}\left(\frac{\delta\alpha}{\delta q}\right)_0 q_0 \{\cos(2\pi[\nu_0 + \nu_{vib}]t) + \cos(2\pi[\nu_0 - \nu_{vib}]t)\} \qquad (2.13)$$

Thus, the induced dipole in a molecular system vibrating with frequency ν_{vib} and irradiated at frequency ν_0, will vary as ν_0 and also $(\nu_0 + \nu_{vib})$ and $(\nu_0 - \nu_{vib})$. Obviously the polar state is more energetic than the relaxed one, i.e. the polarization process is endothermic; hence the spontaneous relaxation must be accompanied by a release of energy. This process is the emission of radiation which we describe as scattering. In Fig. 2.9, even in ultra-filtered liquids where no dust particles are present, illumination by a laser reveals a cylinder of weakly glowing liquid where the laser passes. The 'glow' is the source of the scattered light, and it will appear at frequencies ν_0, $\nu_0 + \nu_{vib}$ and $\nu_0 - \nu_{vib}$.

If we return to Eq. (2.11), we notice that the proportionality constants in α are very different. This point is revealed and emphasized in Fig. 2.9; the values of α_0 and $(\delta\alpha/\delta q)_0 q_0$ are indeed very different. As a consequence we expect to 'see' scatter at ν_0 (arising from α_0) much more intensely than at $(\nu_0 + \nu_{vib})$ or $(\nu_0 - \nu_{vib})$. The former is named after a man who studied the phenomenon extensively in the 19th century, Lord Rayleigh. Rayleigh scatter is, of course 'elastic', i.e. it occurs at the same energy as the incident or source radiation. The inelastic process is named after Raman and is, as we have said, much weaker than its accompanying elastic or Rayleigh relative. One further point is worth emphasizing at this stage: Rayleigh scatter is WEAK, but Raman scatter is even *WEAKER*. Typical Raman/Rayleigh intensity ratios for pure liquids are $\approx 10^{-3}$. Light scattered at a frequency of $(\nu_0 + \nu_{vib})$ is known as anti-Stokes Raman scatter, whereas that scattered at $(\nu_0 - \nu_{vib})$ is termed Stokes Raman scatter. Some texts refer to the anti-Stokes scatter as blue shifted, and the Stokes scatter as red shifted.

If we carry out the experiment described in Fig. 2.9, we observe that the Raman bands appear as 'side bands' close to the Rayleigh scatter. The Stokes Raman feature is always more intense than the anti-Stokes equivalent. To understand this difference we must consider the energy levels involved in the process. Let us assume a very

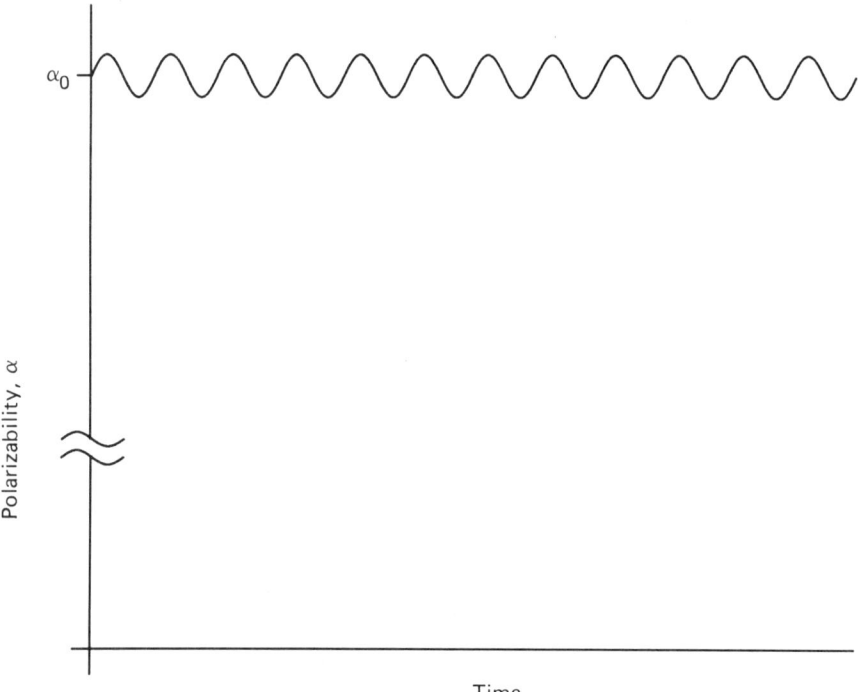

Fig. 2.8 — A graph of the variation in molecular polarizability with time due to vibrations. The ordinate axis is broken to show that the change in α is very small.

simple energy level diagram as applied to chlorine[†], with a single vibrational level at 505 cm^{-1}. This is shown in Fig. 2.10. Polarization of the halogen is caused by interaction with photons of energy $h\nu_0$ momentarily raising the energy of the whole vibrating system by that amount. The polarized condition is not an energy state of the molecule itself; there is no question of the molecule changing its electronic configuration as the interaction occurs[‡]. The polarized state, unlike excited electronic, vibrational or rotational states, is not a fixed energy level and is therefore described as 'virtual'. Relaxation from the virtual state can occur in one of three ways as illustrated in Fig. 2.10.

The most probable relaxation results in elastic scattering, i.e. relaxation with no change in vibrational quantum number, V. Relaxation which occurs with a change in vibrational quantum number produces inelastic Raman scatter. For Stokes scatter $\Delta V = +1$, and for anti-Stokes scatter $\Delta V = -1$. The intensity of Raman scatter is related to the population of the initial state of the molecule. Thus, for Stokes scatter the intensity is normally related to the population of the ground state, $V = 0$, and for

---

[†] For convenience we are ignoring vibrational overtones and rotations.
[‡] An analogous situation to polarizing a molecule is stretching a rubber band. The polarized state of a molecule is like the stretched state of the rubber band. Work is done to produce this state, and in Rayleigh scatter the same amount of energy is released on relaxation.

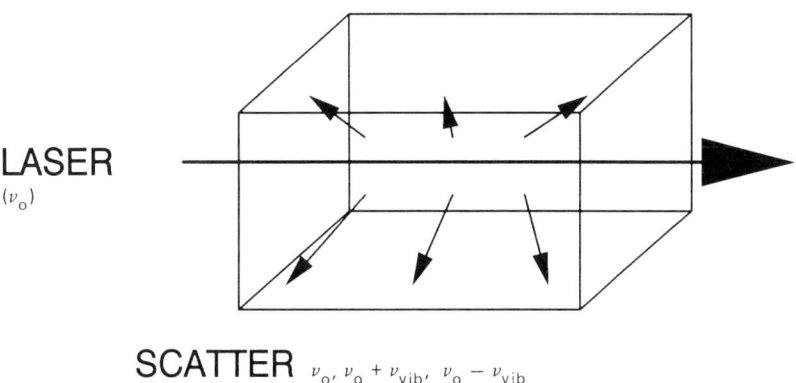

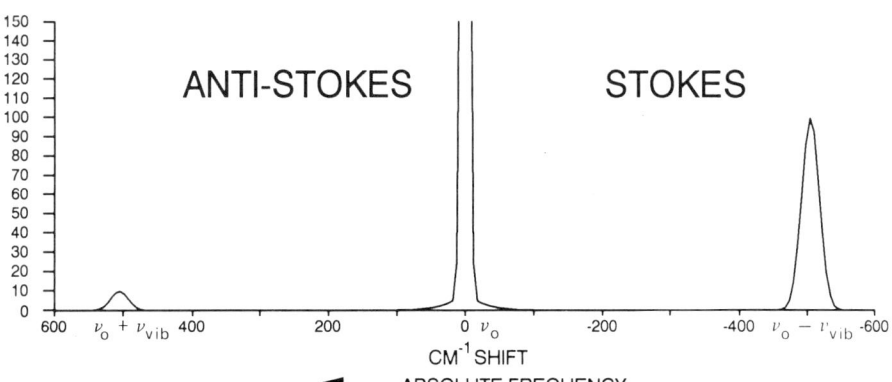

Fig. 2.9 — When an intense monochromatic light source of frequency ν_0 irradiates a sample, the light that is scattered contains frequency components ν_0, $\nu_0 + \nu_{vib}$ and $\nu_0 - \nu_{vib}$. In the case of liquid chlorine the Stokes and anti-Stokes Raman bands are shifted by 505 cm^{-1} from the excitation frequency. If the spectrum is excited with an argon ion laser $\nu_0 = 19436$ cm^{-1}, $\nu_0 + \nu_{vib} = 19941$ cm^{-1} (anti-Stokes) and $\nu_0 - \nu_{vib} = 18931$ cm^{-1} (Stokes). The Stokes:anti-Stokes ratio is approximately 10:1.

anti-Stokes to that of the first excited vibrational level, $V = 1$. These populations are given by the Boltzmann distribution:

$$\frac{n'}{n} = \frac{g'}{g} \exp\left(\frac{-\Delta E}{kT}\right) \qquad (2.14)$$

where n is the population, g is the degeneracy, and ' denotes the upper level. ΔE is the energy gap between the upper and lower levels expressed in J/molecule.

For chlorine gas:

$$\Delta E = 505 \text{ cm}^{-1}$$

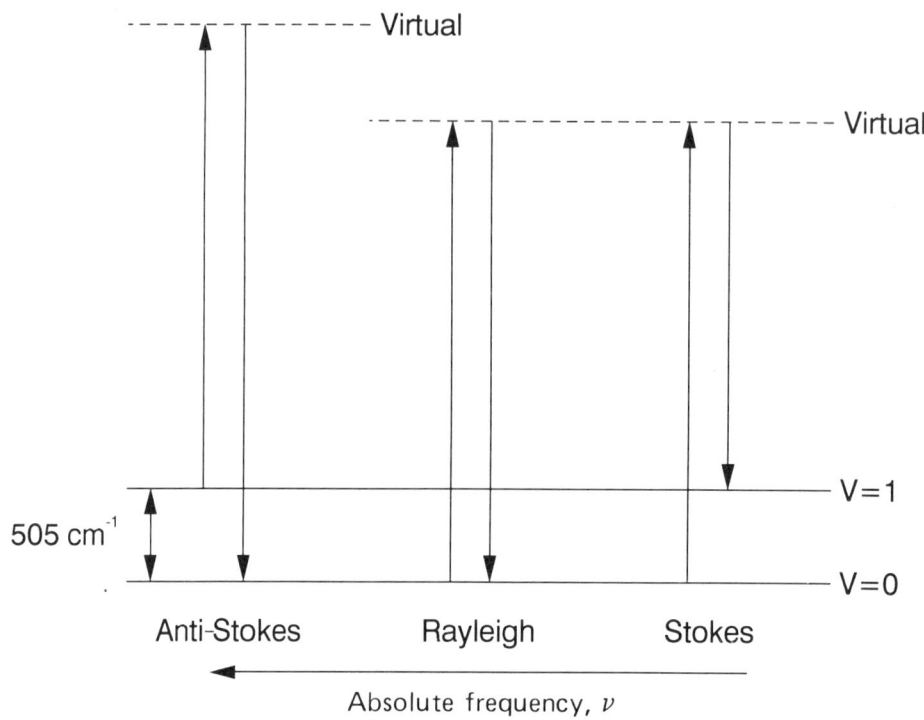

Fig. 2.10 — An energy level diagram showing the origin of the Stokes and anti-Stokes bands in the Raman spectrum of liquid chlorine.

$$\Delta E = h\bar{c} \text{ J} \qquad (2.15)$$
$$= 1.003 \ 10^{-20} \text{ J}$$

where
$\bar{c}$ is the speed of light (cm/sec)
h is Planck's constant (J sec).

Hence the relative populations of the $V = 0$ and $V = 1$ levels can be calculated.

$$\frac{n^{V=1}}{n^{V=0}} = \exp\left(\frac{1.003 \ 10^{-20}}{k \ 298}\right) \qquad (2.16)$$

$$= 0.087$$

However, Raman scattering efficiency depends on the fourth power of the frequency of the light being scattered. Although the intensity of anti-Stokes lines are reduced by the Boltzmann factor, the frequency of the light being scattered is higher and thus scattering occurs more efficiently. The ratio $I_{\text{anti-Stokes}}/I_{\text{Stokes}}$ is given by the following expression.

$$\frac{I_{\text{anti-Stokes}}}{I_{\text{Stokes}}} = \frac{(\nu_0 + \nu)^4 \, n_{V=1}}{(\nu_0 - \nu)^4 \, n_{V=0}} \qquad (2.17)$$

where
 ν_0 is the excitation frequency (cm^{-1})
 ν is the vibrational frequency (cm^{-1})

If the Ar$^+$ line at 19 436 cm^{-1} (514.5 nm) is used as a source, $I_{\text{anti-Stokes}}/I_{\text{Stokes}}$ for the 505-cm^{-1} chlorine vibration should be 0.107. Alternatively, if it is convenient to measure $I_{\text{anti-Stokes}}/I_{\text{Stokes}}$ experimentally, we can use this ratio to determine the temperature of the species under investigation. Although this must be the world's most expensive thermometer, it has been used on several occasions [1].

It must be noted that because the virtual state is not a fixed level, Raman scattered light can be produced from an excitation source with any wavelength, and hence Raman spectra are presented as intensity *vs.* frequency shift from this excitation source. This is why we are now able to record spectra from a near infrared radiation source, whereas conventional experiments use a visible one. The experiment may also be carried out in the ultraviolet. Thus, the Raman spectrum of carbon tetrachloride, for example, will have the same vibrational frequencies whether excited with visible or near infrared lasers. However, as the scattering efficiency varies as $(\nu_0 - \nu)^4$ the relative band intensities will differ depending on the excitation source used. Figure 2.11 shows Raman spectra recorded photographically many years ago, with a mercury discharge lamp as source. The Raman spectrum recorded from a single argon laser line is shown below for comparison.

2.4 INFRARED AND RAMAN ACTIVITY

As we have already discussed, the infrared and Raman spectra should ideally be considered together, since only then do we get a true picture of the vibrational behaviour of a molecule. To do this we must compare and superimpose, in some convenient way, the infrared spectrum plotted as absorbance (or % transmission) *vs.* frequency, and the Raman spectrum as intensity of scatter versus frequency *shift*. An example is shown in Fig. 2.12.

Compression and extension of a bond changes the degree of orbital overlap and may affect the molecular polarizability and the dipole moment. Vibrational modes can be infrared and/or Raman active. From Eq. (2.11), it can be seen that only those modes with a non-zero value of $(\delta\alpha/\delta q)_0$ are Raman active. Infrared spectroscopy only observes vibrations which have a non-zero value of $(\delta\mu/\delta q)_0$. Some vibrations, however, are not observed in either spectroscopy, because they have no effect on either the molecular polarizability or dipole moment. Infrared and Raman activity

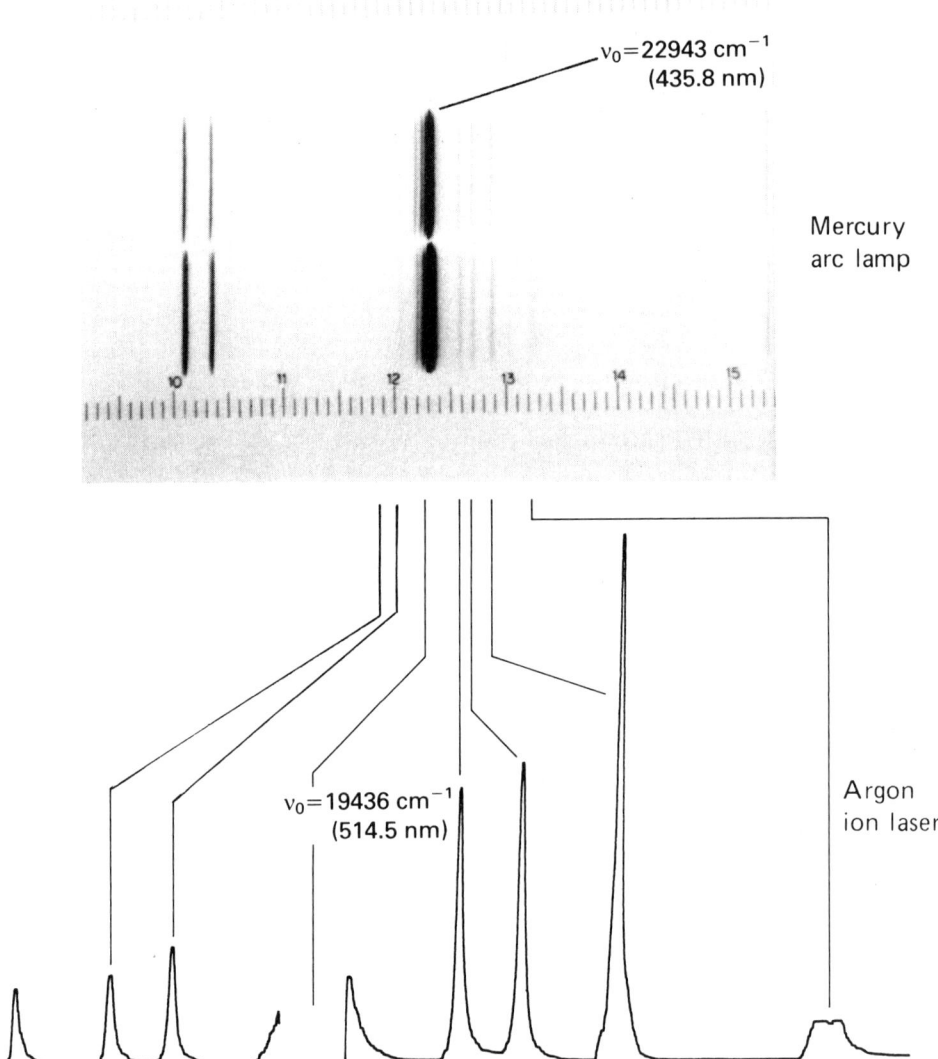

Fig. 2.11 — Raman spectra recorded with different excitation sources are very similar. The Raman spectrum of carbon tetrachloride excited with a mercury arc lamp and recorded on a photographic plate is virtually identical to that recorded using a conventional scanning Raman spectrometer and an argon ion laser.

can be visualized by plotting μ and α against the progress through the vibration, q. The quantity of interest is then the gradient of the plot around the equilibrium position. The normalized coordinate, q, can be expressed in a variety of ways, but for our purpose we will use units of phase angle between 0 and 2π radians to define the progress of the vibration, as shown in Fig. 2.13.

Sec. 2.4]   **Infrared and Raman activity**

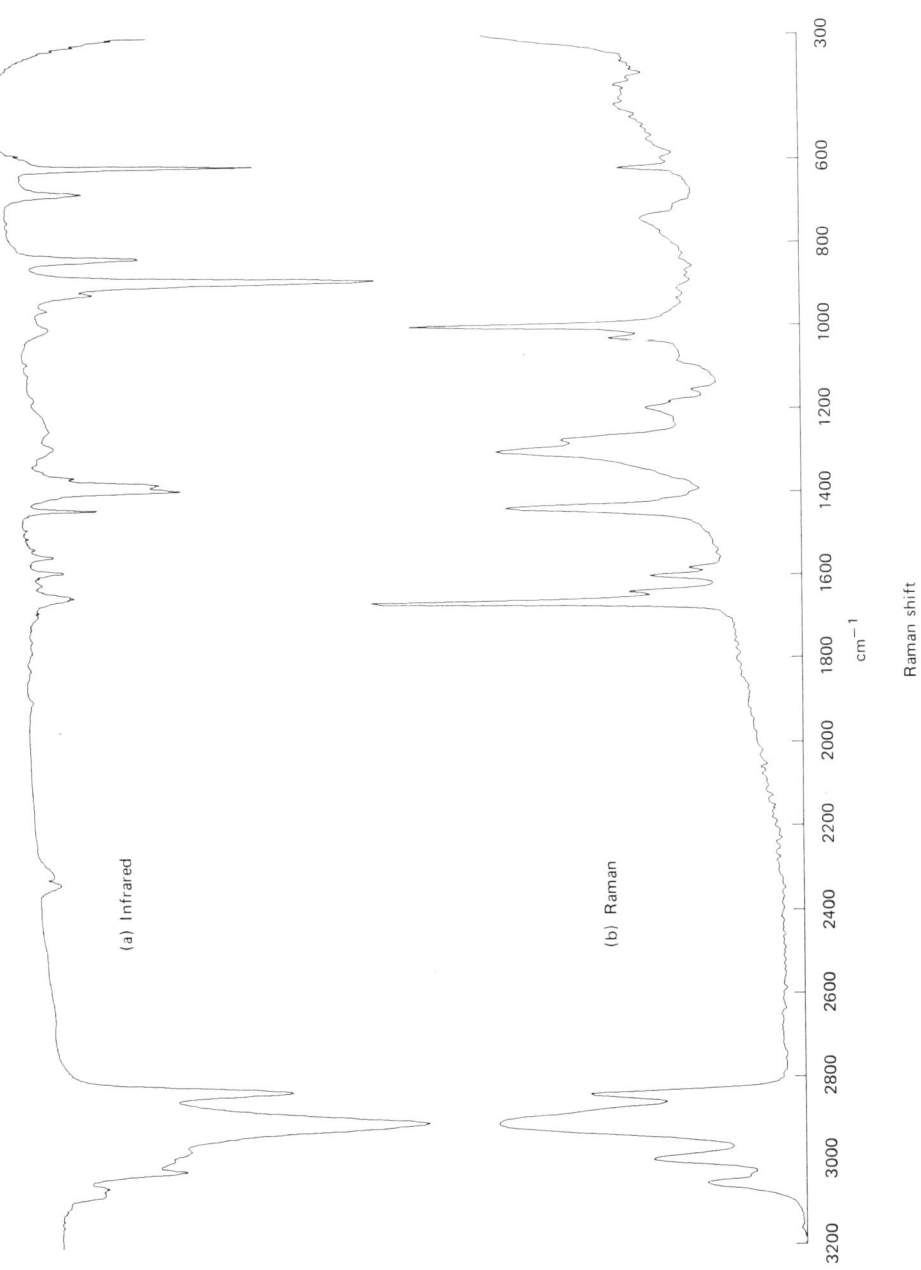

Fig. 2.12 — The infrared and Raman spectrum of styrene/butadiene rubber.

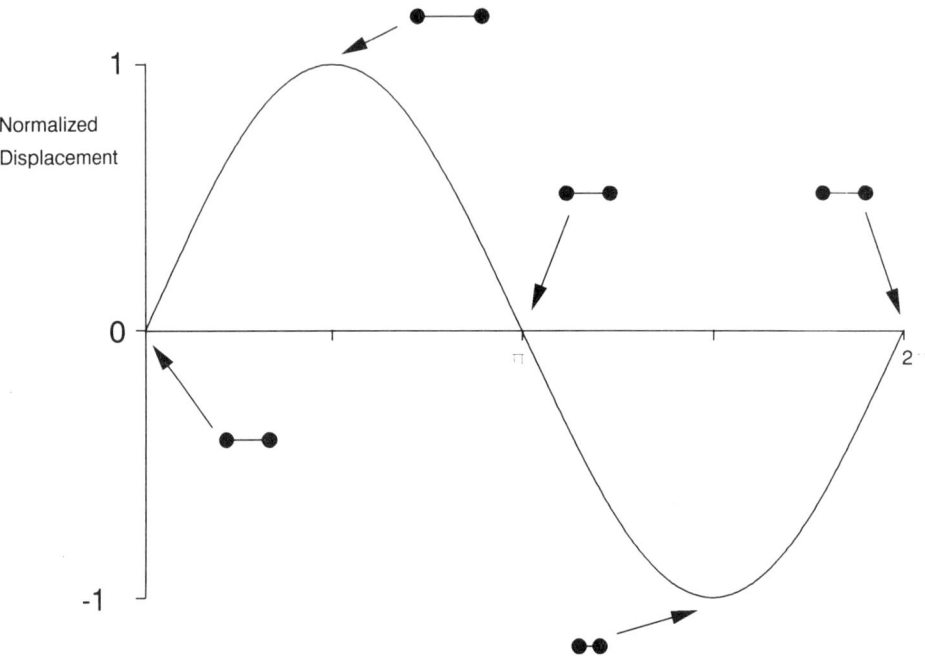

Fig. 2.13 — The progress through a vibration is usually described by using phase angles. This is a graph of displacement from the equilibrium position as a function of phase angle.

Carbon dioxide is a linear triatomic molecule, and hence has four fundamental modes of vibration, $(3n - 5)$. These are shown in Fig. 2.14. There are two linkages and thus two stretching vibrations; one symmetric (v_1) and the other asymmetric (v_2). There are also two bending vibrations, v_3 and v_4, and these are known as a degenerate pair. The two bending vibrations are orthogonal to one another but they are experimentally indistinguishable. The values of α against q for the fundamental vibrational modes of CO_2 are drawn in Fig. 2.15(a). Thus, although in *all* modes, α varies with q, only v_1 is Raman active because it alone possesses a non-zero gradient of $(\delta\alpha/\delta q)_0$.† An alternative representation of molecular polarizability in three dimensions is the polarizability ellipsoid. The polarizability ellipsoid is constructed by placing the centre of the molecule at the origin of a cartesian axis system, and plotting a three dimensional graph of $1/\sqrt{\alpha_i}$, where α_i is the polarizability along the line between the point i and the origin. The ellipsoid can be drawn as a function of phase angle, q, and Fig. 2.16 shows this for the Raman active v_1 mode of carbon dioxide. Turning our attention to dipoles, here we have to consider $\delta\mu/\delta q$ (strictly $\delta\mu_{x,y,z}/\delta q$) for each mode, and in Fig. 2.15(b) an equivalent plot is given, showing that v_1 is silent in the infrared, whilst v_3 and $v_{2,4}$ are infrared active. In centrosymmetric molecules, i.e. molecules which have a centre of inversion, a rule of mutual exclusion applies, wherein no mode occurs in *both* the infrared and Raman spectra.

† A rule of thumb is that if the molecule when fully distorted in one sense is a mirror image of that in the opposite sense, the mode is silent in the Raman effect.

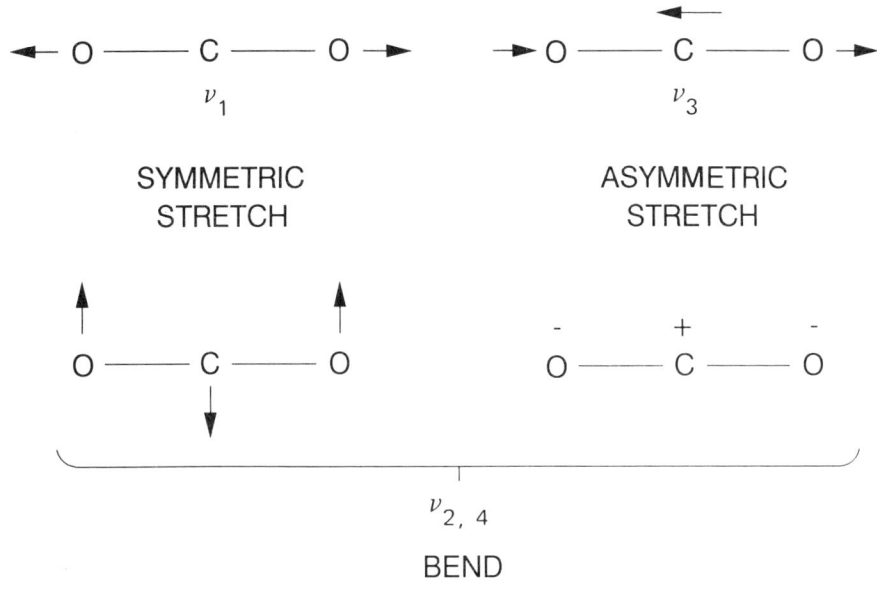

Fig. 2.14 — The fundamental vibrational modes of carbon dioxide.

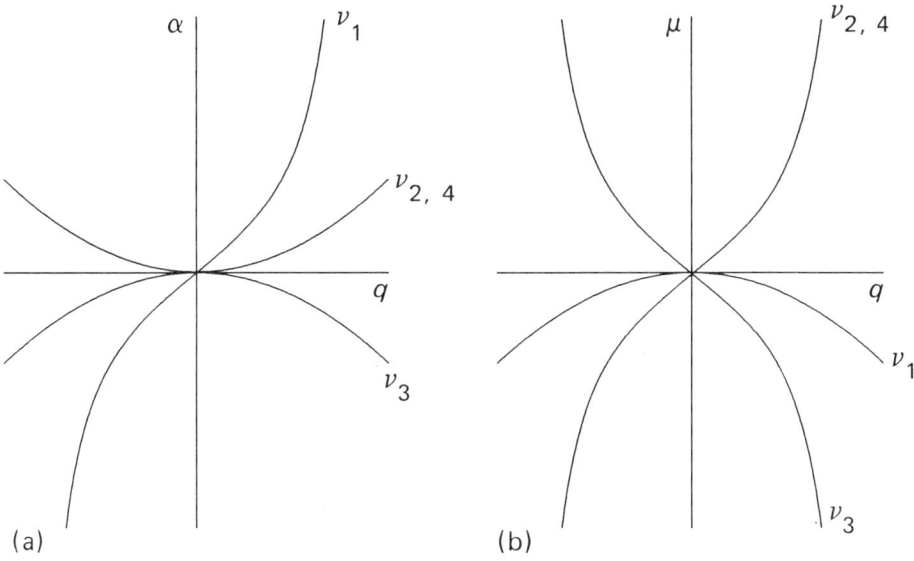

Fig. 2.15 — Graphs of (a) polarizability α, and (b) dipole moment μ for the vibrational modes of carbon dioxide as a function of phase angle.

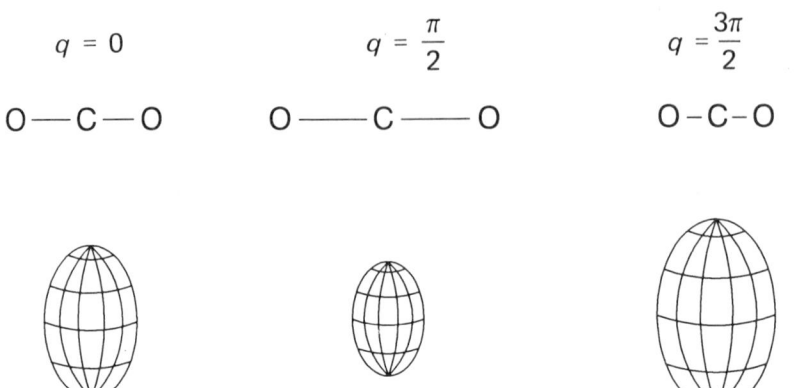

Fig. 2.16 — The polarizability ellipsoid is defined by $1/\sqrt{\alpha_i}$. As α is greater along the bond, the ellipsoid is thinner in this direction but broader orthogonal to it. As the molecule vibrates, the value of α_i in any direction changes and the shape and size of the polarizability ellipsoid varies. Here we show the change in the polarizability ellipsoid with phase angle for the 1 mode of carbon dioxide.

When more complex molecules are involved, an alternative method of classifying fundamental vibrations is required. Fortunately some molecules possess symmetry properties which allow them to be categorized into 'point groups' by using Group Theory. A point group is a collection of symmetry elements such as centre of inversion, reflection planes or axes of rotation. Molecules with the same set of symmetry elements are given the same point group. Water and pyridine both possess C_{2v} symmetry as shown in Fig. 2.17.

Point groups are labelled according to the Schönflies notation, which generally consists of a capital letter, a number and a small letter, e.g. D_{2h}. The number indicates the order of the principal axis of rotation. The capital letter D denotes an n-fold principal axis with n two-fold axes perpendicular to it. If these two-fold axes are absent than the letter C is used. The small letter h shows the presence of horizontal reflection plane, v denotes a vertical plane for C classes and d represents the dihedral plane in the D classes. There are also a few special point groups: tetrahedral T_d, octahedral O_h, and icosahedral I_h.

When the point group of a molecule has been determined, character tables can be used to label fundamental modes of vibration according to their symmetry. A character table lists all the symmetry elements which relate to a particular point group. An example of the C_{2v} character table is shown in Fig. 2.18.

Group theory not only allows us to label the fundamental modes, using irreducible representations, but also to predict infrared and Raman activity. This can be demonstrated by considering the benzene molecule. Benzene has D_{6h} symmetry. Some of its symmetry elements are shown in Fig. 2.19

$E, C_6, C_3, C_2, i, \sigma_h, \sigma_v, \sigma_d$

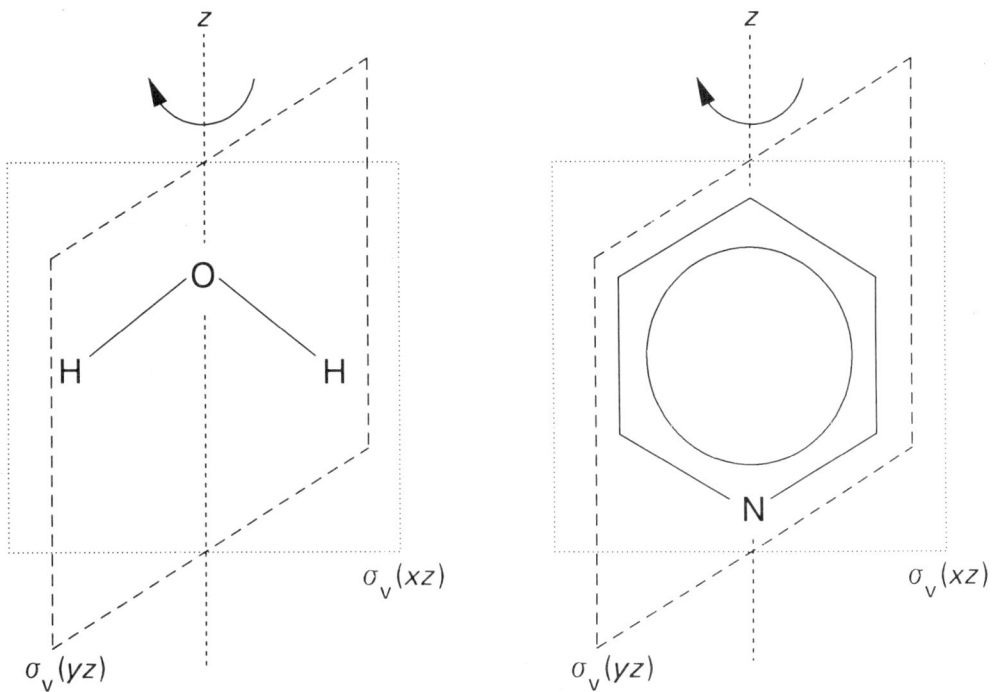

Fig. 2.17 — Water and pyridine possess the same symmetry properties — the identity, E, a two-fold axis of rotation, C_2, and two vertical reflection planes $\sigma_v(xz)$ and $\sigma_v(yz)$. Thus they both belong to the point group C_{2v}.

C_{2v}	E	C_2	$\sigma_v(xz)$	$\sigma'_v(yz)$		
A_1	1	1	1	1	z	x^2, y^2, z^2
A_2	1	1	−1	−1	R_z	xy
B_1	1	−1	1	−1	x, R_y	xz
B_2	1	−1	−1	1	y, R_x	yz

Fig. 2.18 — The C_{2v} character Table.

This molecule has 30 fundamental vibrations, i.e. $3(12) - 6$ and these are shown in Fig. 2.20.

They are labelled v_1, v_2, etc., in order of decreasing frequency within each type of irreducible representation. The character table for the D_{6h} point group is given in Fig. 2.21.

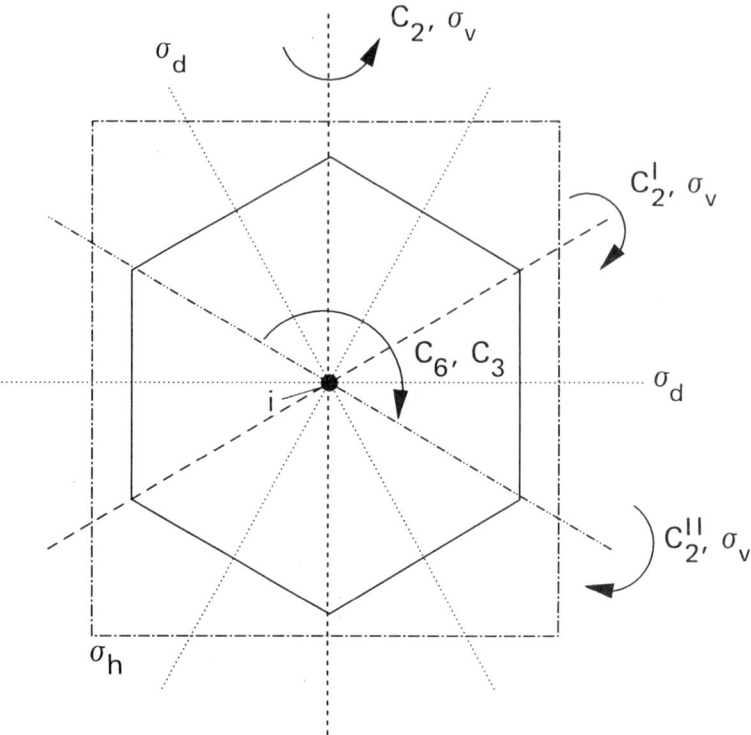

Fig. 2.19 — Benzene belongs to the D_{6h} point group and has many symmetry properties — E, $2C_6$, $2C_3$, $3C_2$, i, $2S_3$, $2S_6$, σ_h, $3\sigma_d$, $3\sigma_v$. The C_6 and C_3 axes are perpendicular to the page, as are the vertical and dihedral reflection planes. The horizontal reflection plane lies in the plane of the paper and the centre of inversion is at the centre of all of these planes.

A vibration is infrared active if it has the same irreducible representation as a component of dipole moment. These are shown by the x, y and z labels in the penultimate column of the character table. Thus, in benzene the A_{2u}, and E_{1u} vibrational modes are infrared active. Similarly, a vibration is Raman active if it has the same irreducible representation as a component of the polarizability. These are represented by x^2, y^2, z^2, xy, xz, yz or a combination of products such as $x^2 - y^2$, indicated in the final column of the character table. The Raman active modes in benzene are A_{1g}, E_{1g}, and E_{2g}. It should be noted that the A_{2g}, B_{1g}, B_{2g}, A_{1u}, B_{1u}, B_{2u} and E_{2u} modes are neither infrared nor Raman active. The principle of mutual exclusion is demonstrated here; the modes which are Raman active are not infrared active, and this can be seen directly from the character table. Figure 2.22 shows the FTIR and FT Raman spectra of benzene. The vibrational modes have been labelled.

Group theory is limited to molecules that are small and symmetric. Most molecules of interest to the analyst are neither. Consider a typical polymer, polytetrafluoroethylene, with a typical molecular weight of 10 000 000. This has around 600 000 atoms per molecule and thus 1 800 000 fundamental modes of vibration. At first sight this may seem disastrous, as we might expect the spectrum to

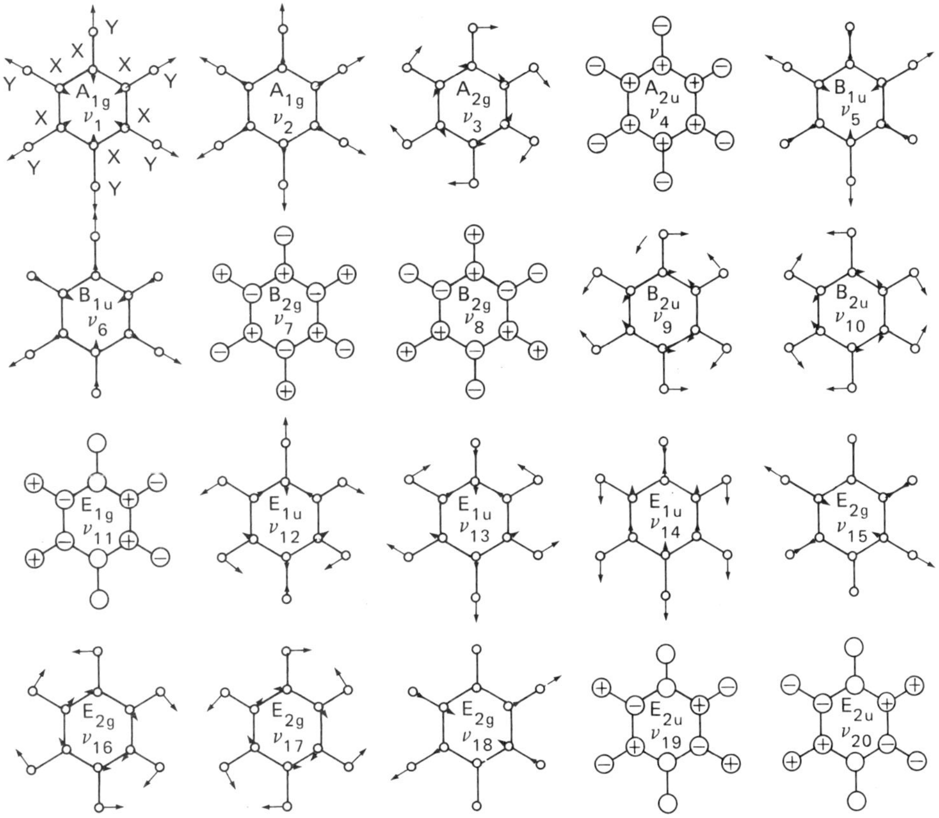

Sec. 2.4] **Infrared and Raman activity** 37

Fig. 2.20 — The fundamental vibrational modes of benzene.

be completely featureless consisting of a similar number of overlapping bands. However, in reality vast numbers of these modes have almost identical frequencies. Figure 2.23 shows a 'density of states' plot for polytetrafluoroethylene [2]. This is a plot of the number of modes of a given frequency *vs.* frequency. It is not a predicted spectrum since it says nothing about infrared or Raman activity. However, it does demonstrate that vibrational spectroscopy can be applied to large and complex molecules, and that discrete sharp spectra can in principle be obtained (rather than a continuum).

When the spectra of molecules containing different functional groups are studied, it is found that they give rise to infrared and Raman bands at specific frequencies. For instance, if several molecules which all contain an −OH group are studied, their spectra show a vibrational band near $v = 3500$ cm^{-1} and another around 1650 cm^{-1}. These are conventionally assigned to −OH stretching and bending vibrations. The motions can be viewed simplistically as shown in Fig. 2.24.

Strictly, all or nearly all of the other atoms within the molecule are involved in the vibration, but their contribution can be ignored because their vectors are very small. The frequency of the vibration is almost exclusively dependent on the magnitude of

D_{6h}	E	$2C_6$	$2C_3$	C_2	$3C_2'$	$3C_2''$	i	$2S_3$	$2S_6$	σ_h	$3\sigma_d$	$3\sigma_v$		
A_{1g}	1	1	1	1	1	1	1	1	1	1	1	1		x^2+y^2, z^2
A_{2g}	1	1	1	1	-1	-1	1	1	1	1	-1	-1	R_z	
B_{1g}	1	-1	1	-1	1	-1	1	-1	1	-1	1	-1		
B_{2g}	1	-1	1	-1	-1	1	1	-1	1	-1	-1	1		
E_{1g}	2	1	-1	-2	0	0	2	1	-1	-2	0	0	(R_x, R_y)	(xz, yz)
E_{2g}	2	-1	-1	2	0	0	2	-1	-1	2	0	0		(x^2-y^2, xy)
A_{1u}	1	1	1	1	1	1	-1	-1	-1	-1	-1	-1		
A_{2u}	1	1	1	1	-1	-1	-1	-1	-1	-1	1	1	z	
B_{1u}	1	-1	1	-1	1	-1	-1	1	-1	1	-1	1		
B_{2u}	1	-1	1	-1	-1	1	-1	1	-1	1	1	-1		
E_{1u}	2	1	-1	-2	0	0	-2	-1	1	2	0	0	(x, y)	
E_{2u}	2	-1	-1	2	0	0	-2	1	1	-2	0	0		

Fig. 2.21 — The character table of the D_{6h} point group.

the force constant, k_{OH}. The value of a force constant depends on the electronic structure of the vibrating bond, which may be slightly perturbed by different side groups. For example, all molecules containing a C=O group have a characteristic frequency in the range 1600–1800 cm^{-1}. The precise frequency, however, depends on whether the C=O is present in a lactam, amide, acid, etc. The force constant $k_{C=O}$ can be related quite elegantly to the electron withdrawing or donating properties of side groups R_1 and R_2 in compounds of the form R_1COR_2. Thus if R_1 is a highly electronegative species such as a chlorine atom and R_2 is a methyl group, the chlorine will induce electrons towards itself, and this increases the π character of the bond. Methyl groups have a weakly donating property and hence when both R_1 and R_2 are methyl groups the bond order is lower than in the previous case. Hence we expect $k_{C=O}$ to be larger, and the band frequency to be higher, in CH_3COCl than in acetone (CH_3COCH_3). The band frequencies obtained experimentally are 1805 cm^{-1} and 1712 cm^{-1} respectively.

The observation that particular fragments of a molecule give rise to vibrations in narrow, clearly identifiable frequency domains is invaluable to chemists as an analytical tool and is usually referred to as the 'Group Frequency Approach'. The use of the group frequency approach is widespread and many tables exist relating functional groups and vibrational frequencies. However, when these correlations are used, it must be remembered that the approach is only an approximation. The observation of a band in a characteristic frequency range can be highly indicative of the presence of a particular chemical group. However, what is, in fact, being

Sec. 2.4] **Infrared and Raman activity** 39

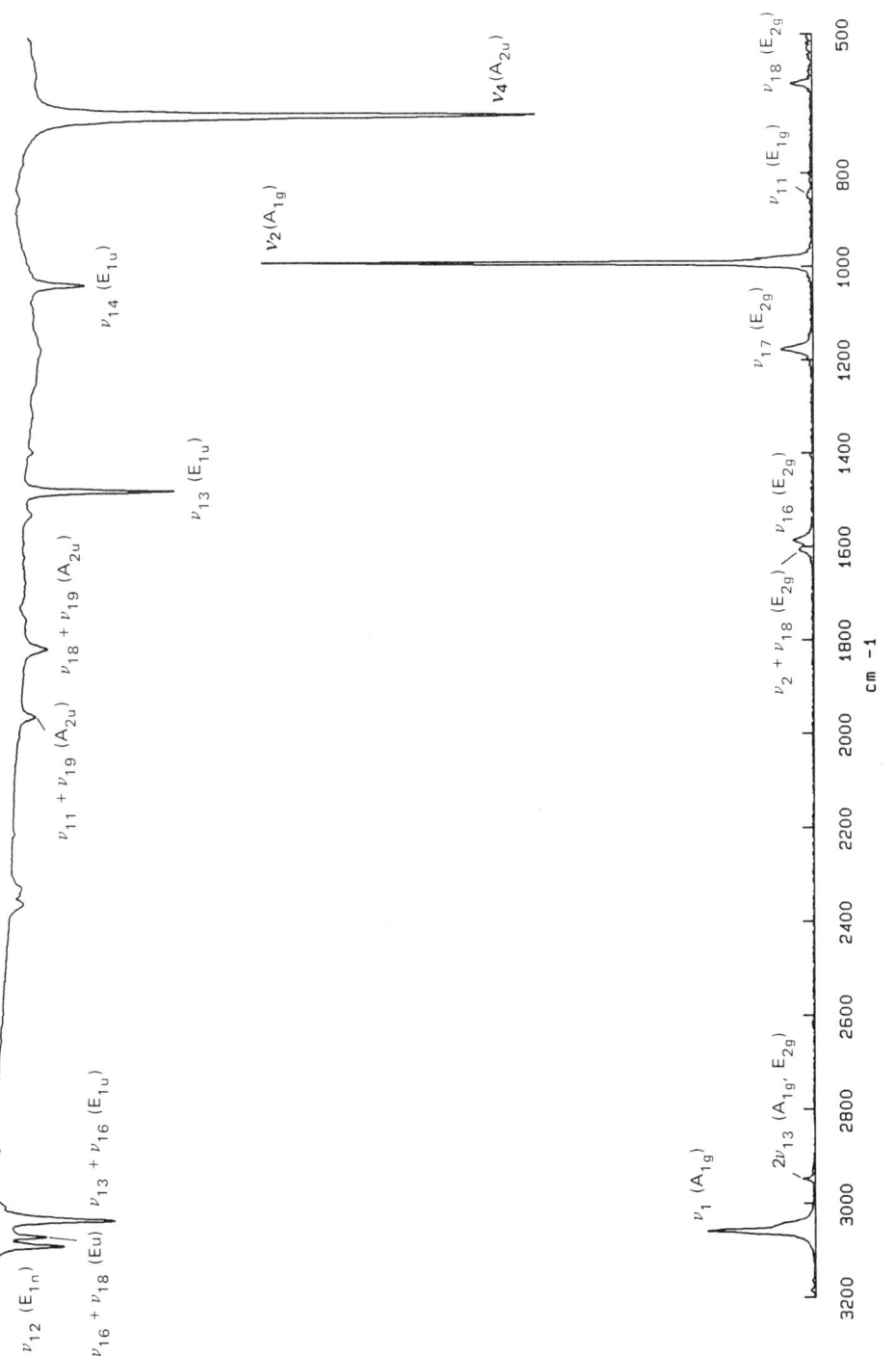

Fig. 2.22 — The infrared and Raman spectra of benzene. The bands are labelled with the corresponding mode number and irreducible representation. The ν_{11} band in the Raman spectrum is very weak and does not appear unless higher laser powers are used.

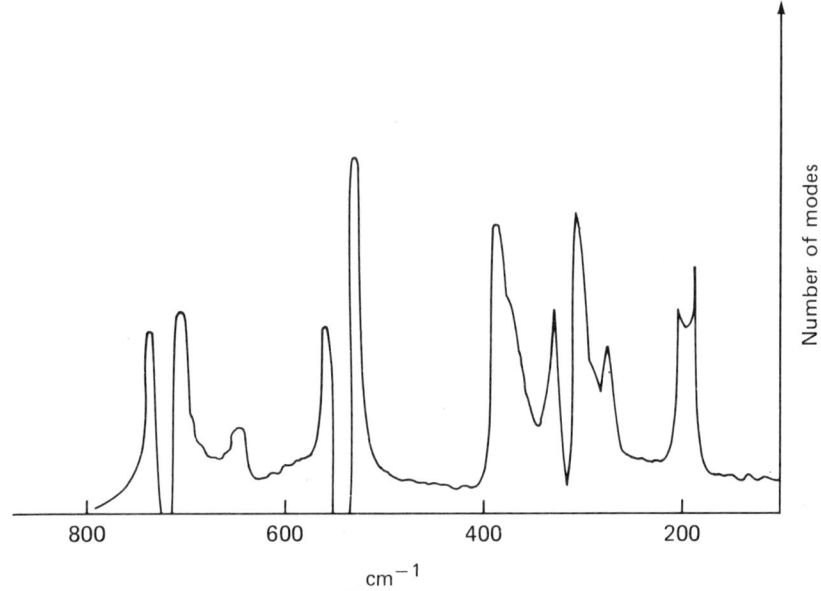

Fig. 2.23 — A density of states plot for polytetrafluoroethylene (PTFE) showing the number of modes of a given frequency *vs.* frequency.

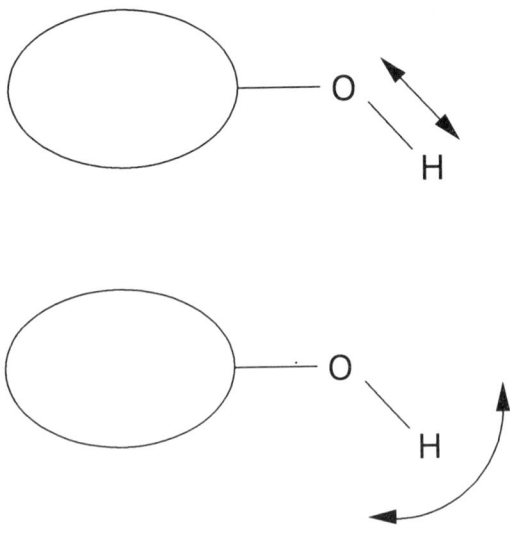

Fig. 2.24 — A simple view of the OH stretching and bending vibrations in a large molecule. The contribution to the vibration from the rest of the molecule is very small.

measured is the frequency of a fundamental mode. If a molecule is composed of atoms of similar masses, a vibration thought to be due to a characteristic group may well be due to a mode involving many atoms in the molecule. As a result, the frequency of the mode may be atypical of the group and the analysis faulty. The solution to this problem lies in looking for more than one indication of the presence of a particular functional group. Thus, for example, in confirming the presence of a benzene ring and its substitution one always checks several regions of the spectrum to see that the correlations all agree. This is where the combined use of infrared and Raman spectra is particularly useful. Figure 2.25 shows the infrared and Raman

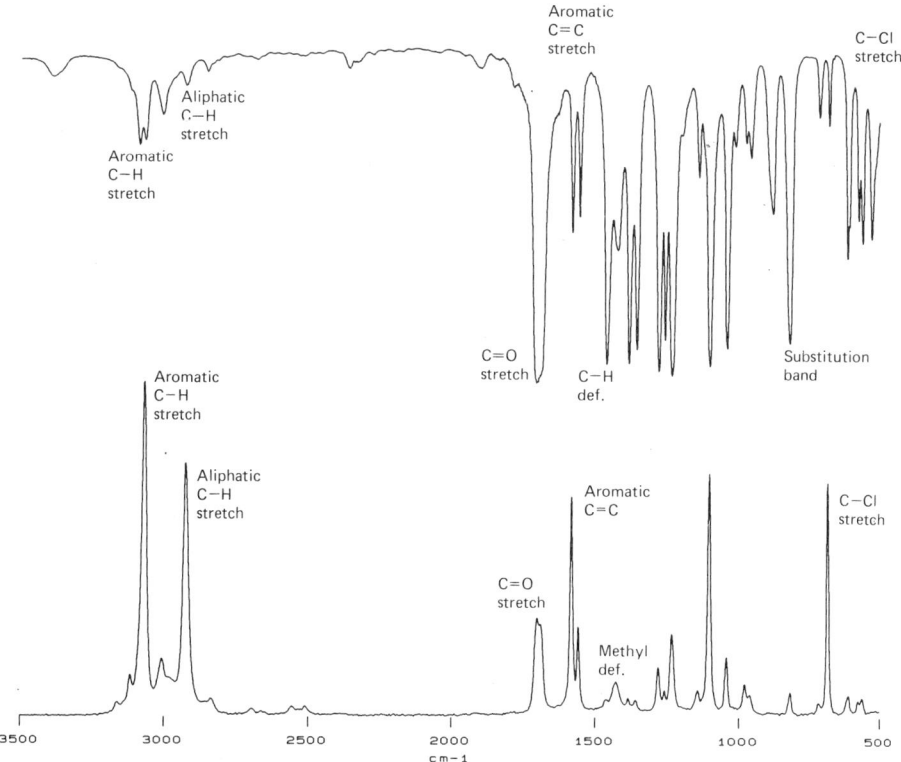

Fig. 2.25 — The infrared and Raman spectra of 2,5-Dichloroacetophenone.

spectra of 2,5-dichloroacetophenone, in which the main functional groups have been assigned.

With the advent of computer-archiving technology, many libraries of spectra have been accumulated and these can now assist with the identification of unknown materials.

The vibrational frequencies of a molecule are not solely related to the molecule in isolation, but are also affected by its environment. Thus, infrared and Raman spectra can provide information on a wide range of physical properties. For instance, evidence relating to intermolecular interactions between solvents and solutes, crystallinity and crystalline structure, orientation, physical state, temperature and pressure can be found in vibrational spectra.

2.5 THE EFFECT OF POLARIZATION AND SAMPLE ORIENTATION ON INFRARED SPECTRA

We have already seen that for a vibration to be infrared active, there must be a change in dipole as the vibration proceeds. Infrared absorption results when the oscillating electric vector of radiation, E, interacts with the oscillating dipole vector, μ, of the molecule. The intensity of the absorption is related to the angle, θ, between these two vectors as shown in figure 2.26.

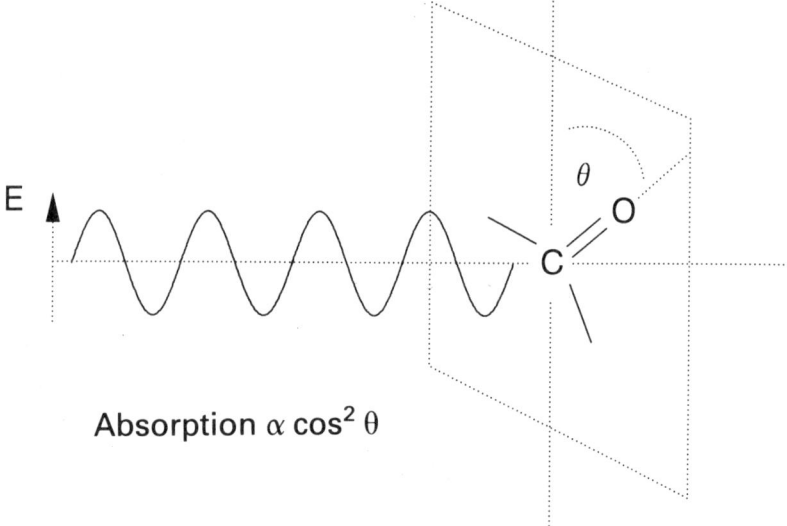

Fig. 2.26 — The angular dependence of the interaction between the electric field of light and an oscillating dipole.

Normally, the dipole vectors of the molecules are randomly distributed and an average absorption is observed. However, when a perfectly oriented sample is studied, such as a single crystal, then the dipole vectors are exactly aligned. If the infrared source is now polarized parallel to the dipole vector, μ, ($\theta = 0$) maximum absorption occurs. If, however, the source is polarized perpendicular to the dipole vector ($\theta = 90°$), no absorption occurs. At intermediate angles of θ, the absorption varies as $\cos^2\theta$. This phenomenon is known as dichroism.

Similar effects can be observed for partially oriented films, such as nylon and polyethyleneterephthalate (PET). Figure 2.27 shows the different PET spectra

Sec. 2.6] **Effect of polarization and sample orientation on Raman spectra**

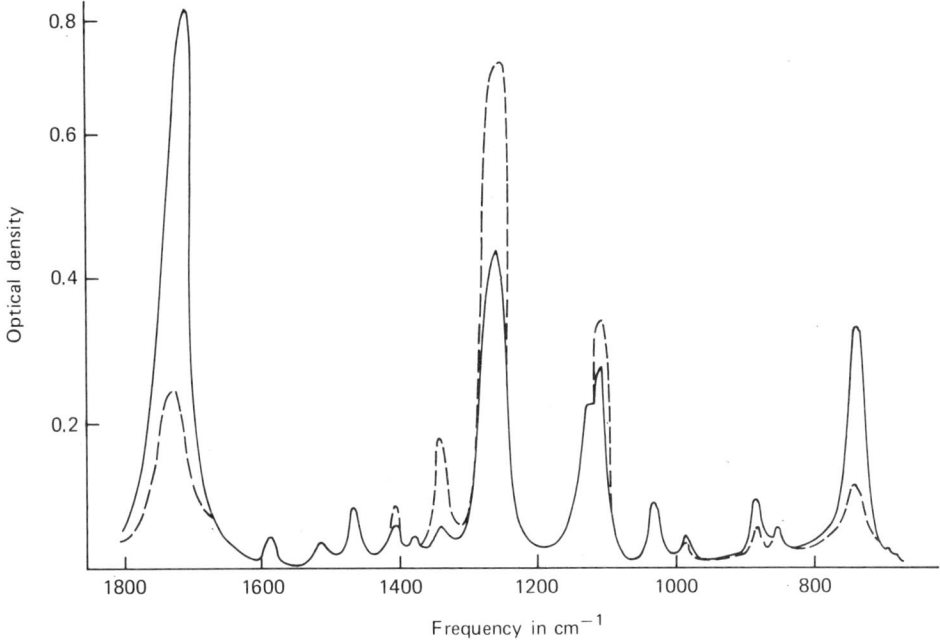

Fig. 2.27 — The infrared spectrum of an oriented PET bottle obtained with polarized infrared radiation. The solid and dashed lines respectively show the spectra obtained with the electric vector of the infrared radiation parallel and orthogonal to the length of the bottle.

obtained when the orientation axis and the polarization axis are (A) parallel and (B) perpendicular. The C=O bond is continually producing an oscillating dipole, but when it is perpendicular to the electric vector of the infrared radiation they cannot mutually interact. If the orientation of the sample was perfect we would expect to see complete extinction of the C=O band. Thus, this method can be used to determine the degree of orientation in a sample [4]. For each band in the spectrum it is possible to determine a 'dichroic ratio', D, i.e. the ratio of the intensities observed in parallel and perpendicular arrangements.

$$D = \frac{I_\parallel}{I_\perp} \tag{2.18}$$

2.6 THE EFFECT OF POLARIZATION AND SAMPLE ORIENTATION ON RAMAN SPECTRA

In the section on infrared spectroscopy, we saw how the intensity of absorption is related to the relative orientation of the electric vector of the infrared radiation and that of the oscillating dipole in the sample. Mathematically this is expressed as the dot product of the radiation electric vector and the dipole vector. The situation is more complex in Raman spectroscopy, but more information can be obtained by performing polarized experiments.

If we use a cartesian axis system to describe three dimensional space, the electric vector of a ray of light can be resolved into the three constituent vectors E_x, E_y and E_z, which point along the x, y and z axes respectively. We can use similar vectors P_x, P_y and P_z to describe the dipole induced by this light ray when it irradiates a sample. These two sets of vectors are related by a 3×3 matrix called the polarizability tensor by the following equation:

$$\begin{bmatrix} P_x \\ P_y \\ P_z \end{bmatrix} = \begin{bmatrix} \alpha_{xx} & \alpha_{xy} & \alpha_{xz} & E_x \\ \alpha_{yx} & \alpha_{yy} & \alpha_{yz} & E_y \\ \alpha_{zx} & \alpha_{zy} & \alpha_{zz} & E_z \end{bmatrix} \qquad (2.19)$$

This matrix equation can be multiplied out to give expressions for P_x, P_y and P_z:

$$P_x = \alpha_{xx} E_x + \alpha_{xy} E_y + \alpha_{xz} E_z \qquad (2.20)$$
$$P_y = \alpha_{yx} E_x + \alpha_{yy} E_y + \alpha_{yz} E_z \qquad (2.21)$$
$$P_z = \alpha_{yx} E_x + \alpha_{zy} E_y + \alpha_{zz} E_z \qquad (2.22)$$

Thus it is clear that the component of the induced dipole in the x direction does not depend solely on E_x, but on E_y and E_z as well. Strictly, Raman scattering should be described by the polarizability derivative tensor where the elements are of the form $\delta \alpha_{nm}/\delta q$. However, when this tensor is multiplied out, the equations derived have the same form, and for the purposes of this discussion we will use the polarizabilities, α_{nm}.

We can define the direction of the x, y and z axes arbitrarily, but it makes sense to utilize the geometry of the Raman spectrometer. With the axes shown in Fig. 2.28, we could direct the laser along the y axis, collect Raman light along the x axis and place the sample at the origin. The analyser, which is a piece of polaroid, is placed before the collection lens and allows light of only a particular polarization to enter the spectrometer. For certain samples, different spectra can be obtained by varying the position of the analyser. This is true even for unoriented samples such as liquids. It is also important to place a polarization scrambler after the analyser, because the spectrometer may be more transmissive to light of one polarization than another. This is less important in interferometers, since they are much less polarization-sensitive than grating monochromators [3].

Now suppose that a non-polarized light source is directed along the y axis. This will have non-zero electric vector components E_x and E_z. E_y will be zero. This will induce electric dipoles in the sample according to the following equations:

$$P_x = \alpha_{xx} E_x + \alpha_{xz} E_z \qquad (2.23)$$
$$P_y = \alpha_{yx} E_x + \alpha_{yz} E_z \qquad (2.24)$$

Sec. 2.6] Effect of polarization and sample orientation on Raman spectra

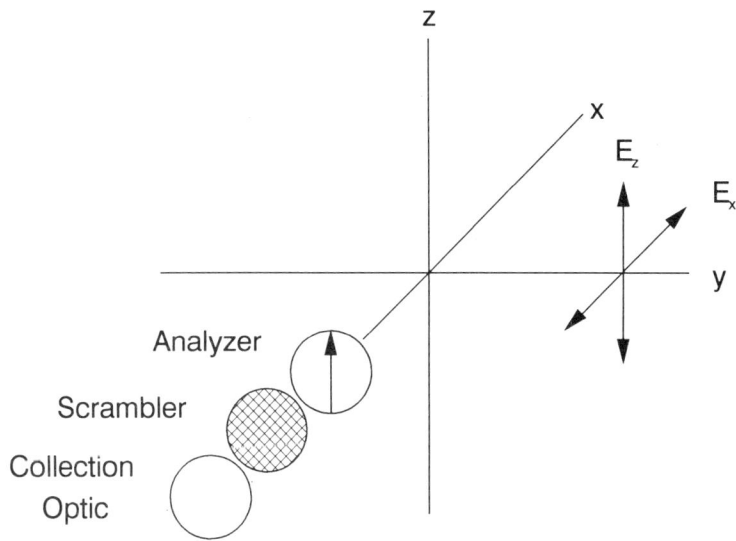

Fig. 2.28 — Apparatus and experimental geometry for 90° polarized Raman experiment.

$$P_z = \alpha_{zx} E_x + \alpha_{zz} E_z \tag{2.25}$$

An oscillating dipole cannot scatter light along the same axis as itself. Thus, if we observe the sample along the x direction, only light due to P_y and P_z will be collected, so we can neglect Eq. (2.23). The intensity of the light scattered is proportional to the square of the induced dipoles. Thus the intensity of light scattered with its electric vector pointing in the y and z axes is given by the following expressions:

$$I_y = \alpha_{yx}^2 E_x^2 + \alpha_{yz}^2 E_z^2 \tag{2.26}$$
$$I^z = \alpha_{zx}^2 E_x^2 + \alpha_{zz}^2 E_z^2 \tag{2.27}$$

We can now choose whether we want to measure scattered light with its oscillating dipole in the y or z axes by rotating the analyser accordingly. In this way, two spectra can be recorded, and information about the symmetry of Raman active vibrations can be obtained by comparing the intensities of the same band in the two spectra. A **depolarization ratio** is calculated for each band according to the following expression:

$$\rho_n = \frac{I_\perp}{I_\parallel} = \frac{I_y}{I_z} \tag{2.28}$$

$$= \frac{\alpha_{yx}^2 E_x^2 + \alpha_{yz}^2 E_z^2}{\alpha_{zx}^2 E_x^2 + \alpha_{zz}^2 E_z^2} \tag{2.29}$$

Here ρ_n denotes that this is the depolarization ratio obtained with a non-polarized excitation source and as such is termed 'natural light'. Hence both E_x and E_z are non-zero. If a polarized laser is used, for example with its electric vector pointing only in the z direction, the depolarization becomes:

$$\rho_p = \frac{\alpha_{yz}^2 E_z^2}{\alpha_{zz}^2 E_z^2} \tag{2.30}$$

since

$$E_x = E_y = 0, \quad E_z \neq 0$$

Expressions (2.24) and (2.25) apply only to an oriented sample where the molecules are fixed in space. For isotropic materials such as a liquid or a gas where the sample molecules are constantly tumbling, an average value of the polarizability has to be used. This is expressed by two quantities; (i) $\bar{\alpha}$, the average value of α_{xx}, α_{yy}, α_{zz} and (ii) β, a quantity which measures the anisotropy in the polarizability. ρ_n and ρ_p can be expressed in terms of $\bar{\alpha}$ and β accordingly to Eqs. (2.31) and (2.32).

$$\rho_n = \frac{6\beta^2}{45\bar{\alpha}^2 + 7\beta^2} \tag{2.31}$$

$$\rho_p = \frac{3\beta^2}{45\bar{\alpha}^2 + 4\beta^2} \tag{2.32}$$

These equations demonstrate a rather surprising result. If an isotropic sample such as a liquid is irradiated with a non-polarized laser, the Raman scattered light can be partially or completely polarized. It is possible to use polarized Raman experiments to assign different spectral bands to different vibrations by comparing experimental results with those predicted by group theory. Raman experiments performed before the advent of lasers used discharge lamps as excitation sources and these produced non-polarized light. Accordingly the depolarization ratios reported in the early literature are ρ_n and, as the following paragraph explains, have a limiting value of 6/7. Polarized excitation sources such as lasers produce depolarization ratios, ρ_p, with a limiting value of 3/4.

By measuring the value of ρ_n or ρ_p for a Raman band, it is possible to derive information about the relative values of $\bar{\alpha}$ and β, and these allow us to deduce symmetry information about the vibration causing the bands. For instance, for a mode which is degenerate or not totally symmetric, $\bar{\alpha} = 0$ and $\beta \neq 0$. These modes give rise to Raman bands with values of $\rho_n = 6/7$ and $\rho_p = 3/4$ and are said to be depolarized (or sometimes 'totally depolarized'). Totally symmetric modes of molecules which have cubic or icosahedral symmetry have $\beta = 0$ and hence $\rho_n = 0$ and $\rho_p = 0$. The polarizability ellipsoids of these modes are perfectly spherical and they are described as being 'totally polarized'. The totally symmetric modes of molecules of lower symmetry produce intermediate values of ρ_n and ρ_p such that $0 \leq \rho_n < 6/7$, $0 \leq \rho_p < 3/4$ since $\bar{\alpha} \neq 0$ and $\beta \neq 0$.

So far we have discussed polarized Raman scattering from isotropic samples. When a perfectly oriented sample, such as a single crystal, is studied it is possible to obtain spectra which are due to just one element of the Raman scattering tensor. In Eq. (2.30), it was shown that if a polarized laser was used as the excitation source, $E_x = E_y = 0$ and $E_z \neq 0$, then

$$I_y = \alpha_{yz}^2 E_z^2 \tag{2.33}$$

$$I_z = \alpha_{zz}^2 E_z^2 \tag{2.34}$$

By varying the experimental geometry, different spectra can be collected each of which 'tunes in' to one particular element of the tensor. Only bands due to vibrations of an irreducible representation with this element in the fourth column of the character table will be recorded. These experiments are described by 'Porto' nomenclature [5]:

$$a\ (b\ c)\ d$$

$a =$ direction of illumination
$b =$ direction of polarization of the laser
$c =$ direction of orientation of the analyser
$d =$ direction of view.

For example, the experiment producing a Raman signal of intensity given by Eq. (2.33) would be labelled $y(zy)x$. Similarly, for a perfectly oriented solid sample with D_{2h} symmetry, a $y(zz)x$ observation would record only the bands arising from the α_{zz} element and only A_g modes would appear in the spectrum. For a partially oriented sample such as a stretched polymer film or a liquid crystal it is often possible to establish the degree of order by using polarized Raman experiments [6,7].

In this chapter we have kept the number of references to original work deliberately small, favouring instead a list of alternative and complementary treatments to those given here.

REFERENCES

[1] M. Lapp, 'Flame Temperatures from Vibrational Raman Scattering', in *Laser Raman Gas Diagnostics*, p. 107, Plenum Press, New York, 1974.
[2] D. J. Cutler, P. J. Hendra and G. Fraser, 'Raman Spectra of Synthetic Polymers: A Review' in *Developments in Polymer Characterization*, Vol. 2, J. V. Dawkins, ed., Applied Science Publishers, 1980.
[3] D. J. Cutler, *Spectrochim. Acta*, 1990, **46A**, 131.
[4] A. Cunningham, I. M. Ward, H. A. Willis and V. Zichy, *Polymer*, 1974, **15**, 749.
[5] T. C. Damen, S. P. S. Porto and B. Tell, *Phys. Rev.*, 1966, **142**, 570.
[6] D. I. Bower and W. F. Maddams, *The Vibrational Spectroscopy of Polymers*, Cambridge University Press, Cambridge, 1989.

[7] D. I. Bower, *J. Polym. Sci. Polym. Phys. Ed.*, 1972, **10**, 2135.

BIBLIOGRAPHY

Molecular vibrations
C. N. Banwell, *Fundamentals of Molecular Spectroscopy*, 3rd Ed., McGraw-Hill, 1983.
N. B. Colthup, L. H. Daly and S. E. Wiberley, *Introduction to Infrared and Raman Spectroscopy*, Academic Press, New York, 1975.

Group Theory
A. Fadini and F.-M. Schnepel, *Vibrational Specvtroscopy: Methods and Applications*, Ellis Horwood, Chichester, 1989.
A. Vincent, *Molecular Symmetry and Group Theory*, Wiley, Chichester, 1977.

3
Conventional laser-Raman spectroscopy

The conventional laser-Raman system consists of a laser source illuminating the sample, followed by collection optics which gather the scattered light and pass it into the spectral analysis and sensitive detection system. Spectral analysis is most frequently carried out by using scanning multiple monochromators [1], but many varied alternatives are frequently found, including spectrographs with position sensitive detectors [2–4].

The fundamental problems in Raman spectroscopy are the very low intensity of the Raman radiation and the relatively high intensity of the light collected at the laser wavelength. As a result, the spectrometer must have superb sensitivity coupled with quite outstanding discrimination properties. In order to maximize sensitivity it is essential to use efficient optical systems carefully matched to each other, followed by sensitive detectors with low noise. The description that follows is not exhaustive, but will enable the reader to understand the logic behind the design of modern conventional instruments.

3.1 LASERS

A conventional Raman spectrometer 'views' a very small patch of the sample. A typical size is of the order of 2000×50 µm, or in a microscope instrument a spot of approximately 1.5 µm in diameter. The exact value depends on the instrument operating conditions. It is optically wasteful if the source is not focused exactly onto the viewed patch. To achieve this it is normal to use gas lasers (most frequently argon, krypton or helium–neon lasers), incorporating, where appropriate, prisms within the cavity to confine lasing to one emission line. The popular lines in current use are listed in Table 3.1, with outputs typically available.

Raman light can be generated by an excitation source of any wavelength. In order to study a wide variety of samples it is desirable to have several laser wavelengths available. This may enable the user to reduce problems of sample heating and fluorescence, by switching from one excitation source to an alternative. Instead of operating two or more lasers one can use a tuneable dye laser†. Although a range of

---

† Recently, a solid-state tuneable laser has appeared, based on a crystal of titanium sapphire. It covers the range from 700 to 1000 nm, when powered with an argon ion laser.

Table 3.1 — Useful laser emission lines for Raman spectroscopy

Laser	Wavelength (nm)	Wavenumber (cm^{-1})	Power (mw)
Argon ion[1]	454.5	22002.2	50
	457.9	21838.8	200
	465.8	21468.4	70
	472.7	21155.1	130
	476.5	20986.4	350
	488.0	20491.8	1000
	496.5	20141.0	400
	501.7	19932.2	200
	514.5	19436.3	1200
Krypton ion[2]	413.1	24207.2	60
	476.2	20999.6	60
	482.5	20725.4	45
	520.8	19201.2	90
	530.9	18835.9	200
	568.2	17599.4	200
	647.1	15453.6	500
	676.4	14784.2	120
	752.5	13289.0	100
	799.3	12510.9	30

1. Spectra Physics Model 2025-3, 3 W all lines, argon ion laser.
2. Spectra Physics Model 2025-11, 0.6 W all lines, krypton ion laser.

dye solutions is available, enabling users to cover a considerable spectral range, most spectroscopists confine themselves to the use of Rhodamine 6G resulting in the output characteristics shown in Fig. 3.1.

A laser does not produce perfectly monochromatic radiation, but rather monochromatic light contaminated with weak non-lasing lines which result from spontaneous emissions in the cavity. This light can be both reflected and Rayleigh scattered by the sample. Since Raman spectrometers are extremely sensitive, these very weak emissions appear as spurious bands in the Raman spectrum. However, since these lines are emitted over 2π steradians as they leave the laser, they fall in intensity rapidly with distance and therefore can be readily reduced by a spatial filter. The alternative or additional precaution is to incorporate a multilayer dielectric rejection filter, designed to pass only the laser line. Where multiple laser lines or a dye laser source are used, the number of filters required becomes impractical and a laser monochromator is employed. In Fig. 3.2 we show several typical filtering devices.

In the vast majority of analytical samples, some background fluorescence is encountered (see Fig. 1.3). In fact, the sensitivity of the instrument in such cases is usually limited by the disappearance of the Raman lines within the noise of the background. Multiple scanning and data averaging are extensively used to improve spectral quality. Since the background varies in intensity with the laser output, it is essential that the laser is stable, i.e. free from short term jitter in intensity. The laser should also be free from long-term drift in output power. Although lasers have steadily improved, they are far from perfect, and background noise derived from the laser can easily limit performance.

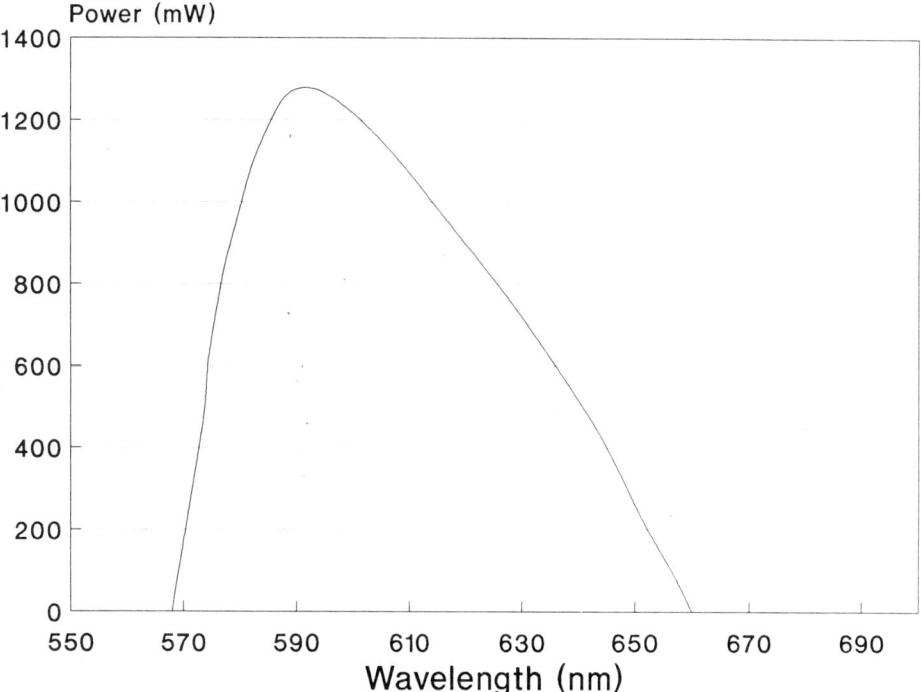

Fig. 3.1 — A graph of power *vs.* wavelength for a Rhodamine 6G dye laser.

3.2 COLLECTION OPTICS

More than 20 years ago, an analysis by Barrett and Adams [5] appeared in the literature concerning the 'front end' of a Raman spectrometer. The authors described the optimum optical arrangement shown in Fig 3.3.

The magnification ratio of the front end optics is usually about ×6. Thus the width of the viewed patch is 1/6 of that of the entrance slit. The height of the viewed patch is typically about 2000 μm, depending on the magnification ratio and the size of the detector. To fill the viewed patch, right-angled, focused laser illumination is used such that the laser beam runs parallel to the entrance slit. The collection optics determine the maximum amount of Raman light that will enter the instrument. This is best thought of in terms of solid angles, and is usually expressed in terms of an *f* number defined by the solid angle of the cone of light collected†. The collection system can be based on a simple or compound lens, or a Cassegrain type mirror system can be used. These alternatives are illustrated in Fig.3.4. The real point worth repeating is that the instrument entrance slit width is demagnified at the sample, i.e. the instrument views only a very small patch.

† Those who use cameras are familiar with *f* numbers. A lens of diameter, *d* and focal length, f_1 is said to have an *f* number equal to f_1/d.

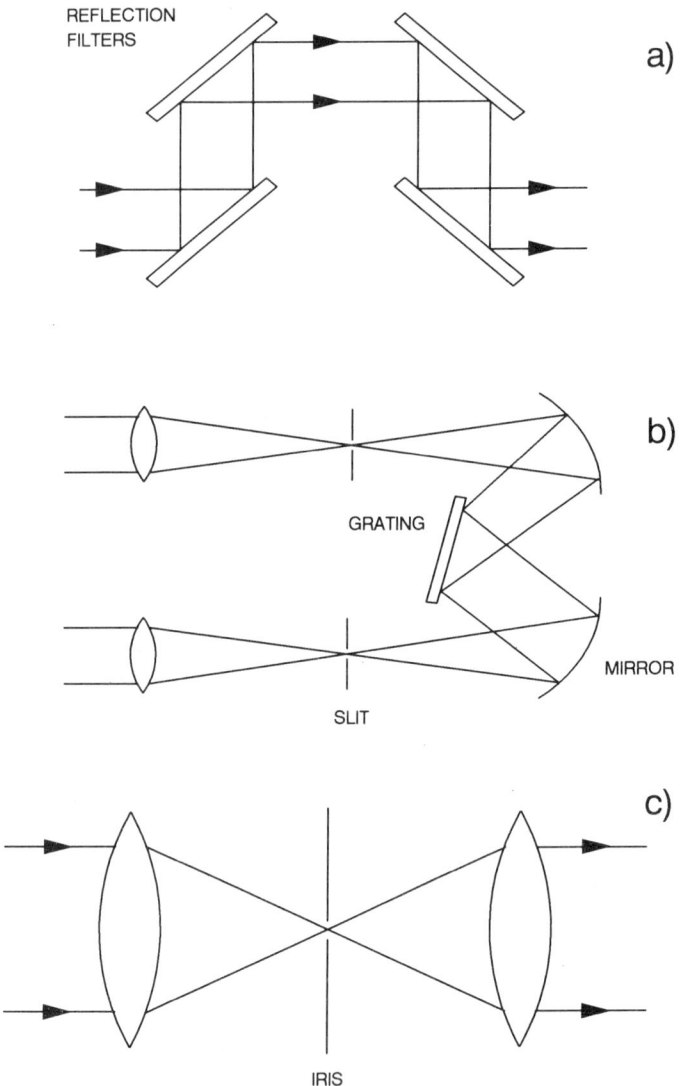

Fig. 3.2 — Typical laser filtering devices: (a) a Chevron arrangement using reflection filters, (b) a laser premonochromator, and (c) a spatial filter.

When a microscope is used at the front of a Raman spectrometer, relatively efficient back-scattering optical systems can be employed. A system frequently encountered is illustrated in Fig. 3.5. A few milliwatts of laser power (more than this would result in immediate incineration of the sample) is passed down the microscope and brought to a focus, confocal with and at the centre of the patch viewed, by the microscope objective. The reflected and scattered light is then passed to the spectrometer for processing. Although the microscope is a high aperture, efficient

Sec. 3.2] **Collection optics** 53

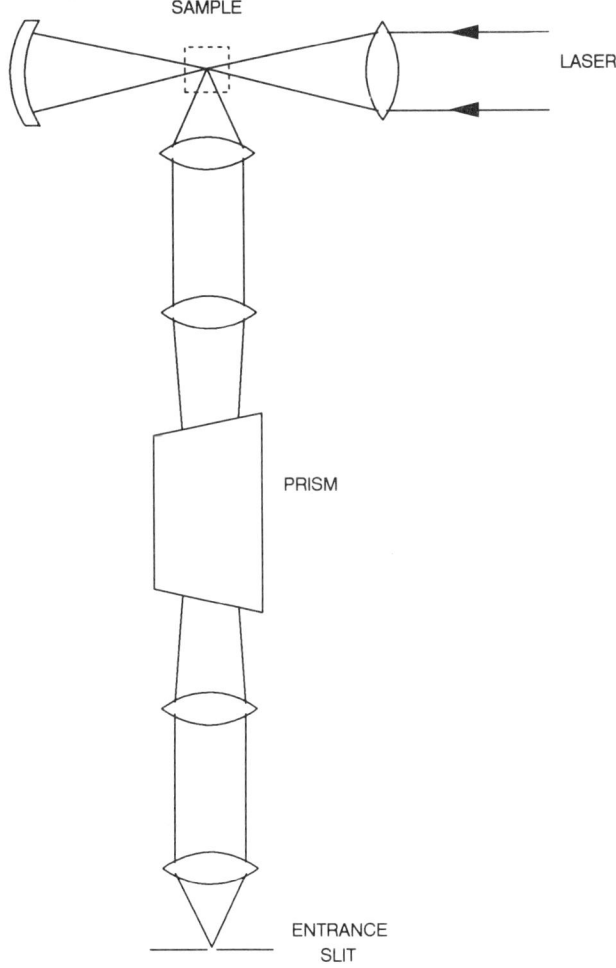

Fig. 3.3 — The Barrett and Adams front-end optical arrangement.

system, the patch of light falling on the entrance slit of the spectrometer is circular and hence does not match the entrance slit unless it is inordinately wide.

In the conventional mode of operation, i.e. incorporating right-angled illumination/collection, the laser spot and sample must be brought precisely to the point viewed by the monochromator. To ease this problem most manufacturers provide either a reverse- or back-illumination lamp, or a 'periscope viewer'. The former provides an image of the entrance slit at the viewed point. This can be located by using a piece of tracing paper, and then the laser is adjusted such that the two are superimposed. The sample is then introduced at this position (easier said than done!). Where a transparent sample is involved, the operator may look towards the slit, and search for the point of no parallax between the projected image of the slit

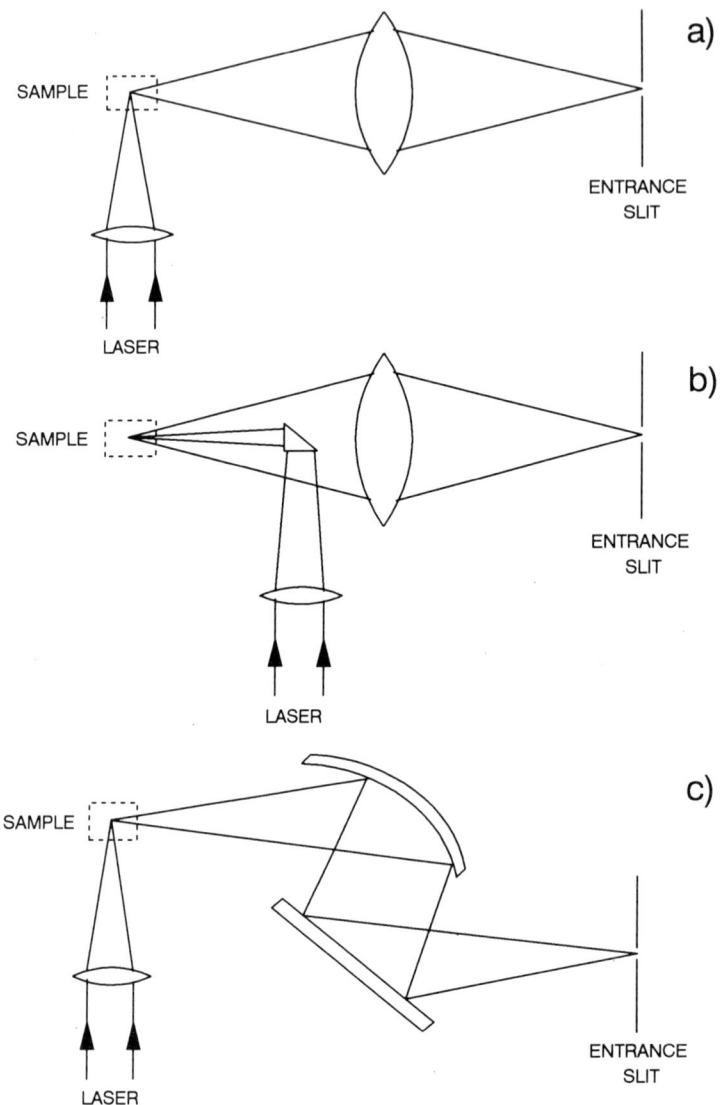

Fig. 3.4 — Possible front-end optics of Raman instruments, (a) 90° scattering arrangement using a lens, (b) 180° scattering arrangement, and (c) an elliptical mirror collection system.

and the focused laser beam. The periscope viewer is a simple 'swing-out' optical system which can be used to look at the viewed patch backwards through the entrance slit of the spectrometer. Alignment of the laser spot and sample can be achieved with the use of cross-wires. Whatever system is available, alignment of the sample requires skill and patience. A sensible routine is to open the slits to a wide setting, adjust the sample until a Raman signal is observed and then gradually close

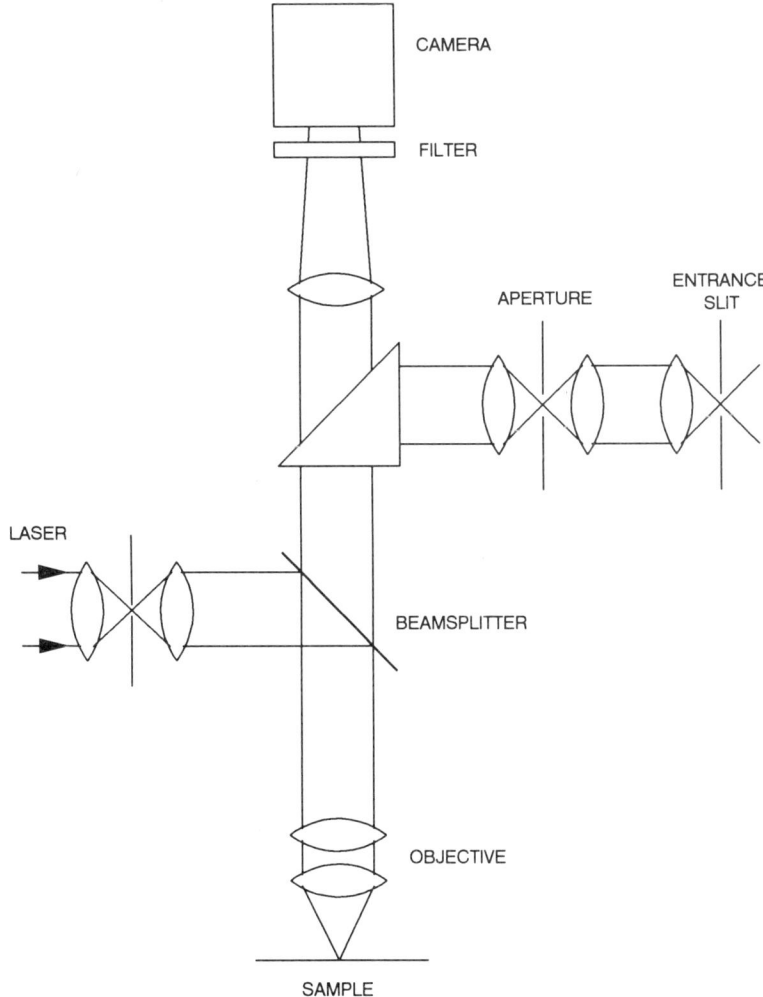

Fig. 3.5 — Microscope collection optics.

the slits to the desired value whilst making any necessary fine adjustments to the sample position. When samples are fluorescent this final step can be most tedious. Since alignment is required to a precision of a few micrometres, the laser and sample adjustment systems have to be of high quality, well maintained and operated intelligently by well trained staff.

3.3 SCANNING MONOCHROMATORS AND SPECTROGRAPHS

In a conventional Raman spectrometer light is analysed by a monochromator. Figure 3.6 shows an optical diagram of a simple monochromator. Light enters the device

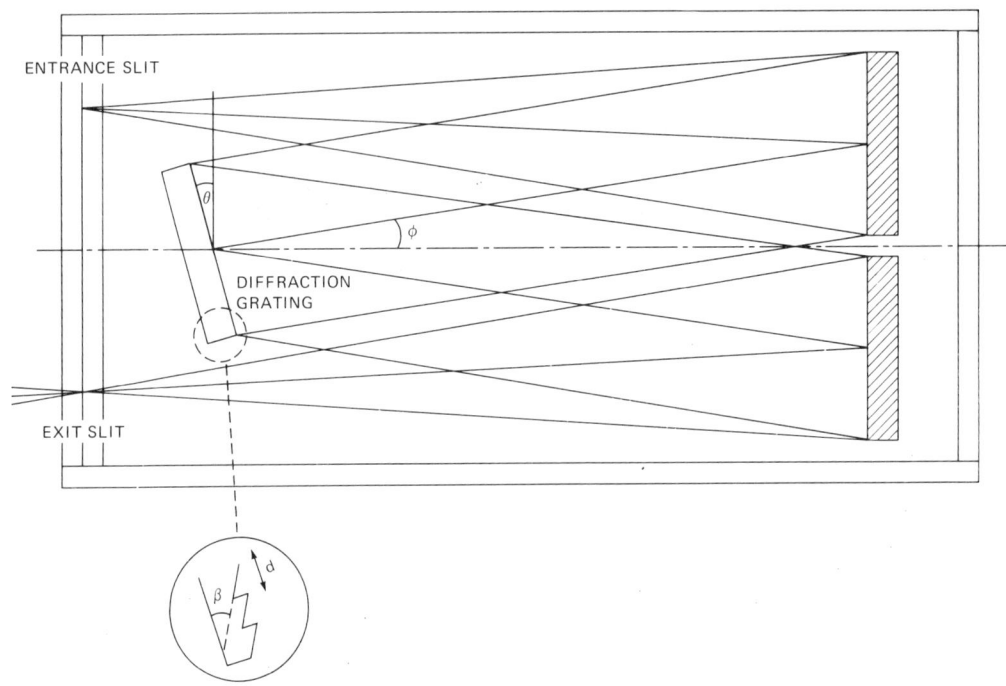

Fig. 3.6 — An optical diagram of a simple single monochromator.

through a narrow slit and is reflected off a collimating mirror to produce a parallel beam. This illuminates a diffraction grating which disperses the light according to its wavelength, according to the grating relationship, Eq. (3.1).

$$n\lambda = 2d \sin \theta \tag{3.1}$$

where n is an integer, λ is the wavelength passed by the monochromator containing a grating ruled with lines spaced d apart, set at an angle θ to the axis of the instrument, as shown in Fig. 3.6. A second mirror then reflects this light onto an exit slit. The grating actually produces a series of images of the entrance slit on the exit slit plane, each displaced by the wavelength of the radiation entering the instrument. However, only one of these images can pass through the exit slit and onto the detector. This image can be viewed either by having a wide exit slit and using a position-sensitive detector (as in a spectrograph) or by rotating the grating and tracking the image across a narrow exit slit to produce a spectrum.

Commercial spectrometers consist of two monochromators coupled together, but several manufacturers have offered triple monochromator instruments. Multiple monochromators are required in Raman work to achieve the essential discrimination between the relatively intense radiation at the laser frequency, and the weak Raman scatter. This is particularly difficult for a polycrystalline solid, or an opalescent sample, since the relative intensities I_{v_0}/I_{Raman} can be as great as 10^{12}, and the Raman lines of interest can be very close to the laser frequency. Perhaps one of the most critical measurements regularly made is that on the longitudinal acoustic mode (LAM) in crystalline polyethylene [6]. Polyethylene is of course a white, opalescent material and the LAM frequency usually lies within $\Delta_v = 5$–25 cm^{-1} shift, (see Fig. 3.7).

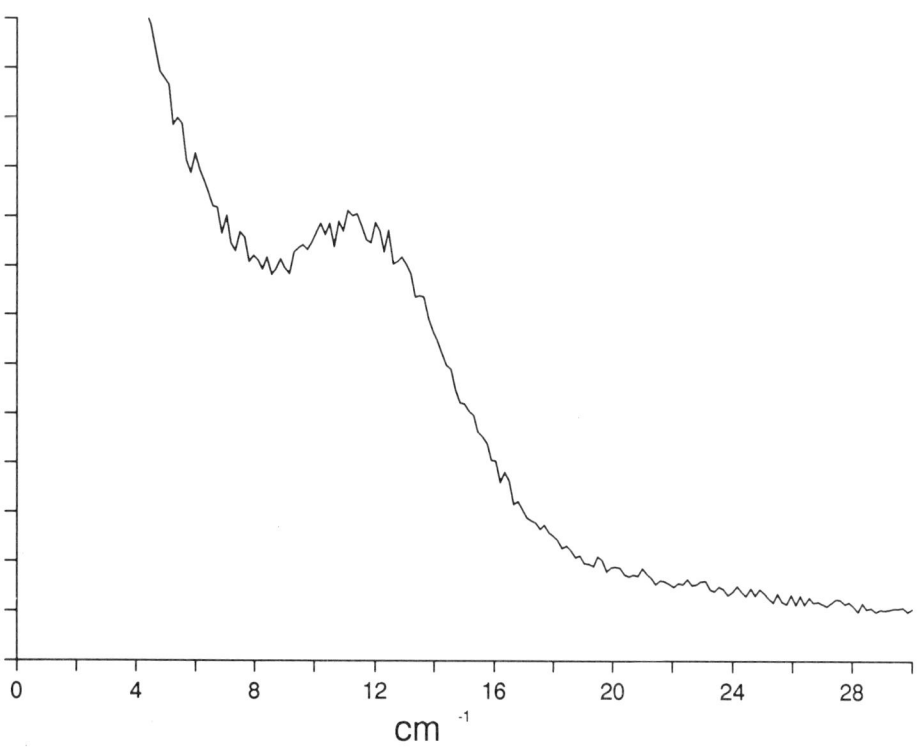

Fig. 3.7 — A longitudinal acoustic mode (LAM) in polyethylene.

If a monochromator is illuminated with perfectly monochromatic light, the output function drawn in Fig. 3.8(a) might be expected. What is actually observed is drawn in Fig. 3.8(b),(c) and (d). Inside the monochromator, light is scattered by dust, the slits, and imperfections in the optical surfaces, especially the gratings. This produces a signal apparently at a wavelength different from that of the source, because the light has taken an abnormal path through the instrument. However, no

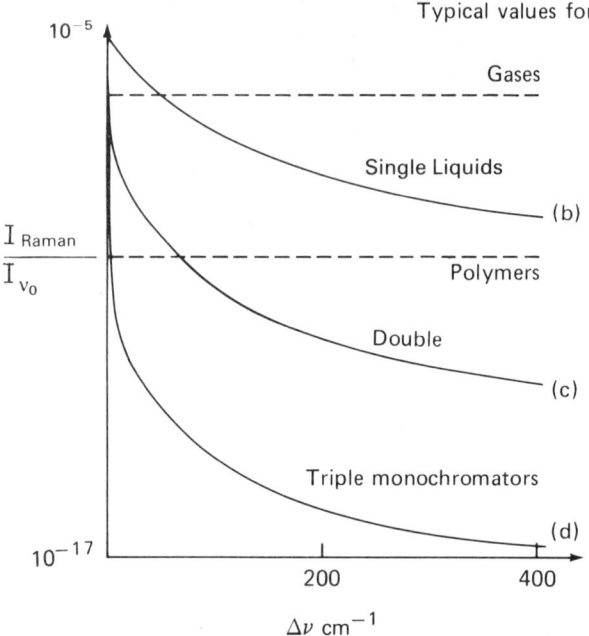

Fig. 3.8 — The effective stray light observed with (a) a hypothetical perfect monochromator, and (b) real single, (c) double and (d) triple devices.

radiation other than that of the source wavelength is actually present. This is a defect of monochromators and is called 'stray light'.

If a second monochromator which is aligned in series with the first diffracts the light leaving the first monochromator, the stray light is reduced, and the addition of a third monochromator will reduce it still further [see Fig. 3.8(b)]. A double monochromator will operate well at larger Raman shifts, but a triple will outperform it particularly very close to the exciting line. However, to synchronize three monochromators mechanically exactly is difficult and hence costly. Various attempts have been made to improve the stray light performance of double monochromators, including re-imaging between the exit slit of the first and the entrance slit of the second monochromator, and the use of holographic rather than ruled gratings. Three commercial solutions to the design of an efficient Raman monochromator are shown in Fig. 3.9.

An alternative approach which has become popular in the last few years is the use of a spectrograph and a position-sensitive detector. As we have already pointed out, a single monochromator does not have an adequate stray-light performance for Raman work, and therefore filters or pre-monochromators are used in these instruments. These instruments can never compare with conventional scanning monochromators as far as stray light is concerned, but they are optically far more efficient because they offer the multiplex advantage which is covered in Chapter 4. Many modern conventional Raman spectrometers incorporate both photomultiplier

Sec. 3.3] Scanning monochromators and spectrographs

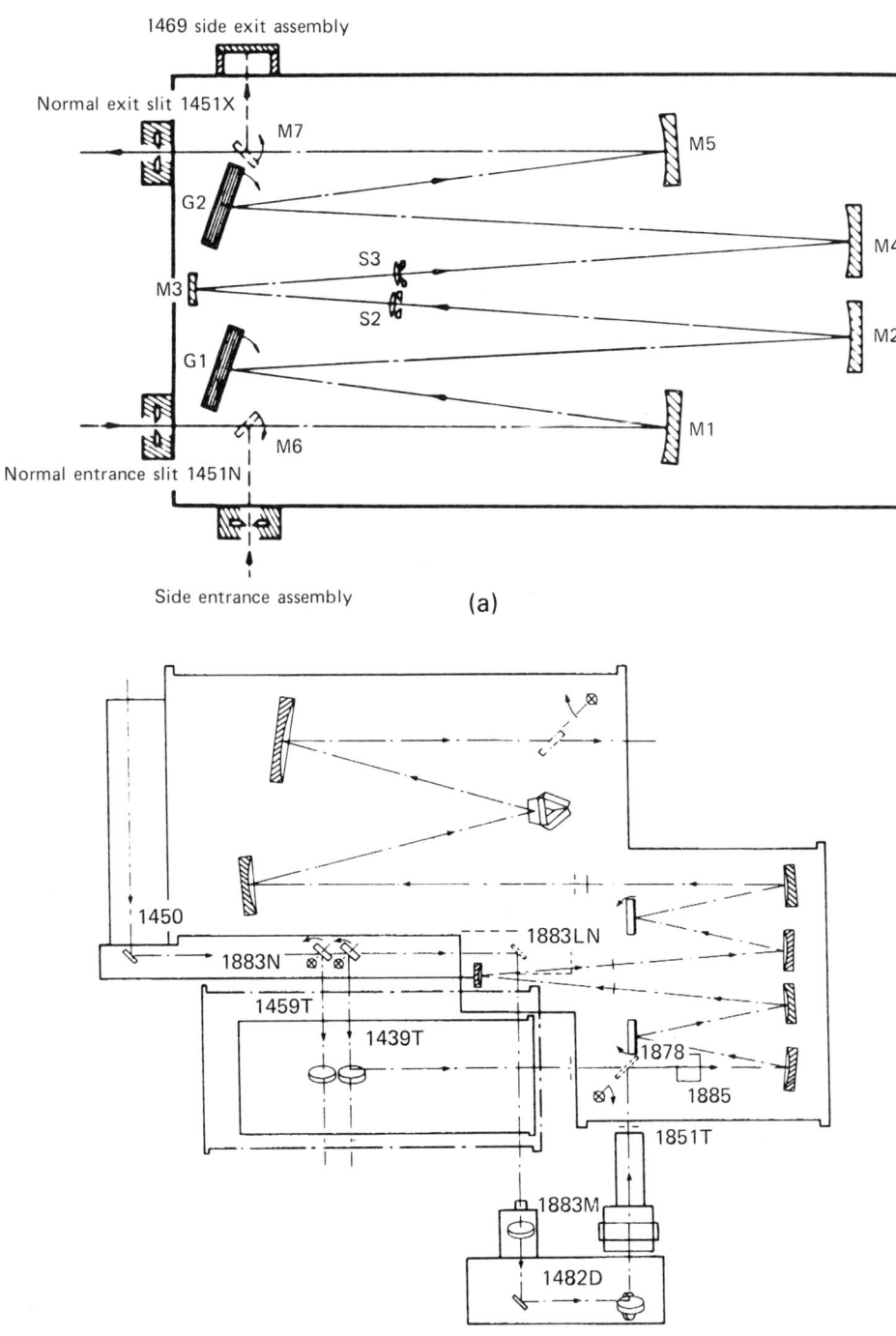

Fig. 3.9 — Three modern conventional Raman instruments: (a) a Spex double monochromator, (b) a Spex triple monochromator.

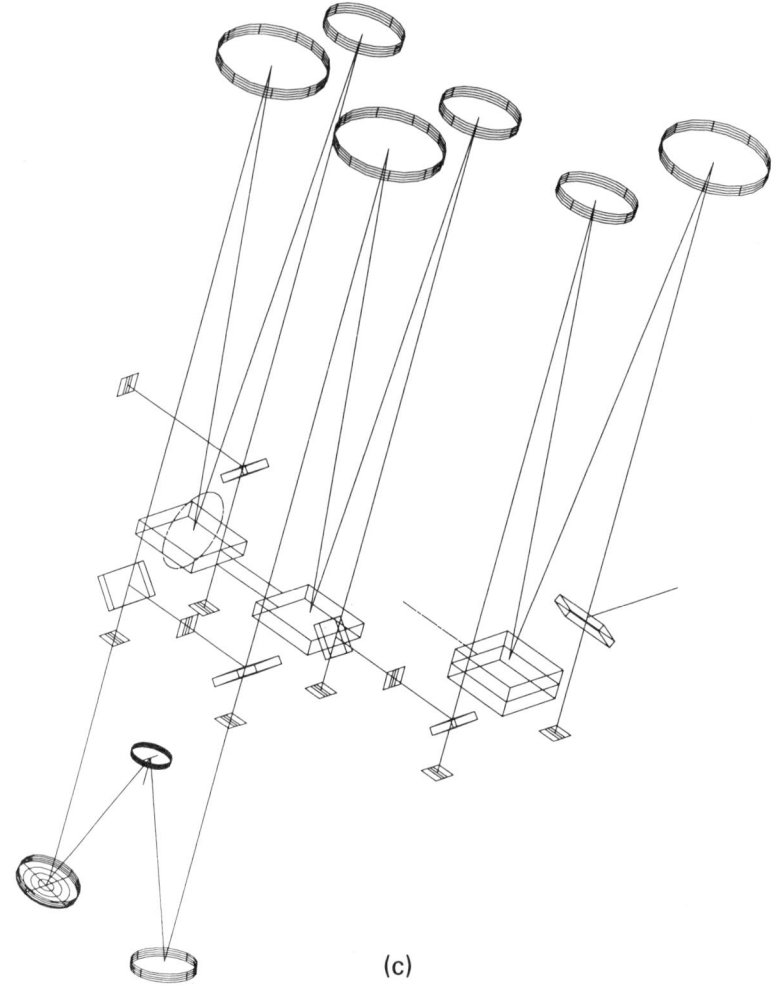

Fig. 3.9 — (c) A Jobin Yvon triple monochromator. Optical diagrams courtesy of Spex Inc. and Jobin Yvon.

tubes and position-sensitive detectors, and can thus operate as both scanning spectrometers and spectrographs.

3.4 SPECTRAL BANDPASS, RESOLUTION AND FREQUENCY PRECISION

A monochromator or spectrograph can operate with, in principle, any slit width. The slit width and the dispersion of the instrument together define the range of wavelengths that will pass through the exit slit. This is called the **spectral bandpass**. If the dispersion of the instrument (a function of the grating ruling, order of the spectrum and focal length of the device) is d cm^{-1}/mm, and the slit width is s mm, then the spectral bandpass will be sd cm^{-1}†. The slit width selected to record a spectrum accurately depends on the width of the spectral lines to be studied. The width of a band is expressed by measuring its full width at half of its maximum height (FWHH). To reproduce a spectral band perfectly, an infinitely narrow bandpass is needed. This is obviously not sensible, since no light would pass through the slits and no spectrum would be recorded at all. In practice it is best to set the spectral bandpass to slightly less than the FWHH of the narrowest band in the spectrum. For crystals, a typical value of the band FWHH is 2 cm^{-1}, whereas liquids have broader bands with FWHHs of 3–5 cm^{-1}. This results in only slight distortions of the true band profile, with the band height being reduced by only a small amount. If the spectral bandpass is greater than FWHH of the band to be measured, the instrument will artificially broaden the band such that its measured FWHH will be equal to the spectral bandpass. If the bandpass is very much greater than the FWHH the band shape can be grossly distorted and appears as a flat bump. In an emission spectrometer such as a conventional Raman instrument, increasing the spectral bandpass will allow more light to fall on the detector and a stronger signal will be measured. Thus the area under a band will vary with spectral bandpass. However, in a ratio recording spectrometer, such as a double-beam conventional infrared instrument, the band area does not change with spectral bandpass. This is because the sample and reference beams are equally attenuated when the slit width, and hence bandpass, is reduced. The equivalent of the slit width in an interferometer is the area of the Jacquinot stop (see Chapter 4). To achieve high resolutions in a FT spectrometer the size of the Jacquinot stop must be reduced, and the throughput of the device falls. Thus, when using an FTIR spectrometer, in order for the band areas to remain constant at different resolutions it is essential to record a new background spectrum whenever the resolution is changed. An FT Raman instrument is analogous to a conventional Raman instrument, and when the Jacquinot stop size is reduced the area of the spectral bands will inevitably decrease. Varying the spectral bandpass will affect the signal-to-noise ratio in spectra recorded regardless of the type of instrument used and this is discussed in Section 3.6.

Although the spectral bandpass determines how well overlapping bands can be distinguished, a more popular concept is resolution. Lord Rayleigh attempted a definition of resolution in the 19th century, though others have been suggested and all are rather arbitrary [7]. According to Rayleigh, two neighbouring monochromatic sources of equal intensity are resolved if the depth of the trough between the two spectral bands is equal to or more than 20% of the band height (see Fig. 3.11). Formally, for example, in a spectrum quoted as being at 5 cm^{-1} resolution this criteria should apply to bands of equal intensity 5 cm^{-1} apart. An illustration of how slit width, and hence spectral bandpass, determines whether overlapping bands are resolved is shown in Fig. 3.12.

As well as the spectral bandpass, there are also several other instrumental factors, any one of which will define the lower limit resolution of the Raman spectrometer. These are the width of the exciting line, errors in the synchronization

---

† The Coderg triple instrument has a dispersion of about 140 μm/cm^{-1} (1200-lines/mm grating and 800-mm focal length). Thus, a slit width of 0.3 mm gives a bandpass of about 2 cm^{-1}.

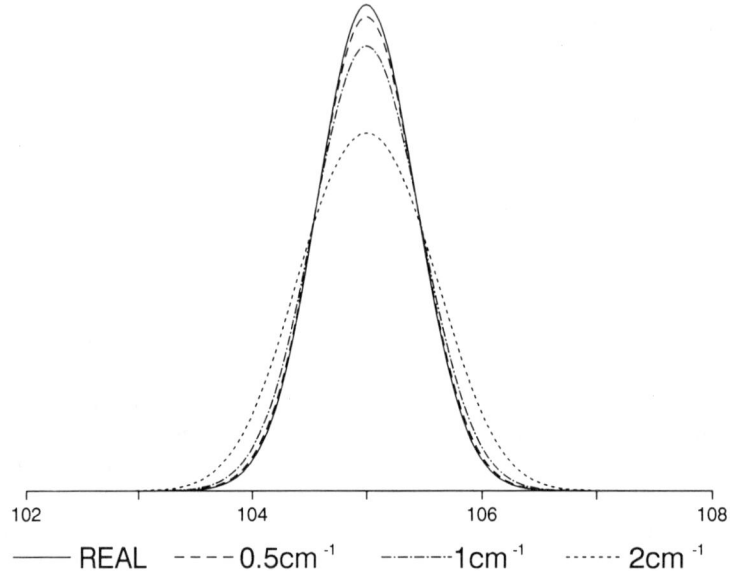

Fig. 3.10 — The variation in band profile with spectral bandpass. The solid line shows the profile of a Gaussian band with a FWHH of 1 cm^{-1}. The dashed lines show how this would be recorded with the spectral bandpass set to 0.5, 1 and 2 cm^{-1}.

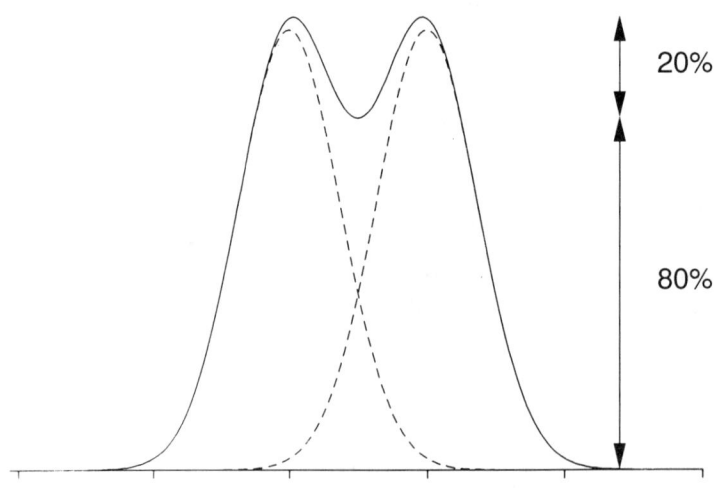

Fig. 3.11 — Two overlapping spectral bands. According to the Rayleigh definition these bands would just be resolved.

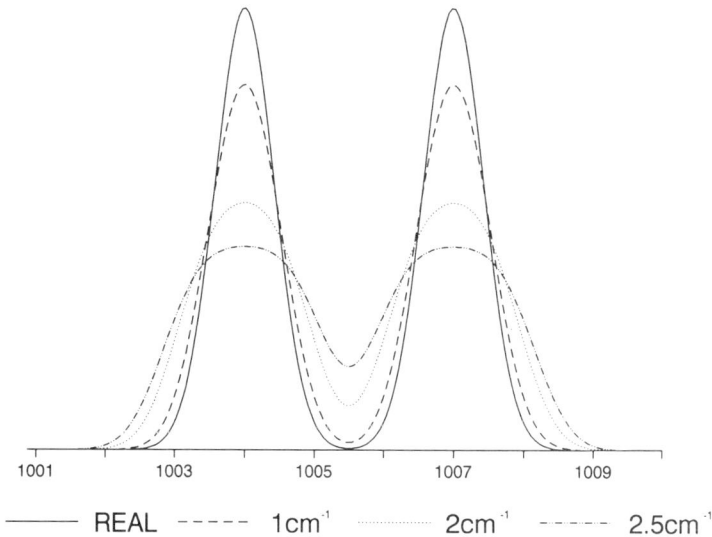

Fig. 3.12 — The solid line shows two Gaussian-shaped bands of FWHH equal to 1 cm^{-1} spaced 3 cm^{-1} apart. The dashed lines show the spectrum that would be recorded with the spectral bandpass set to 1, 2 and 2.5 cm^{-1}.

between multiple monochromators, optical imperfections, and the line spacing of the grating.

Monochromators are scanned and hence potentially inaccurate as far as their frequency calibration is concerned. Further, they are temperature sensitive and hence the calibration may vary wildly. In the Raman experiment, a relatively easy check can be made by determining the apparent position of the laser line. If this is set to $\Delta_\nu = 0$ cm^{-1}, the calibration will be *fairly* accurate. Fortunately, there are standard well characterized spectra in existence, so the frequency calibration can readily be checked [8]. In practice, errors of 2–4 cm^{-1} are by no means unusual in the accuracy with which band heads are quoted in the literature.

3.5 DETECTORS AND DETECTOR ELECTRONICS

Perhaps the ultimate detector for visible and ultraviolet light is the modern high-performance photomultiplier tube (PMT) [9]. Light falls on the front face of the detector, which is an electrically charged photocathode. Electrons are produced as a result of the photoelectric effect, and these are accelerated towards a series of dynodes at which more electrons are produced. Each dynode serves to increase the electron current and the device acts as an 'electron amplifier'. Typically PMTs contain up to 14 dynodes, so the gain of the device is very high. A single photon event at the photocathode produces a large, easily measured pulse at the output. Unfortunately spontaneous emissions occur from the photocathode, so these tubes give an output of a train of pulses even when in the dark. This so called dark current or dark count can be reduced by cooling the tube, and many Raman spectrometers incorporate PMT refrigeration designed to reduce the temperature to approximately −40°C. The total rate of spurious emissions from the photocathode depends linearly on its area. Thus, a small photocathode operated at room temperature will perform as well as a larger cooled detector. This design choice avoids complexity and hence reduces costs.

Many varieties of photocathode surfaces are available, all having different wavelength sensitivities and photon efficiencies, an example of which is shown in Fig. 3.13.

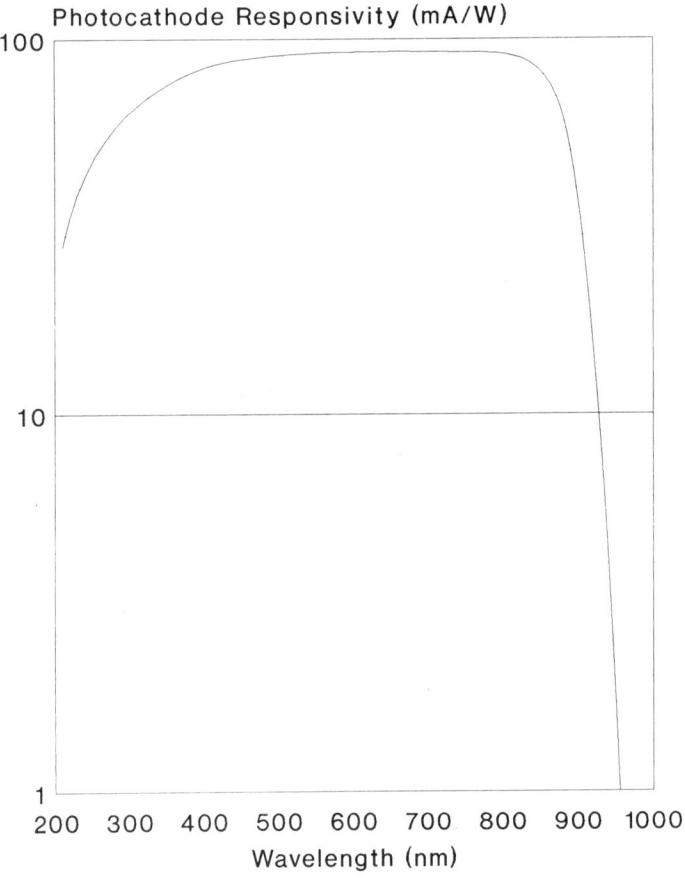

Fig. 3.13 — A typical response curve for a photomultiplier tube.

Unfortunately, tubes which operate well into the red and near infrared invariably seem to have high dark currents.

Solid-state visible detectors, such as silicon photoconductive chips, are common, but they have far higher dark currents than photomultipliers. A good measure of the performance of a device is the level of radiation required to equal the spurious signal produced in the dark, usually called the noise equivalent power (NEP). Since sensitivity varies with wavelength, the value should really be quoted at a single wavelength, but this information is rarely given. Some NEP values for visible detectors are provided in Table 3.2.

Table 3.2 — NEP values for a range of visible detectors

Device	Model	Manufacturer	NEP (W)	Dark count (cps)	Dark current (nA)	Size mm^2
PMT	9863B	EMI	—	40	0.4	6.25
Silicon Photo volatic diode	BPX63	Centronic	2.7×10^{-15}	—	—	1
GaP photo diode	G1961	Hamamatsu	5×10^{-15}	—	—	1.21

The output pulse train for a quantum detector such as a photomultiplier is always analysed in modern Raman spectrometers by using photon counting electronics. These systems consist of an energy discriminator followed by either a ratemeter, a scaler counter, or a suitably interfaced microcomputer fast enough to monitor individual pulses. The discriminator which acts as an energy gate is set to pass counts large enough to originate from electrons leaving the photocathode, but reject those emanating accidentally further along the dynode sequence. Where a counter is used, its output is normally made available for subsequent processing in digital form after each sampling period. A block diagram of a typical photon counter is given in Fig. 3.14.

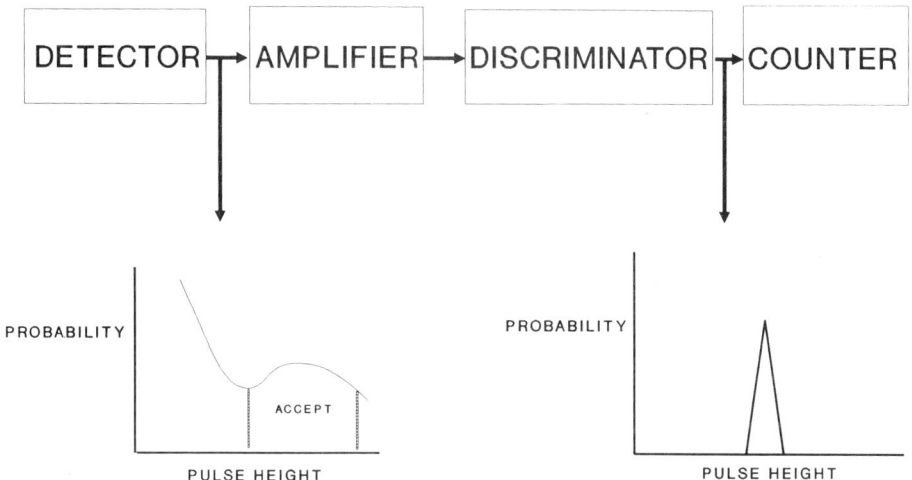

Fig. 3.14 — A schematic diagram of a typical photon counting system.

Photomultipliers are invariably used in scanning instruments, and the radiation leaving the exit slit as the frequency is scanned is focused onto the photocathode by a simple lens. As previously described, a spectrograph can be used in Raman spectroscopy and has the advantage that, whereas the scanning instrument surveys each frequency sequentially, residing over one increment for only a short time, the spectrograph looks at all frequencies within its grasp for the duration of the experiment. This gain in performance is called the multiplex advantage [10]. In this case the detector has to be large and segmented, so that the intensity at each discrete frequency can be monitored separately and continuously.

Recently, several different types of position sensitive detectors have been applied in Raman spectroscopy. Initially, photodiode arrays and sensitive TV cameras (vidicon) were used in conjunction with image intensifiers. Latterly, more sensitive photodiode arrays [11] and charge coupled devices (CCD) have been developed, and these make excellent Raman detectors [2]. A CCD chip consists of a two dimensional matrix of semiconductor detectors known as elements as shown in Fig. 3.15(a). When light falls on the CCD these develop a charge by the photoelectric effect. These charges can be shuffled vertically along the chip by changing the potential applied to an array of transparent electrodes which cover the device. The bottom row of elements can be shuffled horizontally to the corner of the chip where the charge developed is converted into an electrical signal. A CCD detector can be operated in two ways. If the dispersed spectrum covers the entire chip the vertical columns can be added together to enhance the signal to noise ratio. Alternatively, kinetic experiments can be performed by arranging the detector such that the spectral image just illuminates one row of the CCD and collecting a series of spectra as a function of time by shuffling them vertically along the device. This is shown in Fig.3.15(b).

Complex electronic circuits invariably follow all of the position-sensitive detectors, and the data is displayed on an oscilloscope or interfaced to a microcomputer. Although position-sensitive detector spectrographs have the advantage of speed over the more conventional scanning instruments, some of them are more noisy than photomultipliers, i.e. their NEP per channel is relatively high. This restriction does not apply to CCDs which have quite outstandingly low dark currents.

3.6 SOURCES OF NOISE IN SPECTRA

The ideal Raman spectrum consists of a series of emission lines over a completely black background [see Fig. 3.16(a)]. However, as we have already mentioned, Raman detectors have an inherent dark current (count). Thus the spectrum will have a small background, as shown in Fig. 3.16(b). In addition, a real spectrometer has some stray light and this will add to the overall background. The background signal due to stray light increases as the exciting line is approached, as illustrated in Fig. 3.16(c). The dark current and stray light signals are instrumentally derived and are present regardless of the sample. As well as generating Raman light, the laser may also excite fluorescence in the sample. Fluorescence appears as broad spectral emission which adds to the total background as shown in Fig. 3.16(d). Because the fluorescence is a function of the sample, it cannot be reduced by using a better detector or monochromator, as is the case with dark current and stray light.

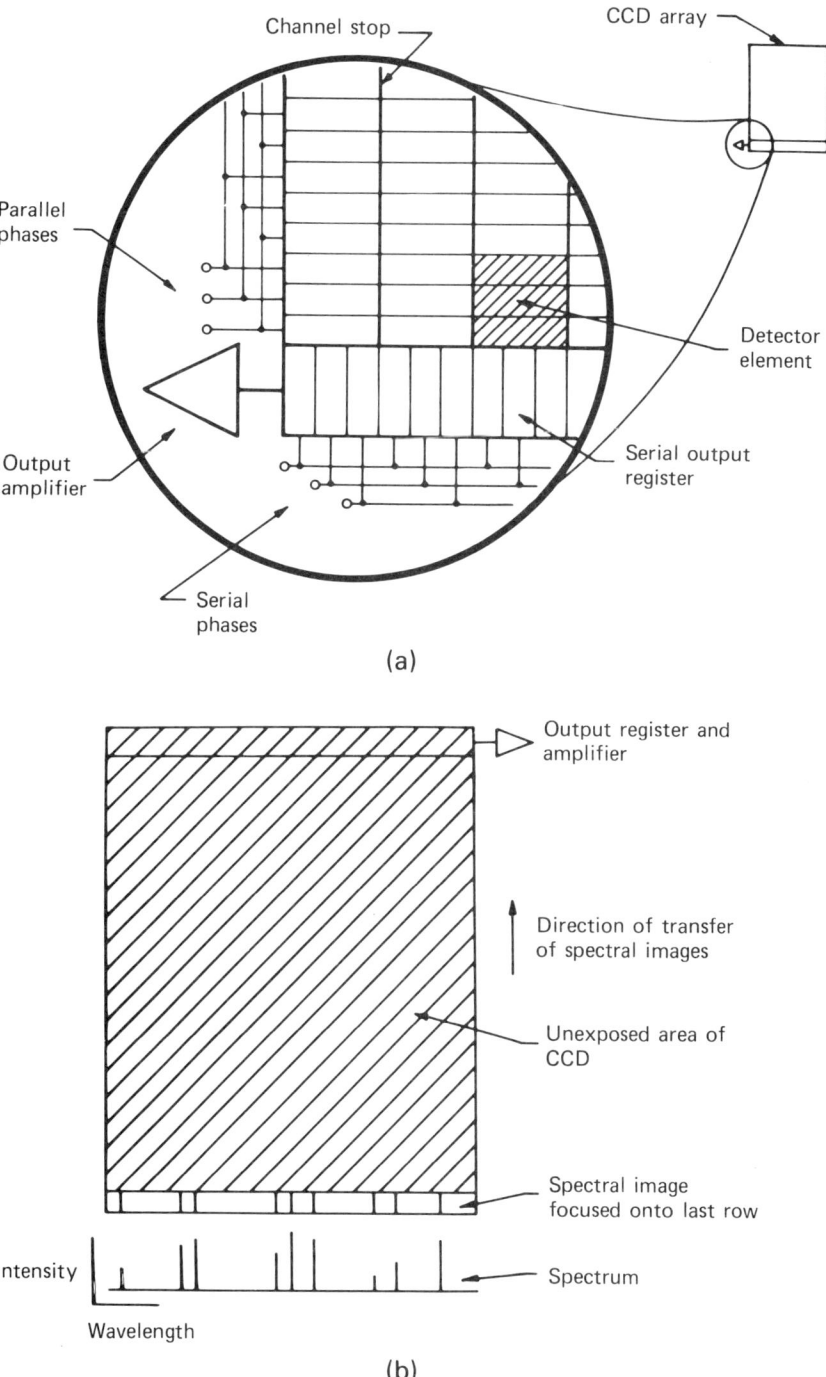

Fig. 3.15 — (a) A schematic diagram of a CCD detector, (b) the use of a CCD in kinetic studies. Both diagrams courtesy of Professor D. Batchelder.

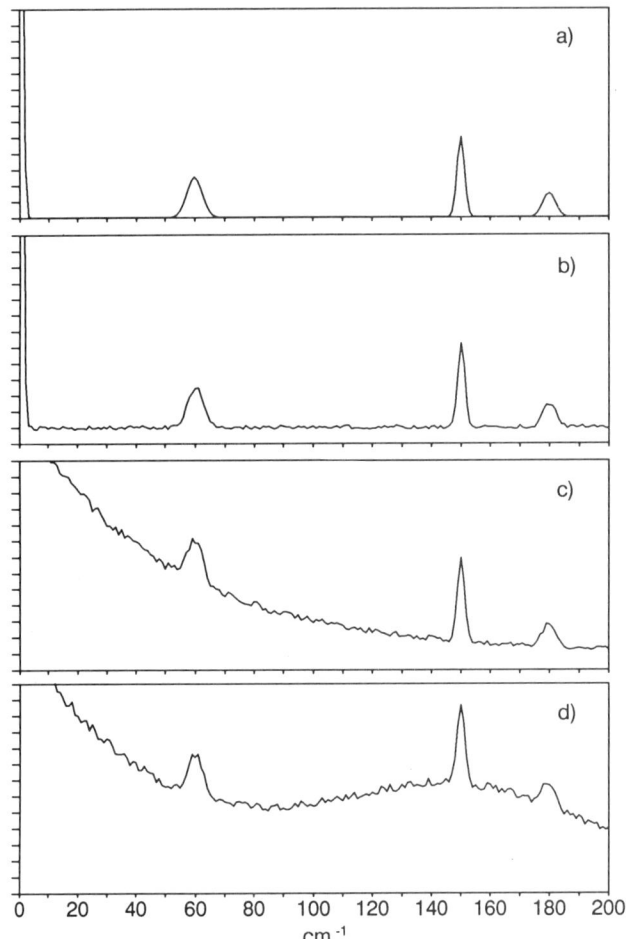

Fig. 3.16 — (a) Shows an idealized Raman spectrum with no background and no noise. In (b) the whole spectrum is slightly displaced owing to the detector dark current, and a random noise component has been added. When stray light is introduced (c) the whole spectrum appears over a sloping background and the noise level increases. Finally, if the sample is fluorescent the background may continue across the spectrum (d) and the signal to noise ratio is reduced still further.

When a real spectrum is examined, invariably it is found that the spectrum has a random component, both in its background and over its peaks. This randomness or jitter is called noise. The acid test of an instrument's performance is the ratio of the signal to the noise at its output, since this defines the lower limit below which the Raman lines cannot be measured. Although it is quite feasible to see a weak band against a constant background of any intensity, when it appears against a noisy one there are severe problems. Spectroscopists tend to be suspicious of spectra where the signal to noise ratio is below three.

Sec. 3.6]   **Sources of noise in spectra**   69

If the Raman spectrum consists of a series of lines against a black background (this never really happens, see Fig. 3.16) the Raman signal will be S_{Raman} photons detected per second†. The signal itself will, however, be noisy owing to statistical variations. This phenomenon is called 'shot noise' and is due to the fact that photons do not arrive at the detector at a constant rate, but rather as a random stream. The statistical variation of any signal S is given by $\sqrt{S}$. Thus the Raman signal S_{Raman} has an associated statistical noise, $N_{Raman} = \sqrt{S_{Raman}}$. Similarly, the signal due to the detector dark count, S_{Dark}, also has a statistical variation, $N_{Dark} = \sqrt{S_{Dark}}$ (see Fig. 3.17).

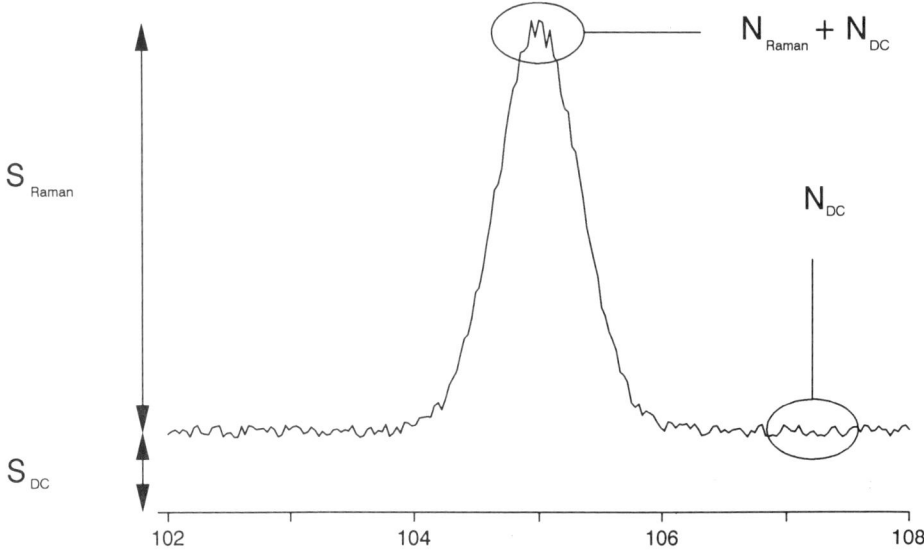

Fig. 3.17 — A spectral band showing associated noise.

A monochromator allows us to select the narrow range of wavelengths of light that fall on the detector. Thus, the noise varies with wavelength as the spectrometer is scanned, because of the different levels of radiation it passes. In a region of the spectrum where there is no signal other than the dark current, the noise level is N_{Dark}. When a Raman line is scanned the associated noise is given by $\sqrt{(S_{Raman} + S_{Dark})}$. Thus, the noise at the band head will invariably be greater than that of the surrounding black background, (see Fig. 3.17). Hence, the ratio of the signal to noise is given in Eq. (3.3).

$$\frac{S}{N} = \frac{S_{Dark} + S_{Raman}}{\sqrt{S_{Dark} + S_{Raman}}} \qquad (3.2)$$

where $S_{Raman} \gg S_{Dark}$

† Since we know the wavelength of the light, S_{Raman} can be expressed in units of power.

$$\frac{S}{N} \approx \sqrt{S_{\text{Raman}}} \qquad (3.3)$$

The signal-to-noise ratio at the head of a band-limits the accuracy with which the band-height and band position can be determined, and also reduces the ability to resolve overlapping bands.

We now move on to more realistic spectra, where we have a spectrum superimposed on a background emanating from some source such as fluorescence or stray light. If the level of the background, beneath a Raman band of intensity S_{Raman}, is S_{Back}, where $S_{\text{Back}} = S_{\text{Dark}} + S_{\text{Stray}} + S_{\text{Fluor}}$, then the signal-to-noise ratio becomes:

$$\frac{S}{N} = \frac{S_{\text{Raman}} + S_{\text{Back}}}{\sqrt{S_{\text{Raman}} + S_{\text{Back}}}} \qquad (3.4)$$

However, we are only interested in S_{Raman}, so the effective signal-to-noise ratio for the Raman band is:

$$\frac{S}{N} = \frac{S_{\text{Raman}}}{\sqrt{S_{\text{Raman}} + S_{\text{Back}}}} \qquad (3.5)$$

For a given Raman band intensity there is a limit to the background intensity that can be tolerated before the signal to noise ratio falls to a useless level.

All the above analysis presupposes that the source of radiation is of a nominal stability. However, in an emission experiment such as ours, the intensity of the total spectrum is a function of the laser power and of the alignment of the sample. Both the output of a laser and other factors are subject to short term variations and long term drift. It is the short-term fluctuations which are particularly troublesome, since they contribute to the noise in the background and also in Raman bands. Fortunately the lasers normally used by Raman spectroscopists are relatively highly stabilized (typically better than 1% rms).

Several methods of improving signal and hence the signal-to-noise ratio exist. However, there are always compromises which result in either increasing the time required for the experiment or degrading the resolution. The value of the spectral bandpass can be increased by opening the slits of the monochromator so that more light falls on the detector, but this will degrade the resolution. The incident laser power can be increased, but at the risk of damaging the sample. In a scanning monochromator, the total number of photons detected can be increased by scanning more slowly. To illustrate these points we show a simple spectrum recorded under a variety of conditions in Fig. 3.18. If a spectrograph and a position-sensitive detector are used, the S/N is improved by increasing the sampling time, i.e. the total duration of the experiment. In both systems the spectrum can be recorded repetitively and the data averaged.

When a monochromator is repetitively scanned, the spectrum is recorded quickly (fast enough to remove the possibility that laser instabilities or instrument variations

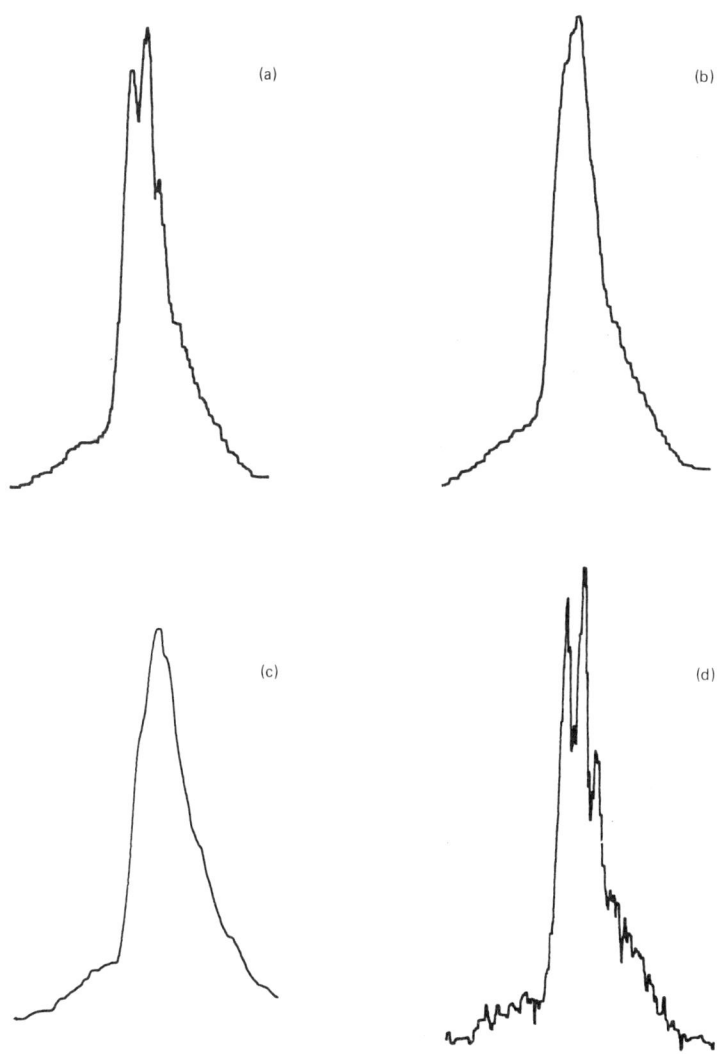

Fig. 3.18 — The Raman spectrum of carbon tetrachloride recorded between 435 and 485 cm^{-1} on a Coderg T800 scanning spectrometer under a variety of instrumental conditions. (a) 1 cm^{-1} bandpass 25 cm^{-1}/min scan rate, (b) 2 cm^{-1} bandpass 25 cm^{-1}/min scan rate, (c) 1 cm^{-1} bandpass 250 cm^{-1}/min scan rate and (d) 0.5 cm^{-1} bandpass 25 cm^{-1}/min scan rate.

will become problematic) over and over again, and a computer is used to sum the spectra produced. Since the effect of carrying out the operation x times is to increase the sampling time by this amount, the gain in the S/N value approximates to $\sqrt{x}$. Again, as with slow scanning, in principle scanning can be continued until the result is satisfactory. However, there are two limitations. If the random variations in the total signal are not truly random, the $\sqrt{x}$ factor will not be achieved. Further, if the spectra recorded are not perfectly superimposed in the computer, band position and shape are perturbed and resolution is degraded. In real scanning instruments, these limitations place a restriction on the maximum number of scans worth accumulating, typically about a hundred.

Fig. 3.19 — A typical Raman gas cell. The laser illuminates the sample from above through a lens. A concave mirror is placed beneath the sample so that the laser operates as an external resonator, considerably increasing the power density at the sample. The scattered light is collected by the lens on the left hand side. Courtesy of Dr C. J. Vear.

3.7 SAMPLING

Perhaps the biggest advantage the Raman spectroscopist possesses is the ease and versatility of sampling. Solids, liquids and gases need only be enclosed in a vessel transparent to the laser and Raman light. As a result, a host of simple cells for innumerable experiments have been devised, and a few are illustrated in Figs 3.19–3.27.

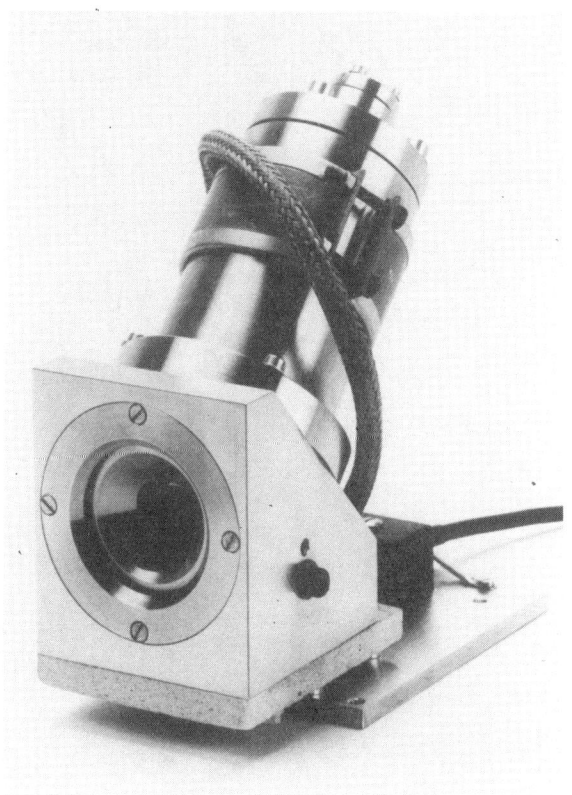

Fig. 3.20 — A high temperature reaction Raman cell, courtesy of Ventacon Ltd.

Heating (Fig. 3.20) and cooling (Fig. 3.21) the sample poses no problems unless temperatures are so high that incandescence occurs. Water is an excellent solvent, so aqueous solutions are accessible, and working electrochemical cells are readily studied (Fig. 3.22). Although care must be exercised in the construction of cells that are strong enough, high pressure studies are also possible (Figs 3.23 and 3.24).

As pointed out in Chapter 2, samples frequently absorb in the visible. Since the excitation source in Raman spectroscopy is very intense, catastrophic heating of the sample can occur. To overcome this problem it is possible to move the laser spot rapidly across the surface of the sample or simply to move the sample under a stationary laser spot by spinning it with a small electric motor (see Fig. 3.25). An alternative approach for strongly absorbing liquids is to flow them through an appropriately transparent cell.

Raman cells have been constructed to study reactions *in situ*, such as flowing polymer melts [12] (see Fig. 3.26), whilst other measurements have been made remotely using either a microprobe [13] or a telescope [14] to collect the Raman light.

Remote Raman measurements have also been made by using fibre-optics to couple experiments with Raman spectrometers [15]. The experimental arrangement

Fig. 3.21 — A low temperature Raman cell. Liquid nitrogen can be used as a refrigerant and the temperature controlled by using a small heater at the sample position. Courtesy of Ventacon Ltd.

is shown in Fig. 3.27. The use of fibre-optics has a number of advantages, particularly versatility. Samples need not be placed in the spectrometer and this is particularly attractive for studying samples under high temperature or pressures, and in chemically hazardous environments. In addition, one spectrometer may be interfaced with many fibre optic probes.

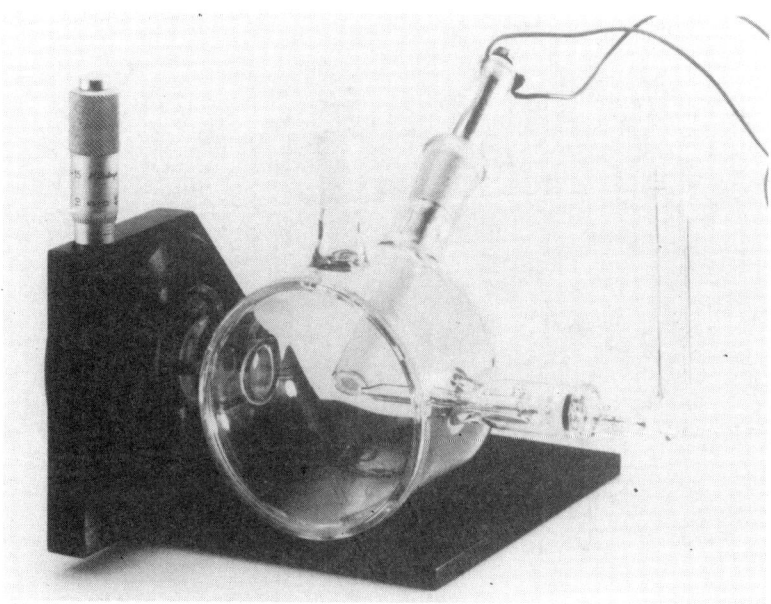

Fig. 3.22 — An electrochemical Raman cell, courtesy of Ventacon Ltd.

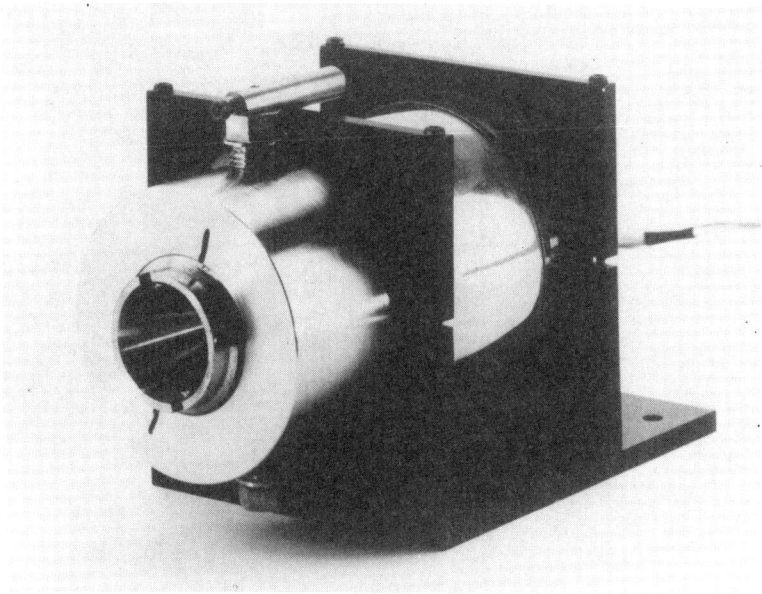

Fig. 3.23 — A high-pressure Raman cell used for making measurements up to 3500 Bar, courtesy of Ventacon Ltd.

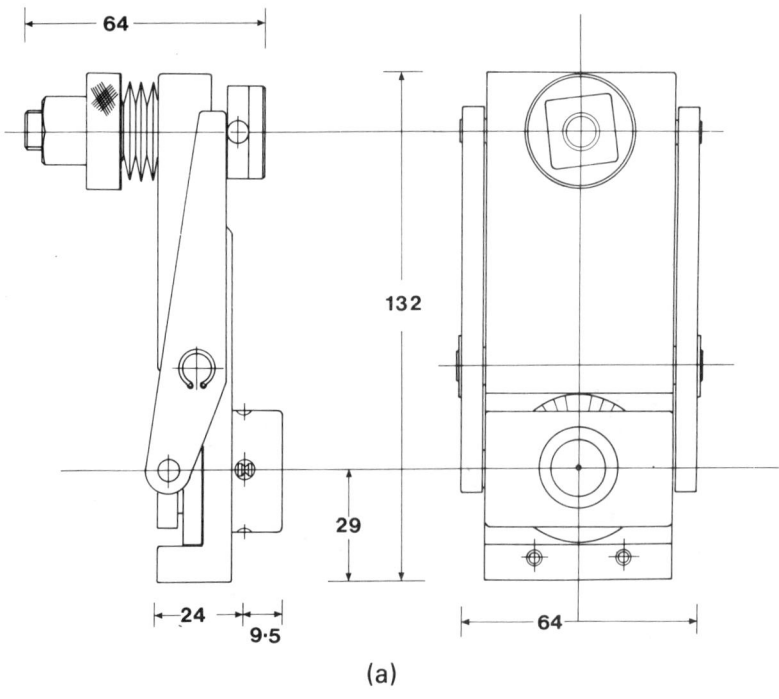

Fig. 3.24 — A diamond anvil Raman cell for achieving very high pressures, courtesy of Diacell Ltd.

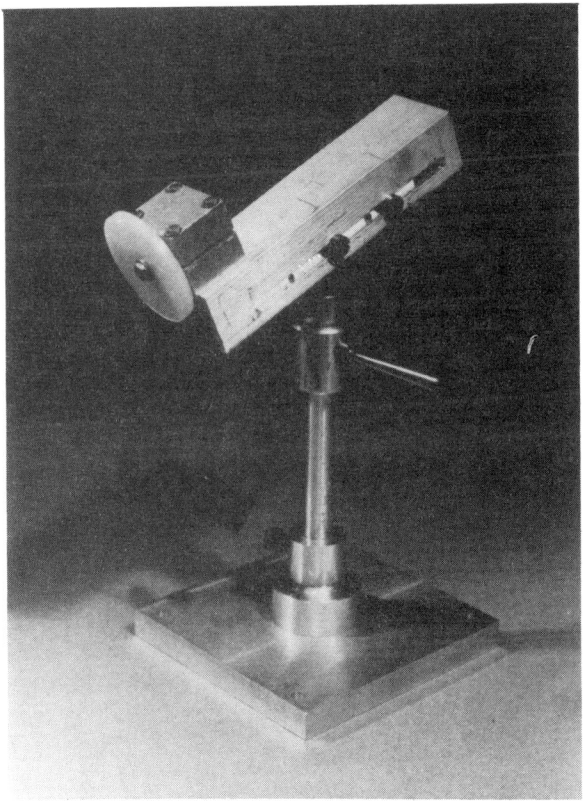

Fig. 3.25 — Raman spinning sample cell, courtesy of Dr I. D. M. Turner.

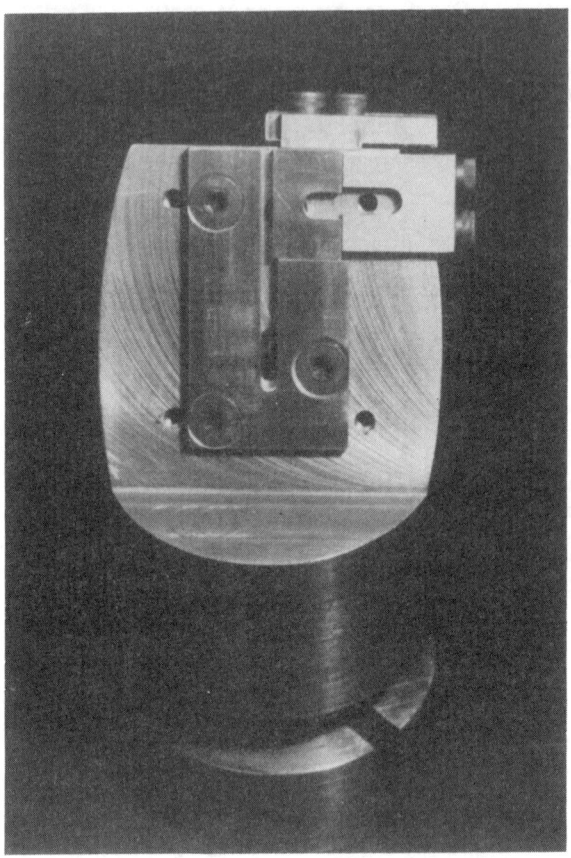

Fig. 3.26 — A Raman arrangement for recording spectra of flowing polymer melts, The unit fits onto the rose of a laboratory extruder and is heated. A thick glass plate is screwed down over the central vertical groove and the polymer is forced to flow upwards behind the glass. Spectra are excited at any position along the groove. (Courtesy of Dr D. B. Morris.)

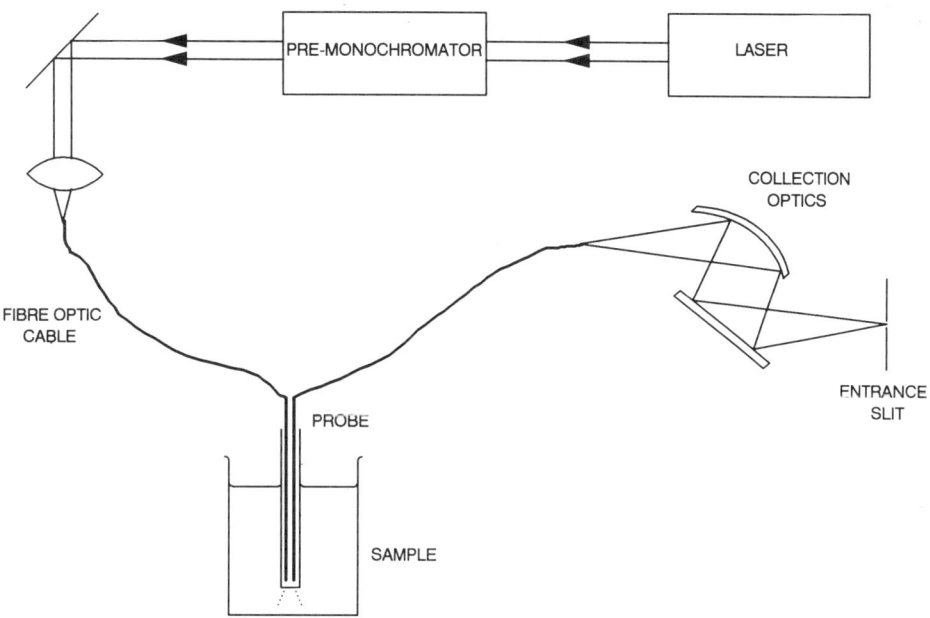

Fig. 3.27 — A diagram showing a typical fibre optic coupling system to record Raman spectra (after McCreery).

REFERENCES

[1] D. P. Strommen and K. Nakamoto, *Laboratory Raman Spectroscopy*, Wiley, New York, 1984.
[2] D. N. Batchelder, *European Spectroscopy News*, 1988, **80**, 28.
[3] H. O. Hamaguchi, *Appl. Spectrosc. Rev.*, 1988, **24**, 137.
[4] M. Bridoux and M. Delhaye, in *Advances in IR and Raman Spectroscopy*, Vol 2, R. J. H. Clark, and R. E. Hester, eds., Heyden, London, 1976.
[5] J. J. Barrett and N. I. Adams, *J. Opt. Soc. Am.*, 1968, **58**, 311.
[6] D. I. Bower and W. F. Maddams, *The Vibrational Spectroscopy of Polymers*, Cambridge University Press, Cambridge, 1989.
[7] A. P. Thorn, *Spectrophysics*, Chapman & Hall, London, 1988.
[8] P. J. Hendra and P. J. Loader, *Chem. Ind.*, 1968, 718.
[9] W. Budde, *Physical Detectors of Optical Radiation* Vol. 4, of Optical Radiation Measurements, Academic Press, New York, 1983.
[10] P. B. Felgett, *J. de Physique et le Radium*, 1958, **19**, 187.
[11] D. K. Veirs, V. F. K. Chia and G. M. Rosenblatt, *Appl. Optics*, 1987, **26**, 3530.
[12] P. J. Hendra, D. B. Morris, R. D. Sang and H. A. Willis, *Polymer*, 1982, **23**, 9.
[13] J. D. Louden, 'Raman Microscopy', in *Practical Raman Spectroscopy*, D. J. Gardiner and P. R. Graves, eds., Springer-Verlag, Berlin, 1989.
[14] D. A. Leoñard, 'Measurement of Aircraft Turbine Engine Exhaust Emissions' in *Laser Raman Gas Diagnostics*, M. Lapp and C. M. Penney, eds., Plenum, New York, 1974.
[15] S. D. Schwab and R. L. McCreeny, *Anal. Chem.*, 1984, **56**, 2199.

4

Fourier transform vibrational spectroscopy

4.1 INTRODUCTION

The conventional Raman spectrometers surveyed in Chapter 3 are exclusively dispersive instruments. Dispersive instruments held a dominant position in infrared spectrometry until about 20 years ago. However, current sales of infrared dispersive instruments are totally eclipsed by those of Fourier transform interferometers. Since this change from spectrometer to interferometer is upon us in Raman spectroscopy, it is essential that we consider the reason for abandoning dispersive instruments.

4.2 PRINCIPLES OF INTERFEROMETRY

The radiation which enters an interferometer is split into two beams of (ideally) equal intensity. One beam is subjected to an optical delay, and the two beams are then recombined. If the delay is zero or equal to an integral number of wavelengths of the radiation, constructive interference will occur. If the delay is $n\lambda/2$, where n is an integer and λ the wavelength, destructive interference results. When an ideal monochromatic source passes through the interferometer, and the optical delay is varied, the signal detected is a cosine wave [see Fig. 4.2(a)]. Now consider the signal obtained when two ideally monochromatic sources of different wavelength pass through the interferometer. The signal is the sum of the signal that would be obtained for the two sources if they were analysed separately [see Fig. 4.2(b)]. As the source becomes increasingly polychromatic, the beat frequency becomes longer and longer. Thus, when a polychromatic source illuminates the interferometer, the beat frequency is so long that we observe only a single characteristic burst, as shown in Fig. 4.2(c). This trace is called the interferogram and the point of maximum intensity is known as the 'centre burst'. This occurs when the optical path difference is zero.

The interferogram is the sum of the cosine waves for all of the wavelength elements in the polychromatic source, multiplied, in each case, by a factor reflecting

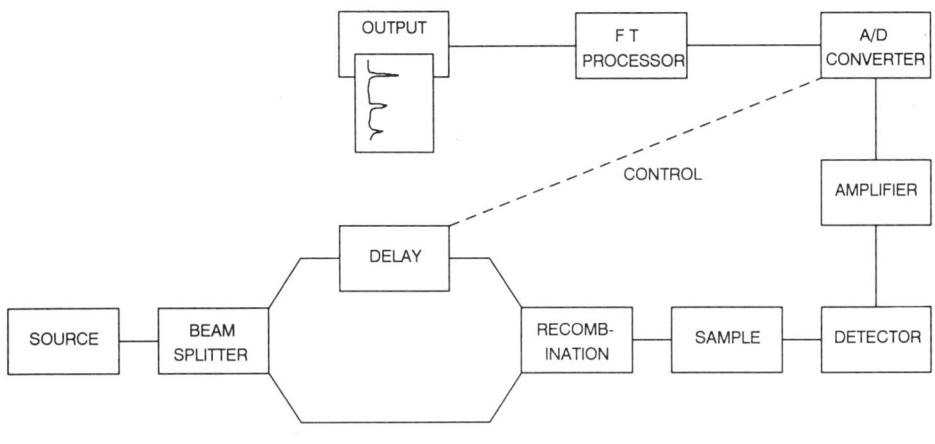

Fig. 4.1 — (a) A schematic diagram of an interferometer, and (b) a simple Michelson interferometer.

their intensity. The signal generated by varying the optical delay when monochromatic radiation of wavelength λ and intensity I passes through the interferometer is

$$I(x) \propto I_\lambda \cos\left(\frac{2\pi x}{\lambda}\right) \qquad (4.1)$$

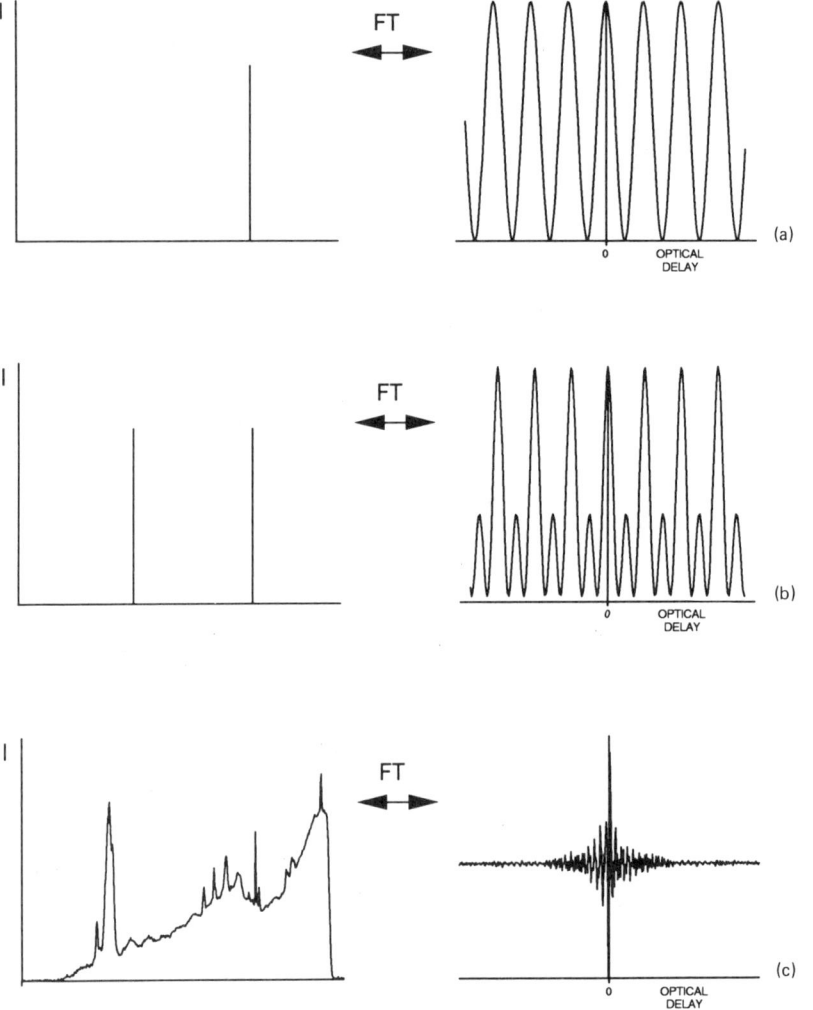

Fig. 4.2 — Interferograms produced by (a) a monochromatic source, (b) two monochromatic sources, and (c) a polychromatic source.

Where x is the optical delay.
Therefore, the interferogram from a polychromatic source has the form

$$I(x) \propto \sum_{\lambda=0}^{\lambda=\infty} I_\lambda \cos\left(\frac{2\pi x}{\lambda}\right) \qquad (4.2)$$

As we have seen, the interferogram is a function of intensity *vs*. optical delay. The movable mirror in the interferometer is moved at a fairly constant rate, and thus the interferogram can be expressed as a function of intensity *vs.* time. The interferogram can be resolved into its constituent cosine waves, and expressed in terms of intensity *vs.* frequency (reciprocal time). The frequency analysis process is that named after its originator, Joseph Fourier (1768–1830) [1,2]. The mathematical process which is used to turn the interferogram (in the time domain) into the spectrogram (in the frequency domain) is called the Fourier transform. Thus, the interferograms and stick spectra shown in Fig. 4.2 are in fact Fourier transforms of each other. The Fourier transform is carried out by using a computer algorithm. For p data points, the number of calculations required varies as p^2. In 1965, Cooley and Tukey [3] developed an algorithm such that the number of calculations required varies as $p \log(p)$. This results in a large reduction in computation time at realistic values of p, and is known as fast Fourier transformation (FFT); it is used on all commercial instruments [4,5]. A more detailed description of the Fourier transformation process is given in Appendix 1.

4.3 DATA PROCESSING IN FT INTERFEROMETERS

Light entering the interferometer is analysed by recording the intensity trace *vs.* optical delay, digitizing, and transforming the spectrogram from the interferogram. The interferogram is sampled at regular intervals and digitized by an analogue-to-digital (A–D) converter. The number of data points recorded is related to the length of the interferogram (i.e. the range of optical delay over which the machine scans), and the periodicity of data sampling. The data sampling is triggering by the use of a reference laser which passes through the interferometer and is subject to the same optical delay as the light being analysed. A separate detector records the interferogram (a cosine wave) from this reference laser. Generally, at the sign changes of this cosine wave, a pulse is sent to the A–D converter which triggers it to sample the main interferogram. Thus, two data points in the main interferogram are collected for every cycle of the reference laser interferogram. Some spectrometers sample more than two data points per cycle, but the general principle is the same for all. Helium–neon lasers are almost exclusively used as the reference, because although they do not have a very short wavelength (which restricts the sampling frequency), they are compact and cheap. The wavelength of the He–Ne laser is 632.8 nm, so that for every centimetre of optical path difference an instrument can collect 31 606 data points if it samples two data points per cycle of the reference laser interferogram.

Since the optical delay is achieved in an interferometer by using a relatively slowly moving mirror system, the interferogram appears at the detector as an AC signal in the audio-frequency range. If the delay is changing at 1 cm/sec, radiation of wavelength 10 μm (absolute frequency 1000 cm^{-1}) will appear as an AC signal, at the detector, with frequency 1000 Hz. The resulting electronic amplification and detection problems are akin to those found in compact disc (CD) audio systems.

The Fourier transformation process cannot yield more data points than those input. In practice, the number of data points in the spectrogram is half that in the interferogram. Thus, unlike a conventional dispersive instrument, the spectrum obtained is discrete rather than continuous. Interferometer-based spectrometers produce a finite data-point density in the computed spectrogram. In the previous example, there would be 15 803 data points in the spectrogram. For a 4000-cm^{-1} spectral range, this corresponds to 4 data points per wavenumber. The spectrum produced is actually a histogram with straight lines drawn between the data points. For most purposes this is not a problem, but when small regions of a spectrum are expanded, the discrete nature of the data becomes obvious, particularly at coarse data point pitching. This is shown in Fig. 4.4.

This now presents somewhat of a philosophical problem. How can we define the resolution of a spectrum when it is composed of discrete data points? As we have seen, Lord Rayleigh attempted to define resolution quantitatively in the 19th century (see Section 3.4), but his definition assumes that the data is continuous. Figure 4.3 shows two overlapping bands separated by 1 cm^{-1}. To resolve this profile, a bare minimum of two data points per wavenumber are required, if the data points happen to fall in the 'correct' place. Thus, it is clear that many more data points per wavenumber are needed to define band profiles adequately. Most infrared manufacturers try to convince their customers that two data points are always adequate to define a resolution element. Several have in the past floated the idea that *one* is sufficient! However, we can safely assume that one centimetre of optical delay can be expected to produce enough data points to resolve bands about one wavenumber apart (or for narrow bands produce a half-width of *about* one wavenumber).

This then brings us onto frequency accuracy. There is a myth that all interferometers are precise in frequency. This erroneous view arises from the precision of the He–Ne wavelength, which is indeed reproducible. The frequency of the calculated points in the spectral histogram is therefore precise. However, the shape of the interferogram is dependent on the alignment of the radiation and the He–Ne laser as they pass through the interferometer. As the radiation is moved off-axis, these two beams will no longer follow exactly the same path, and will experience slightly different optical delays. This results in the frequency of spectral bands shifting in the spectrogram by as much as 1 or 2 cm^{-1} in most commercial instruments. If the user records IR absorption spectra, in transmission, and never varies the manufacturer's alignment or the thickness of the sample across the patch studied, then there should be no frequency errors. If off-bench measurements are made *or* accessories involving reflections are used, the beam passing through the system is unlikely to be aligned perfectly with the factory conditions and as a result band shifting will inevitably occur. This can be a problem with ATR and similar measurements, particularly where spectral subtractions or ratioing are attempted. FT Raman instruments may suffer from absolute frequency inaccuracies, but because any displacement will be consistent across the spectrum, the frequency shifts measured should be correct.

The peak-picking routines which have been built into the software of many spectrometers list the frequencies of spectral bands. For each band, these quote the frequency of the data point which has the greatest intensity. They then display this 'peak' frequency with quite indecent precision (three decimal places is not uncommon). In a well aligned interferometer the frequency of the data points is known very accurately. However, as seen in Fig. 4.3, unless this data point coincides exactly with the real band head, the accuracy of the 'peak' frequency quoted is meaningless. Thus, peak picking routines are a useful indication of band position, but it must

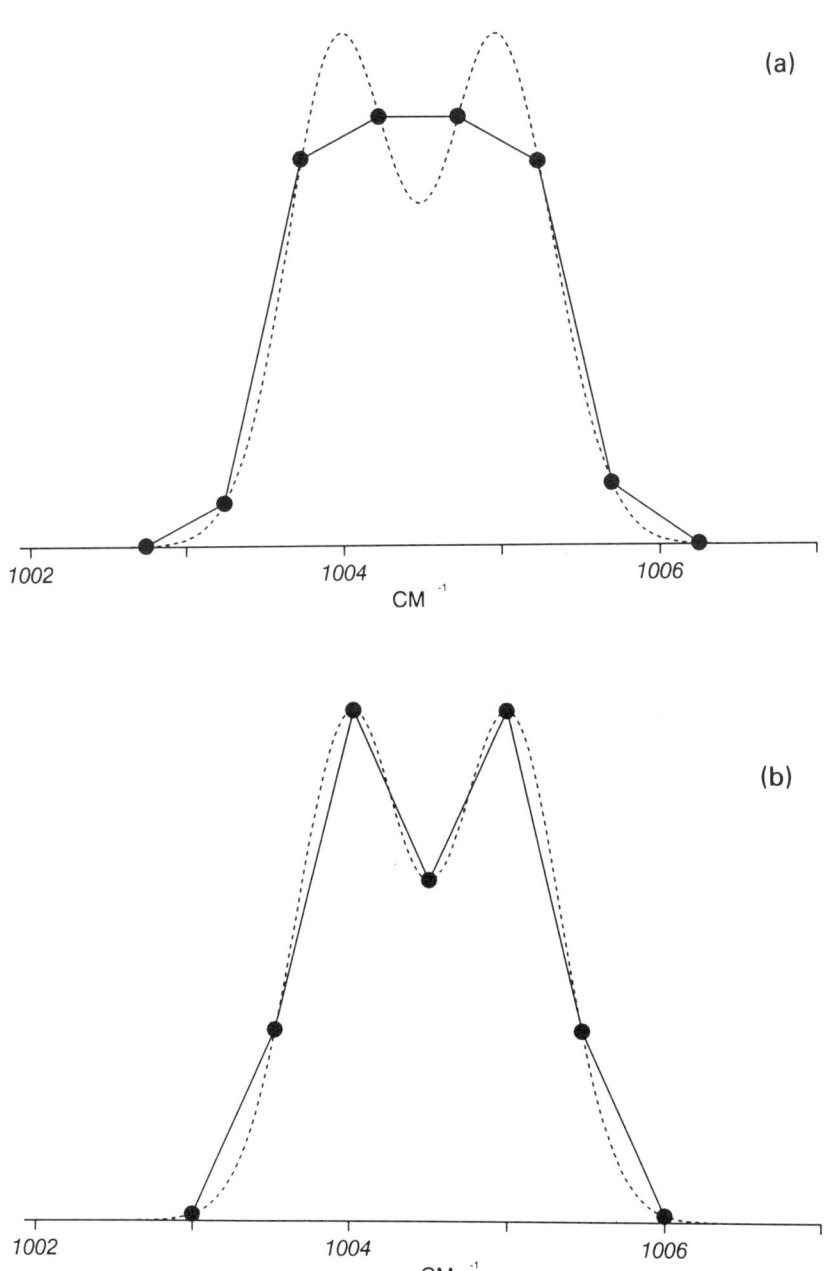

Fig. 4.3 — Two overlapping bands showing the effect of data-point position. In both (a) and (b) the dashed line shows the true profile and the solid line shows the digitized spectrum. The data-point pitching is the same in both (a) and (b), but in (b) they are displaced.

always be remembered that it is the data point rather than the band position that is being quoted. When it is desired to measure small band shifts it is, therefore, essential to have a high density of data points in the spectrogram.

There are several methods of increasing the data-point density. The most legitimate method is to scan the movable mirror further and collect more data points in the interferogram. However, this is time consuming and the interferometer mechanism will have a limited range of optical delay. Alternatively, there are two data processing techniques that can be applied. Perhaps the simplest is to insert extra data points (of appropriate intensity) between the existing data points in the spectrogram, and use a computer routine to smooth the spectrum. However, this does degrade the resolution of the spectrum. Another technique frequently encountered is 'zero filling' [5]. Additional data points of value zero are added to the ends of the interferogram, and the FFT algorithm transforms these to produce extra points in the spectrogram. This is not quite as ludicrous as it sounds, because the values of the data points in the wings of the interferogram are in fact very close to zero anyway. As a result, the density of the data points in the spectrogram is greater, but again the spectral resolution is degraded. It is not unusual to add seven times as many zeros as real data points actually collected, but this will increase the computing time. The benefits of zero filling are illustrated in Fig. 4.4.

The interferogram, in theory, continues out to an infinite distance on either side of the centre burst. Therefore, in a practical spectrometer the interferogram must be deliberately curtailed (or truncated). This inevitably leads to errors, and the calculated spectrum does not perfectly represent the light entering the interferometer. For most purposes these errors are insignificant, but there are a few situations where problems may occur. If a single sharp monochromatic source illuminates the interferometer, truncation errors appear in the FT spectrum as 'ringing'. The characteristic side lobes produced around the main band are shown in Fig. 4.5. Ringing is frequently encountered with samples which have very narrow bands, such as gases. Ringing may also be observed in spectra which have been excessively zero filled. The problem may be reduced by a process called apodization (see Appendix), but this once again degrades the spectral resolution. Spurious side lobes may well obscure weak real spectral features and/or cause confusion.

Another problem associated with the truncation of the interferogram is the accurate representation of both the intense centre burst and the much weaker side features (see Fig. 4.2(c)). In order that both can be recorded meaningfully, high resolution A–D conversion is essential. Usually, 16-bit systems are used, so that the centre burst can be up to almost 33000 times greater than the weakest feature digitized, whilst still accurately depicting both.

4.4 THE ADVANTAGES OF THE INTERFEROMETER IN RAMAN SPECTROMETRY

It is generally accepted that the interferometer has several advantages which make it superior to the monochromator when used in infrared spectroscopy. It is the purpose of this section to explain these advantages and to illustrate their relevance in interferometer-based Raman spectrometers.

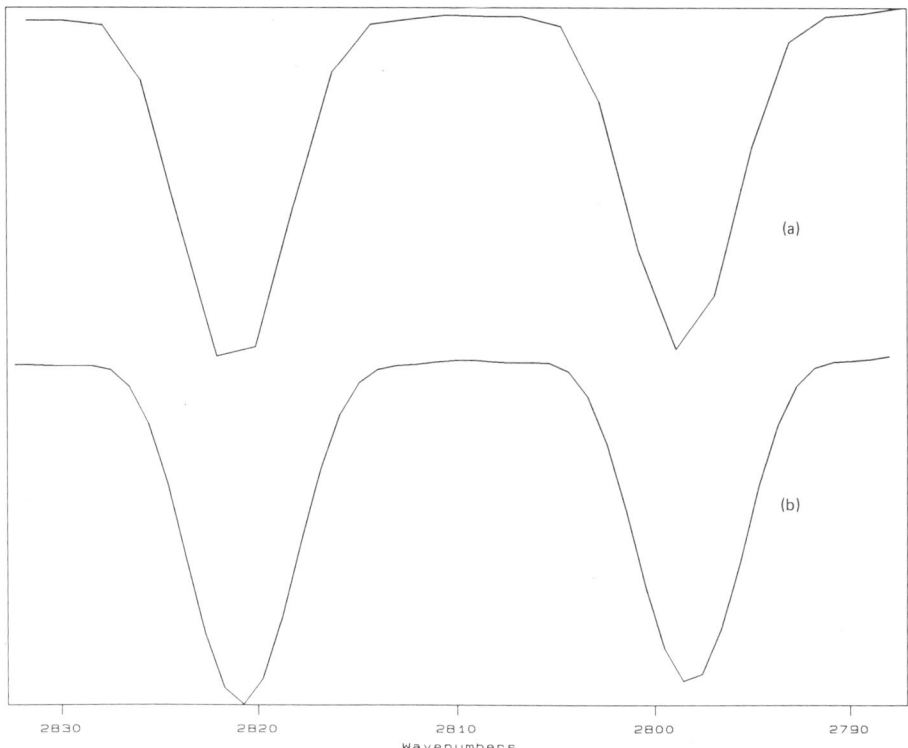

Fig. 4.4—Part of the infrared spectrum of HCl recording with (a) no zero filling clearly showing the spectrum as a crude histogram, and (b) two times zero filling to double the data-point density.

4.4.1 The Jacquinot advantage

The amount of Raman light produced in an experiment is very small, so it is important to collect and analyse this light efficiently. The monochromator and interferometer differ fundamentally in the way they perform this task, but both have restricted optical efficiencies. Monochromators have an inherently low throughput. The diffraction grating produces a series of images only one of which is passed through the exit slit. In a scanning instrument the efficiency is further reduced because it samples only a small fraction of the Raman light at any given time. If a multiple monochromator system or narrow slit-width (high resolution) is used, the fraction of Raman light passing the entrance and exit slits is further diminished. The situation is made even worse by reflective losses from mirrors and diffraction gratings. The interferometer is also optically inefficient, because half of the radiation entering the device is reflected back to the source. As in the monochromator, there are also losses resulting from imperfections in the optical components.

To compare the optical efficiency of instruments quantitatively, we use a measure called the étendue. This parameter is the product of the half angle of the cone of light collected [$\tan^{-1}\{1/(2 \times f$ number of the collection optics)$\}$] and the area of the

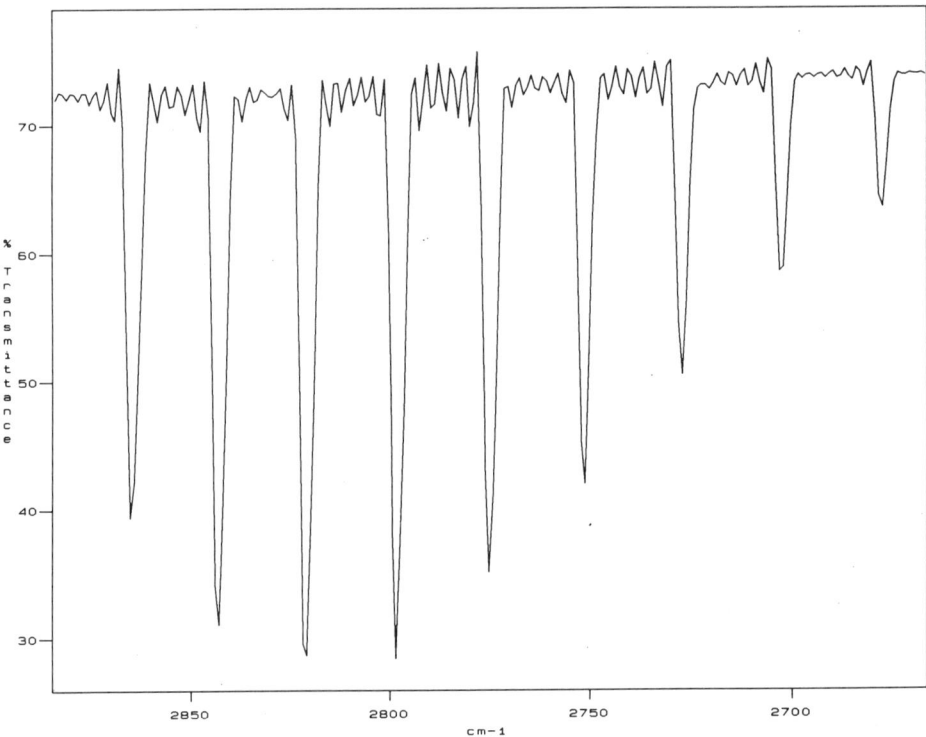

Fig. 4.5 – The infrared spectrum of hydrogen chloride showing the effect of 'ringing'. Each band in the spectrum is accompanied by characteristic side lobes.

entrance aperture. The collection optics of conventional and FT Raman spectrometers have similar f numbers. However, whereas the light entering a monochromator must pass through a narrow slit (typically 60 µm per cm^{-1} spectral bandpass), the entrance to an interferometer is a large circular hole, known as the Jacquinot stop, which is typically 8 mm in diameter.† The area of an 8-mm Jacquinot stop is $\approx 5 \times 10^{-5}$ m^2, compared with $\approx 7 \times 10^{-7}$ m^2 for a slit 600 µm wide and 1.2 mm high. Thus, for spectra run at 10 cm^{-1} resolution, the étendue of an interferometer is approximately 70 times greater than that of a monochromator. The advantage at 5 cm^{-1} resolution is even greater. In FT spectroscopy this improvement is known as the Jacquinot advantage [6].

4.4.2 The Felgett advantage

A scanning monochromator makes very inefficient use of the light entering it, as only a small range of wavelengths fall on the detector at any given time. The rest of the light is simply wasted. A spectrograph on the other hand is a far more efficient device

† In an interferometer with collimating optics of focal length 250 mm ($f7$), an 8-mm diameter Jacquinot stop will restrict resolution to 4 cm^{-1}. For higher resolutions, the diameter of the Jacquinot stop must be reduced and hence the étendue falls.

because the entire spectrum, or a major section of it, is recorded simultaneously and continuously. Thus for a given measurement time, a spectrograph will detect far more photons and have a vastly superior signal-to-noise ratio than a scanning instrument. This is called the multiplex advantage.

An interferometer shows a similar advantage first described by Felgett [7]. In effect, all wavelengths of light entering the device are detected simultaneously. Because the interferogram detected is the sum of the interference patterns generated by each wavelength, each data point contains some information about every wavelength. However, to obtain the interferogram, the moving mirror must be scanned for a finite time (usually a few seconds). The spectrogram can be calculated only once a complete interferogram has been recorded. In this sense the interferometer is like a very rapid scanning instrument, and the measurement time is an integer multiple of the time for one scan. This differs from a spectrograph, where the measurement time is a continuous variable. The Felgett advantage applies only if the performance of the instrument is limited by detector noise, otherwise it may be lost completely or become a disadvantage if the instrument is shot-noise limited [8].

4.4.3 The Connes advantage

In theory, all spectrometers can show an improved signal-to-noise ratio if spectra are averaged. However, this relies on the fact that the spectra can be exactly superimposed. Any displacement error between spectra will cause band shapes to be distorted and the signal-to-noise ratio will, as a result, fail to improve. Scanning monochromators are subject to mechanical wear, which may cause significant displacement errors. Interferometers have the advantage that the frequency scale is generated from the He–Ne laser, the wavelength of which is invariant and known very precisely. This enables spectra to be exactly superimposed and co-added. This is called the Connes advantage [9].

4.5 CONCLUSION

The relative performances of monochromator based instruments and those based on interferometers can be compared in two ways; the time taken to record spectra of equivalent signal-to-noise ratio and resolution, or the signal-to-noise ratio recorded in a given measurement time. Under the same recording conditions, because an interferometer will always have the advantages outlined previously, it will perform considerably better than a monochromator.

REFERENCES

[1] J. B. J. Fourier, *Théorie Analytique de la Chaleur*, Firmin Didot, Paris, 1822.
[2] R. P. Wayne, *Chem. Brit.*, 1987, **23**, 440.
[3] J. W. Cooley and J. W. Tukey, *Mathematics of Computation*, 1965, **19**, 297.
[4] P. R. Griffiths, *Chemical Infrared Fourier Transform Spectroscopy*, Wiley, New York. 1974.
[5] E. O. Brigham, *The Fast Fourier Transform*, Prentice Hall, New Jersey, 1974.
[6] P. Jacquinot, *J. Opt. Soc. Am.*, 1954, **44**, 761.
[7] P. Felgett, *J. Phys. Radium*, 1958, **19**, 187.
[8] J. Chamberlain, G. W. Chantry and N. W. B. Stone, *The Principles of Interferometric Spectroscopy*, Wiley, New York, 1979.
[9] J. Connes and P. Connes, *J. Opt. Soc. Am.*, **56**, 896 1966.

5

The development of FT Raman spectroscopy

5.1 PIONEERING EXPERIMENTS

The origins of the FT Raman technique, like those of so many phenomena familiar to scientists, date back to pioneering experiments made many years ago. As early as 1964, Gebbie, Chantry and Hilsum demonstrated, as part of their long-term research into the use of interferometry in spectroscopy, that Raman measurements were feasible by using near infrared laser sources [1]. Their observation was overwhelmed by their invaluable work in the far infrared and was ignored. In the mid 1980s, interest revived because the experimental difficulties faced by Gebbie and his colleagues had largely been overcome. Several scientists became involved at that time, and a number of preliminary papers were published [2–5]. Their efforts produced a Raman system that has formed the basis of all subsequent work. Chase [2,3] found that the most successful illumination source was the neodymium-doped yttrium aluminium garnet (Nd^{3+}:YAG) solid-state c/w laser operating at 1.064 µm. Scattered light was collected and filtered to remove the laser radiation, and the remainder was passed into a near-infrared interferometer. Output from the interferometer was collected, and focused onto a cooled germanium detector, the electrical signal of which was processed by conventional FTIR hardware. Preliminary observations showed that, even though their instruments lacked the sensitivity of conventional Raman setups, many coloured and fluorescent materials hitherto inaccessible became easy to study. For example, a spectrum produced early in the programme was that of sodium fluorescein showing *no* fluorescent background.

On the experimental side, the pioneers made rapid and significant progress. A variety of sampling systems and light collection optics were devised and evaluated. The filtering process was thoroughly investigated, several interferometers were studied and detectors of various types assessed. In parallel with the early work, Weber, whose reputation in the study of gas-phase spectra is considerable, devised a spectrometer best suited to vapour-phase spectroscopy and showed excellent results [5]. Perhaps the most helpful contribution we can make at this stage is to describe Chase's instrument as it was in operation in 1985 (see Fig. 5.1).

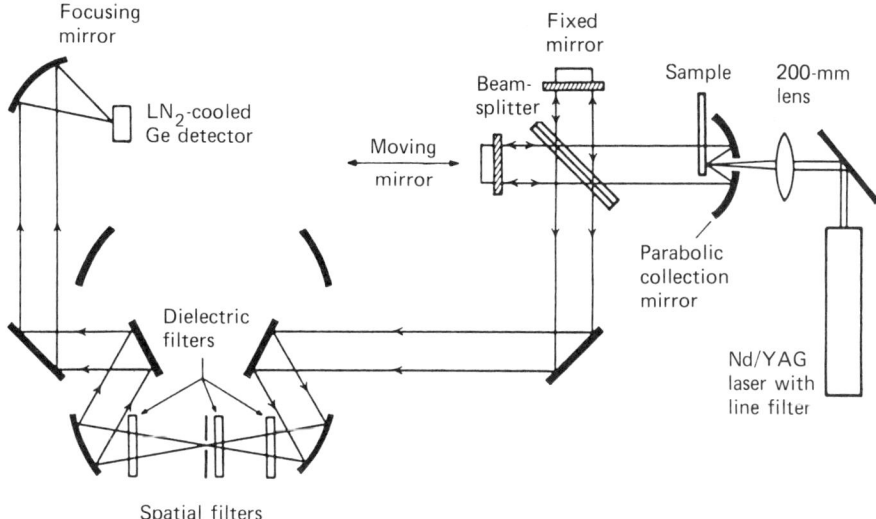

Fig. 5.1 — The spectrometer used for Chase's early FT Raman experiments. The Raman light was collected with a parabolic mirror, passed through a Nicolet 7199 interferometer and detected with a Judson germanium detector at 77 K. Reproduced with permission from D. B. Chase, *J. Am. Chem. Soc.*, **108**, 7485. (1986) Copyright American Chemical Society.

At that time, the favoured method of illumination for the sample was that typical of most conventional spectrometers, i.e. the 90° approach. The inability to view the laser beam without the appropriate aids (near infrared viewer and phosphor cards) makes alignment of the laser sample and collection optics incredibly difficult. The problem was solved by projecting output from a small He–Ne laser along the optical axis of the Nd^{3+}:YAG device. This was achieved either by selectively picking up the He–Ne beam off a removable prism, or by passing the beam through the optical cavity of the near-infrared laser. Obviously, once alignment of the sample was approached, the He–Ne laser is turned off, laser safety goggles were donned and the Nd^{3+}:YAG laser was switched on. Precise location and adjustment of the sample position was achieved by using a high quality micropositioner. Scattered light was collected either by mirror or lens optics, and passed to the Jacquinot stop of the interferometer. Chase's preference for 90° illumination in the mid 1980s probably resulted from his use of interferometers with small Jacquinot stops. The patch viewed at the sample was small (typically a disc of diameter 300 µm). Chase's arrangement was manufactured at this time as a commercial unit by Aries Inc. of Concord Massachusetts. This was normally fitted by the customer to either the very excellent Bomem research-grade FTIR, fitted with a calcium fluoride or quartz beam-splitter, or to the large Nicolet 6700 series instruments.

In 1986, the detector of choice was the cooled germanium device, although it was known that its limited long wavelength sensitivity made detection of the Raman bands at shifts beyond 3000 cm^{-1} inaccessible (see Fig. 5.2). Chase and Rabolt performed a considerable service by encouraging manufacturers of the indium

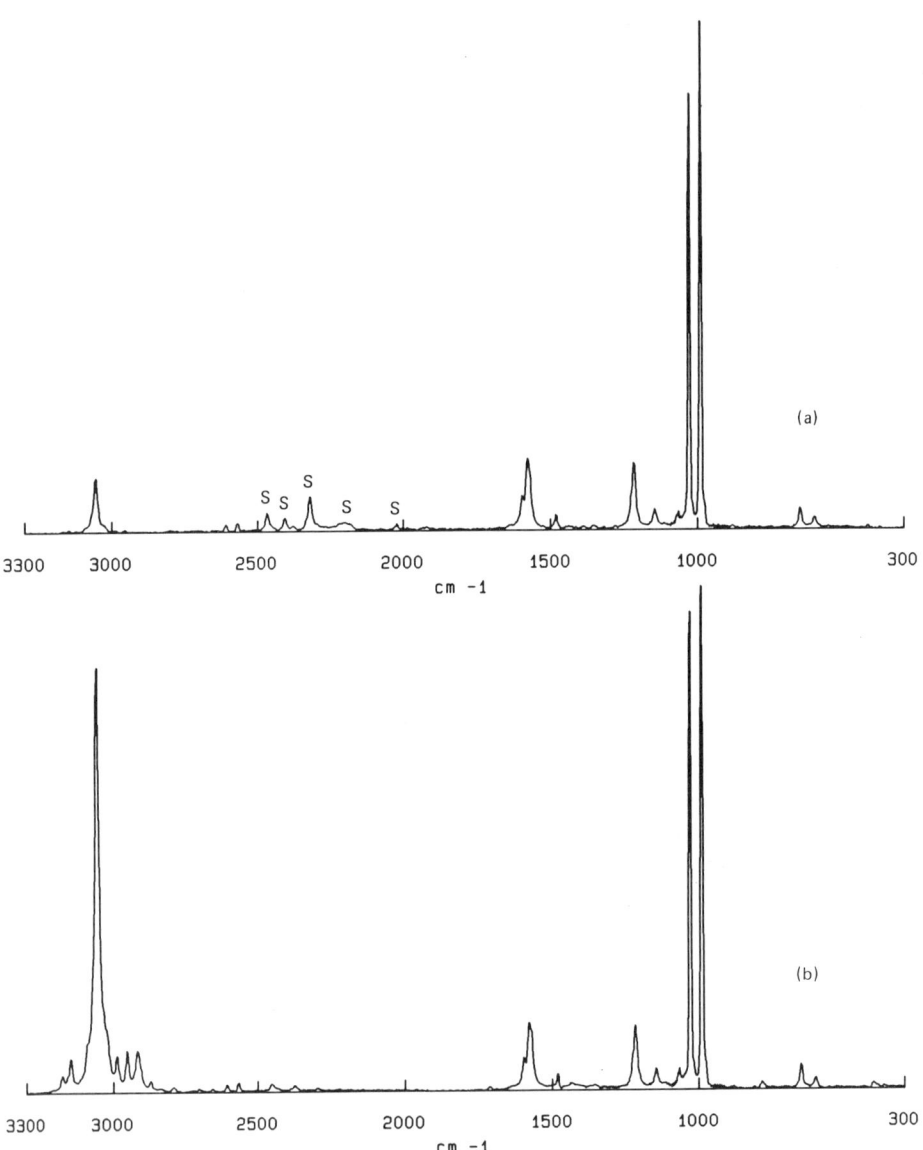

Fig. 5.2 — (a) The FT Raman spectrum of pyridine recorded using a Judson germanium detector at 77 K. The C–H stretching bands around 3000 cm^{-1} are very attenuated. The bands marked S are spurious and are due to plasma lines from the Nd^{3+}:YAG laser. (b) The FT Raman spectrum of pyridine recorded using an Epitax InGaAs detector at room temperature which covers the full spectral range. Uncorrected data.

gallium arsenide detector to refine their products, and hence laid the foundation for all the commercial instruments reported later in this chapter. The InGaAs detector operates to slightly longer wavelengths (and hence higher shifts) than the germanium, but is noisier.

Sec. 5.2]   **The development of analytical FT Raman spectrometers**   93

So far we have not described the crucial filter arrangements, because in the mid 1980s Chase was actively involved in assessing several different approaches. His instrument was configured to allow him to use two methods–reflection and transmission. To reiterate, the collected light is almost entirely composed of elastically reflected and scattered radiation at 1.064 µm. The Raman contribution is minute. To remove the 'laser' radiation, there must be very efficient rejection at that frequency. On the other hand, the filter system must 'cut in' as steeply as possible to allow the user to approach the laser line and hence see Raman bands at small shifts. In the mid 1980s this was (and still is) very hard to do. Chase's solution was to use either four rejection filters in a Chevron arrangement (see Chapter 3, Fig. 3.2) or three or four multilayer dielectric transmission filters. In Fig. 5.3 we show the transmission

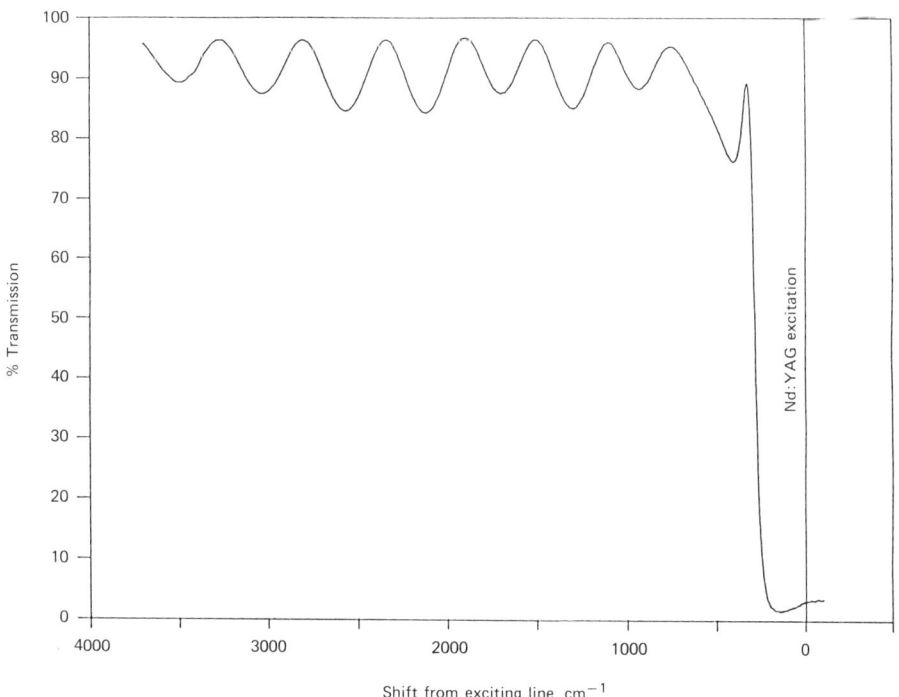

Fig. 5.3 — The transmission characteristics of a Barr Associates multilayer rejection filter as a function of wavenumber shift from the Nd^{3+}:YAG excitation line.

properties of a modern transmission filter suitable for Raman work. This is certainly superior to those available to the pioneers.

5.2 THE DEVELOPMENT OF ANALYTICAL FT RAMAN SPECTROMETERS

After the publication of reports by Chase and Rabolt, it was agreed between one of the authors (P. J. H) and Perkin Elmer UK Ltd. that it would be worthwhile attempting to build an analytical-grade FT Raman instrument. This was an important step in the development of the technique, since all previous experiments had used very high quality and hence expensive research-grade instruments. A successful series of feasibility trials was completed in the Spring of 1987 using a Perkin Elmer model 1710 FTIR fitted with a quartz beam-splitter, germanium detector and several borrowed lasers [6]. The results demonstrated that it was possible to produce good quality spectra with a low cost interferometer. After this work, a fully operational FT Raman spectrometer was developed at Southampton University, so as to explore the potential of the technique as a routine analytical tool. In collaboration with many industrial colleagues, many varied applications in a wide range of areas were studied, and a series of publications has resulted [7–17]. The spectra that were recorded at this time were of excellent quality and adequate for most applications. Consequently, a decision was made to improve the versatility and ease of use of the instrument, at the expense of a slight reduction in performance. The instrument was fully enclosed and the laser interlocked with the sample-area lid so that it could be used in the open laboratory [7]. Originally, a mirror collection arrangement was used (see Fig. 5.4) but since this severely constrained the size of the sample that could be studied, it was replaced by a compound lens system. This permitted samples to be mounted on

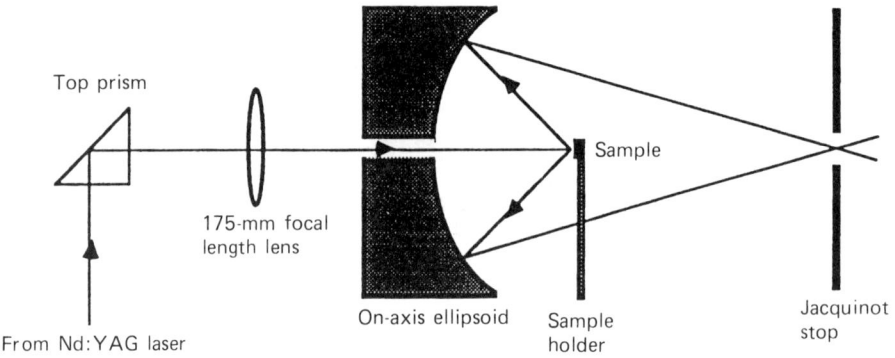

Fig. 5.4 — A diagram of the original on-axis mirror collection optics used on the initial Southampton prototype.

standard 3×2-in. infrared sample cards which slotted easily into the spectrometer at a prealigned position. As a result, no sample alignment was required, a marked improvement over the classical technique. An InGaAs detector operating at room temperature was used, so that the full spectral range was available.

In the original prototype at Southampton the Nd^{3+}:YAG laser was coupled to the spectrometer through an optical fibre. However, as soon as a smaller laser was available from Spectron Lasers Ltd., it was incorporated inside the spectrometer case. A full optical diagram of the prototype is shown in Fig. 5.5, and along with a photograph of the instrument as it was shown at the International Conference on Raman Spectroscopy held in London in September 1988 (see Fig. 5.6). The fully

Sec. 5.3]   **The essential elements of an FT Raman spectrometer**   95

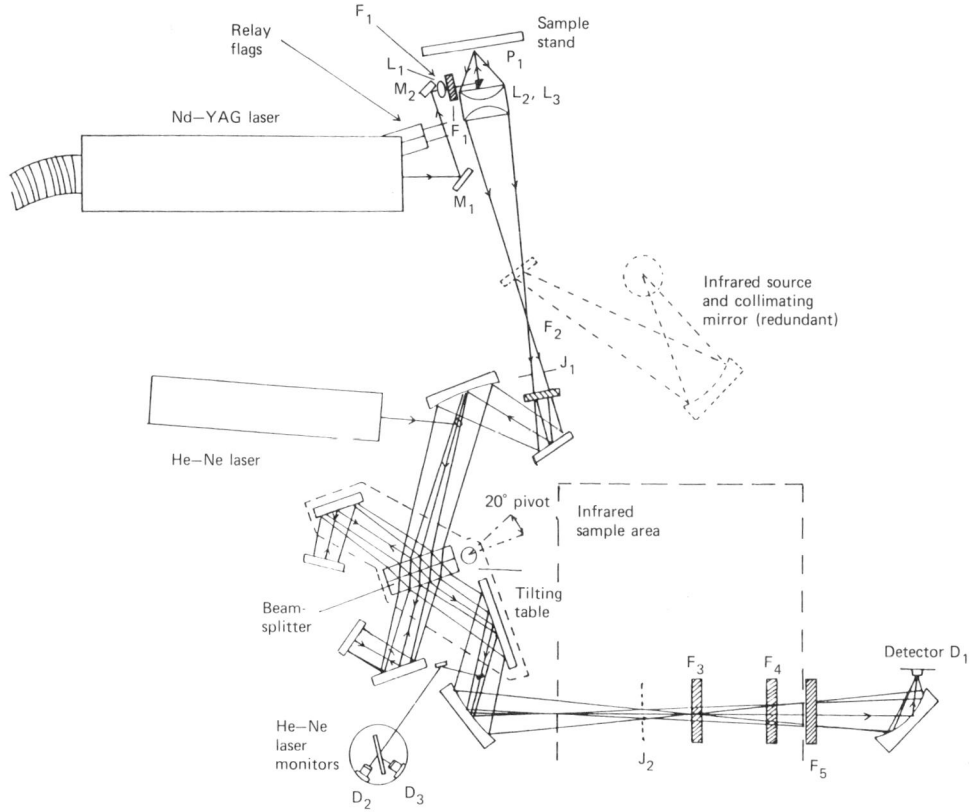

Fig. 5.5 — A complete optical design of the 1710 prototype FT Raman spectrometer.

developed Southampton machine formed the basis for the production version of the Perkin Elmer system which has recently been announced and is described later in this chapter.

In the following two years, work continued to expand the applications of FT Raman. In April 1990 a special edition of *Spectrochimica Acta* (**46A** No. 2) appeared, devoted exclusively to the instrumentation and applications of the technique, many of them developed at Southampton. This was used as the work book for a summer school held in April 1990.

5.3 THE ESSENTIAL ELEMENTS OF AN FT RAMAN SPECTROMETER

Here we describe the basic components of our prototype spectrometer. Later, we will describe the commercial instruments currently available, highlighting the different solutions employed by different manufacturers.

It can be seen in Fig. 5.5 that radiation from the laser is reflected off mirror M_1 and through two metal 'flags' mounted on solenoids which are controlled by

Fig. 5.6 — A photograph of the 1710 prototype as exhibited at the 11th International Conference on Raman Spectroscopy in London, September 1988.

proximity switches on the sample-area lid. When the lid is closed, the solenoids are energized and the flags are pulled out of the beam, allowing the radiation to pass to M_2 through lens L_1 and onto filter F_1. The filter passes the 1.064 μm laser radiation, but blocks the white light emitted by the high-pressure xenon lamp which powers the laser. The radiation then passes through prism P_1 onto the sample. The illuminated patch at the sample is about 0.5 mm in diameter. If L_1 is removed, and this is easy to do, the patch is expanded to about 1.5 mm in diameter.† P_1 can be rotated or tilted, permitting us to centre precisely the laser beam at the sample along the viewed axis of the spectrometer.

Reflected and scattered light is then collected by the lens pair L_2 and L_3. To minimize reflection losses, L_2 is plano-convex and L_3 biconvex. The collected radiation is focused onto the Jacquinot stop J_1 through a filter F_2. The filtered radiation then passes through the interferometer which is of modified Michelson design, involving two fixed mirrors and a rocking beam-splitter/mirror assembly. The combined beam passes into the infrared sample area, where it forms an image at a virtual Jacquinot stop J_2. Just downstream of this position, two more filters identical to F_2 are placed. Filters F_2, F_3 and F_4 remove the radiation at the laser wavelength and permit us to record shifts down to $\Delta v = 300 \text{ cm}^{-1}$ on a routine basis. The filter characteristics are drawn in Fig. 5.3, from which it will be clear that, although

† Where samples are damaged by heating by the laser beam, the consequent reduction in brightness can be of value.

effective, they are far from ideal in that their transmission varies greatly with wavelength.

Radiation leaving the infrared sample area then passes through filter F_5 and then onto a pair of mirrors. The second of these is an off-axis ellipsoid, and this focuses the radiation upwards onto the vertically mounted detector D_1. In Perkin Elmer 1700 series interferometers, a small He–Ne laser illuminates a miniature mirror, at the front of the interferometer, which passes the radiation along the optical axis. The transmitted He–Ne light is collected on a pair of detectors D_2 and D_3 after reflection off a polarizing reflector. The sine wave generated on D_2 and D_3 indicates the optical delay and the phase relationship between D_2 and D_3 indicate the direction of scan. The He–Ne radiation continues along the optical axis and appears at the sample as a small red spot. It also would fall on the main detector D_1 and interfere with the recording of the interferogram, but F_5 removes it.

The main detector D_1 is a 2-mm diameter InGaAs device, specially selected to produce a low level of dark noise, and is operated at room temperature. On occasion, we may mount the detector in a cryostat and cool to either -80 or $-190°C$. The effect is to reduce the dark noise by a significant amount ($\times 2$ or slightly better) but the 'absorption edge' of the device moves to the blue and severely attenuates the performance of the detector at high shifts. The performance characteristics are illustrated in Fig. 5.7. The effect of cooling on the spectrum is illustrated in Fig. 5.8. We consider that the use of detectors which operate over restricted ranges is highly undesirable. In an analytical environment, it is inevitable that over the years incomplete spectra derived from cold detectors will become confused with complete ones, and misleading diagnoses are inevitable.† Since perfectly adequate performance can be obtained with room-temperature devices, and these cover the range out to shifts of $3500 \, cm^{-1}$, we consider that it is irresponsible for commercial manufacturers of analytical equipment to promote cooling facilities until improved detectors are fitted as standard. A liquid-nitrogen-cooled, extended-range germanium detector is offered by North Coast Optical Systems Inc. (Santa Rosa, California). This has a very low dark current and is sensitive out to 3500-cm^{+1} shift. However, this detector is very sensitive to cosmic radiation, and this currently limits its wide-scale application.

One or two other points are worthy of explanation. It will be noticed that we place F_2 in front of the interferometer and F_3 and F_4 after it. F_2 prevents most of the near infrared radiation at the laser wavelength from entering the interferometer. This attenuation prior to the interferometer is essential, because otherwise the He–Ne detectors D_2 and D_3 become 'blinded' and lose information about the optical delay of the interferometer. The problem could be solved if the appropriate filter were placed in front of D_2 and D_3. However, better rejection is achieved if the filters are separated, rather than placing them all in the infrared sample area. Therefore, we place F_2 at the Jacquinot stop J_1. Filters change their characteristics if they are tilted with respect to the optical axis [18]; the effect is shown in Fig. 5.9. It can be seen that for our Barr Associates filters, if the angle of tilt is a little more than $10°$, the radiation

---

† v_{CH} in aromatic compounds frequently produces very weak bands, but in aliphatic materials these bands can be the strongest in the spectrum. If the detector is 'dead' in this region, assignments may easily be confused.

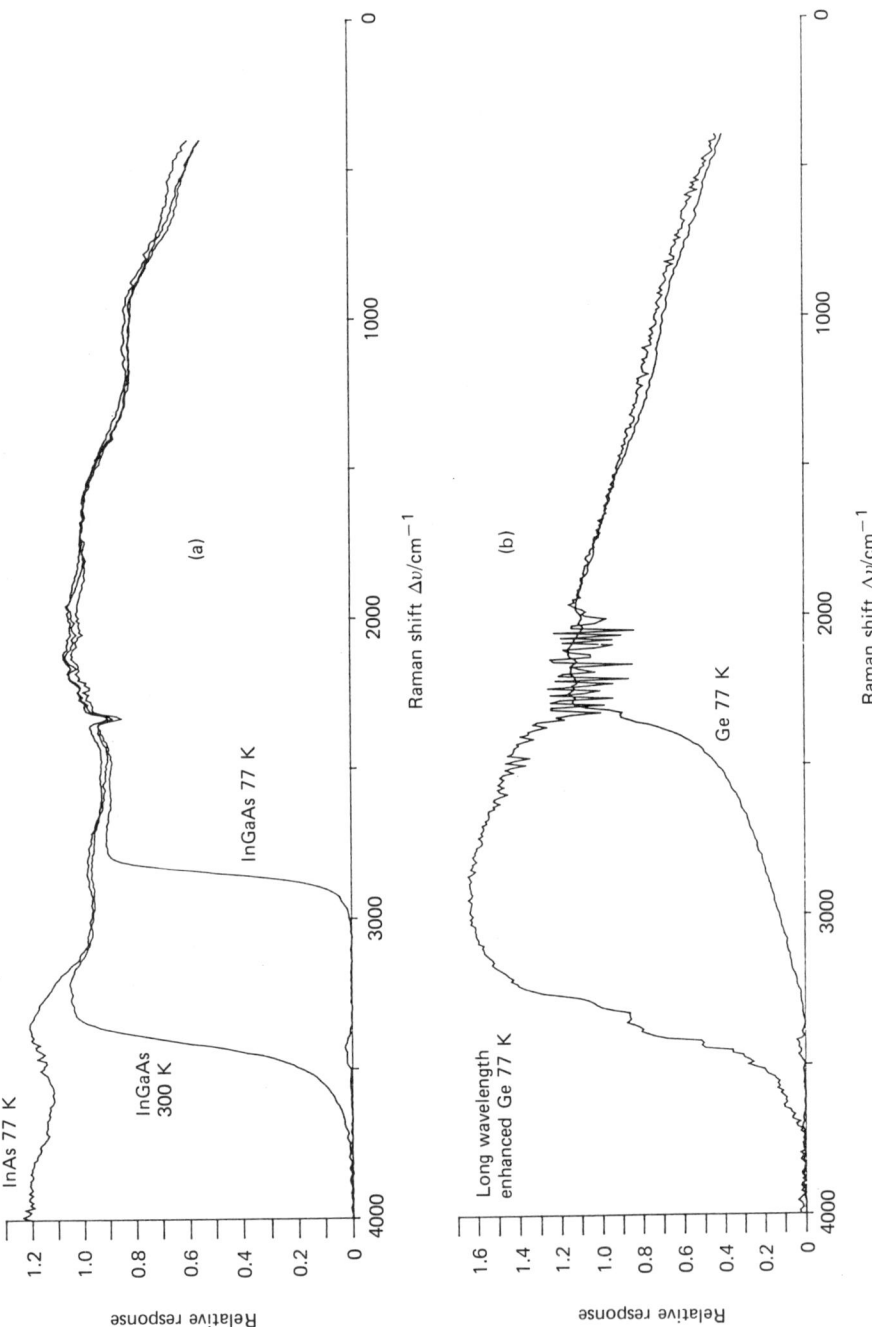

Fig. 5.7 — The characteristics of different detectors suitable for FT Raman spectroscopy; (a) InAs at 77 K, InGaAs at 300 K and InGaAs at 77 K; (b) an extended-range germanium detector at 77 K compared with a Judson germanium detector at 77 K. Detector responses normalized to 1.3 μm. Courtesy of Dr D. J. Cutler.

Sec. 5.3]　　　**The essential elements of an FT Raman spectrometer**　　　99

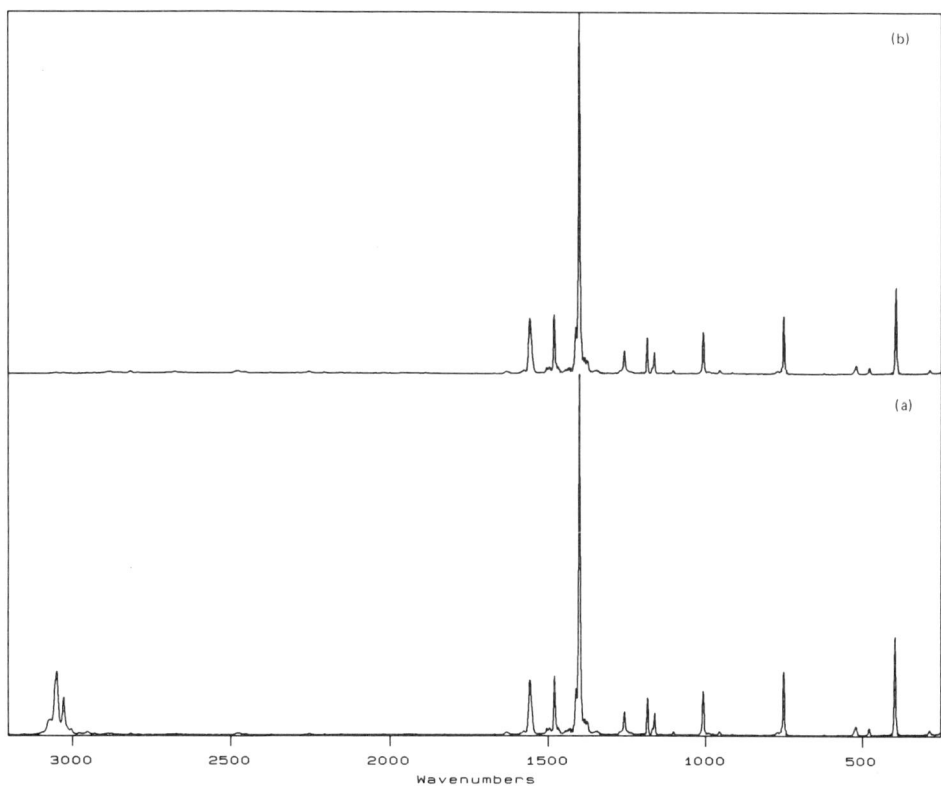

Fig. 5.8 — The FT Raman spectra of anthracene recorded using an InGaAs detector, (a) at room temperature, and (b) cooled to 77 K. Uncorrected data.

at the laser wavelength will still be adequately attenuated, but it will be possible to approach closer to the exciting line than when normal incidence is used.† An example of the successful use of filter tilting to observe low frequency Raman bands is given in Fig. 5.10.

The InGaAs detector has a very high impedance (typically in excess of 15 MΩ) and thus a specially designed low-noise preamplifier has to be employed to amplify the signal and feed it into the A–D converter. This is triggered by the He–Ne fringes, and the apodization, scan speed etc. are controlled by the normal FTIR electronics. The data is displayed in standard format (with high Raman shifts to the left). The detectors and photographic emulsions used in conventional Raman spectroscopy have always been less sensitive at the red end of the visible spectrum. As a result, Raman bands appearing at high shifts have been artificially attenuated when compared with those at lower shifts. The variation in the instrumental sensitivity with wavelength was a smooth featureless curve and as such was largely ignored.

† It is important that even when the filters are tilted that the rejection at the laser wavelength of each filter is in excess of 98%.

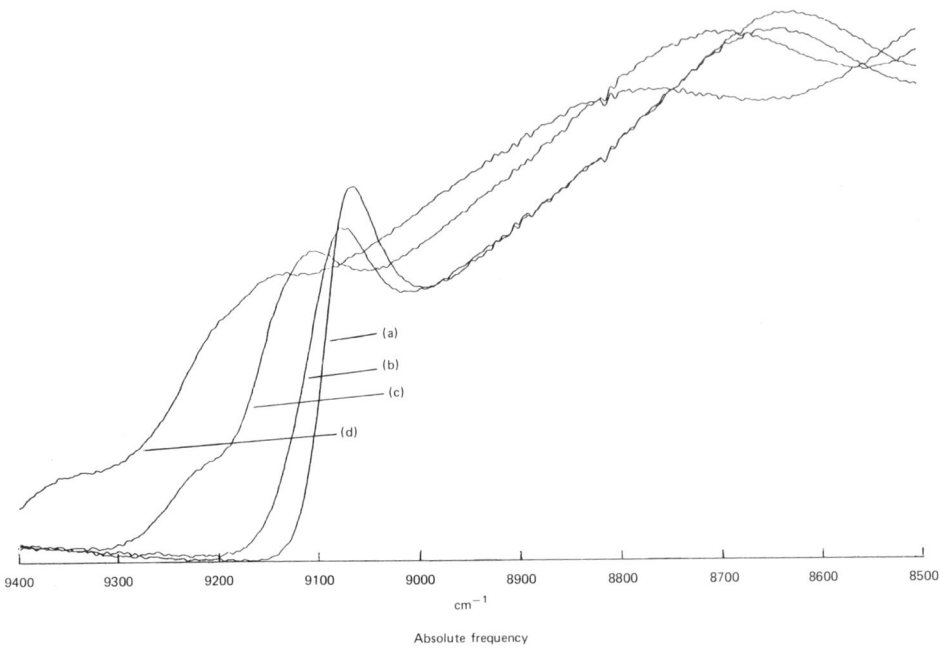

Fig. 5.9 — The transmission characteristics of a typical rejection filter when tilted with respect to the optical axis at (a) 0°, (b) 6°, (c) 12° and (d) 18°.

However, the detector and filter combinations in FT Raman spectrometers give rise to an oscillating sensitivity profile, so the spectra recorded are considerably distorted. Different filters and detectors all have markedly different sensitivity characteristics, and thus the overall sensitivity profile will vary from instrument to instrument. Fig. 5.11 shows raw FT Raman spectra of sulphur recorded by different instruments. It can be seen that the relative band intensities are very dissimilar. However, it is relatively simple to produce a correction function which compensates for the filter and detector characteristics (see Fig. 5.12) [19]. Multiplying raw data by this curve will adjust relative band intensities and remove artificial oscillations in fluorescent backgrounds (Fig. 5.13). The preparation of correction curves and the variation in Raman spectra obtained with different excitation wavelengths are discussed in Chapter 6.

Users of FT Raman instruments will inevitably compare their spectra with those in the literature, which have been recorded almost exclusively with visible lasers. At first sight the spectra may appear quite similar, but this is deceptive. As a rule, the efficiency of Raman scattering is a function of v^4, where v is the frequency of the light being scattered, so the relative band intensities in Raman spectra will vary with the wavelength of the excitation source. In conventional Raman spectroscopy this has only a slight effect, since the visible laser sources have similar wavelengths. In the near infrared, the values of v are halved, so Raman spectra excited in the near infrared are fundamentally different from those recorded in the visible. Bands at

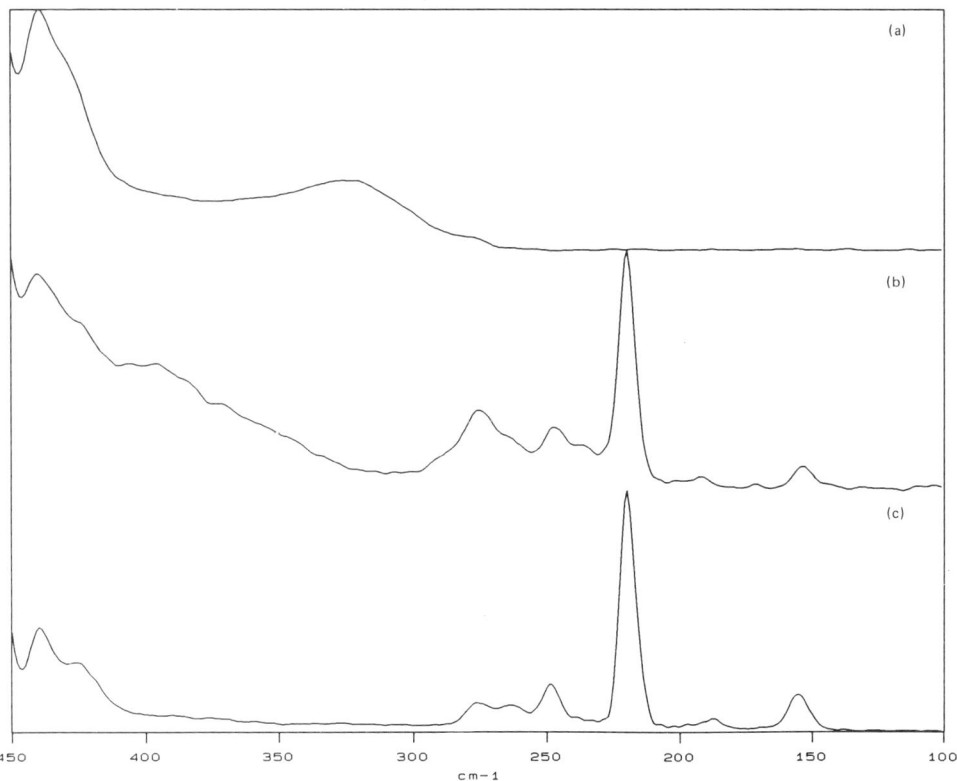

Fig. 5.10 — The FT Raman spectrum of sulphur recorded with the rejection filters tilted with respect to the optical axis at (a) 0°, (b) 6° and (c) 13°. Uncorrected data.

high Raman shifts recorded in the visible should appear to be more intense than those recorded in the near infrared. However, spectra recorded in the visible are not generally corrected for the falling detector sensitivity at high shifts, and quite by chance spectra excited by the visible and near infrared look the same.

The laser we have used to date is relatively unstable and its intensity fluctuates about 1.5% at full power and by considerably more at lower levels. The effect of this variation in source intensity is to multiply the interferogram by a completely random noise function. This then transforms to give spikes and increased noise in the spectrogram. Fortunately, the instability occurs at the same frequency of the mains electricity supply (50 Hz in the UK) and its first few harmonics. This is a much lower frequency than the beat of the Raman interferogram, and fortunately most of the spurious effects of the laser instability appear outside the FT Raman range in the frequency domain. However, if the interferometer scanning speed and hence the data acquisition rate were reduced, laser instability could be far more problematic.

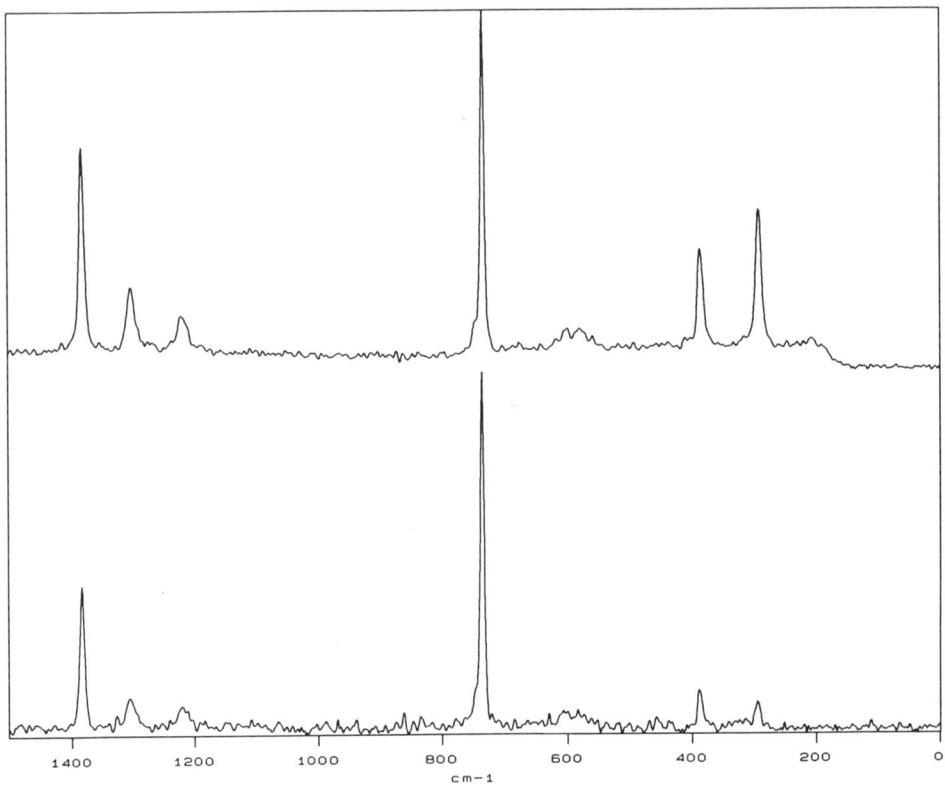

Fig. 5.11 — Uncorrected FT Raman spectra on PTFE recorded of different instruments using different detectors and filters.

The signal-to-noise performance of a spectrometer is always limited by one facet of its design. In present FT Raman spectrometers the quality of spectra obtained is constrained by the inherent noise of the detection system. As detector technology is improved another element of the spectrometer will limit its performance, and this is most likely to be the stability of the excitation source. This is certainly the case in conventional Raman spectroscopy where the detection systems are very quiet indeed. The fluctuation in the output of diode-pumped Nd^{3+}:YAG lasers is very much less than that of their discharge-lamp-pumped predecessors, often better than 0.1% RMS. These devices are very much smaller than the laser technology traditionally used as they do not require water cooling or bulky power supplies. As the power output of diode-pumped lasers increases, and their cost falls, they will be perfect excitation sources for FT Raman spectrometers. Indeed, instrument manufacturers are beginning to offer them as an option in their FT Raman systems (see Section 5.6).

As the sensitivity of FT Raman instruments improves, spectroscopists will attempt to record smaller and smaller Raman signals. Somewhere in the noise they may find a few surprises! They are several well known effects that can generate

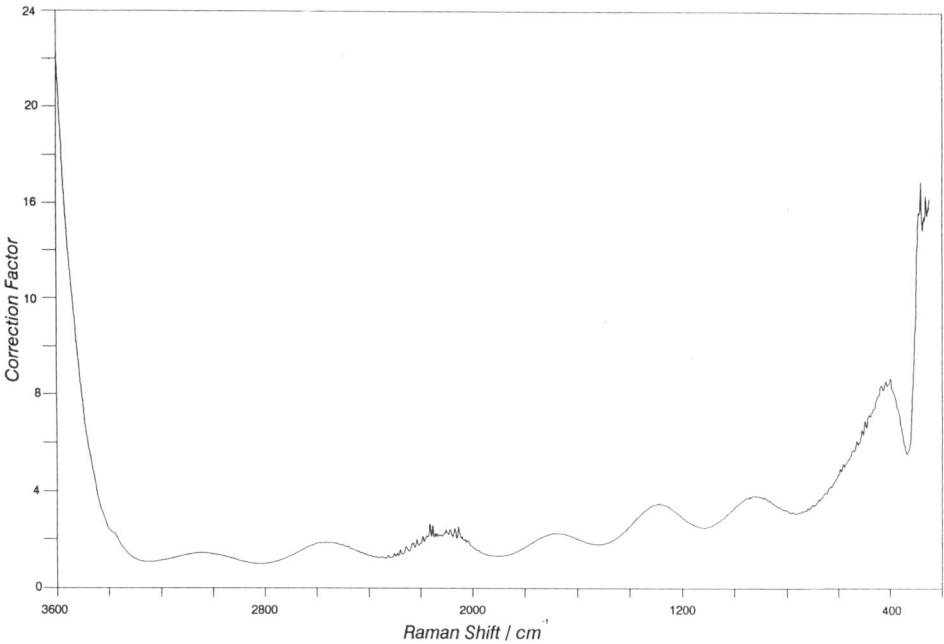

Fig. 5.12 — A typical correction function used to compensate for the instrumental sensitivity curve.

entirely spurious lines in FT Raman spectra, some of which can be reasonably intense. The xenon discharge lamps used to pump current Nd^{3+}:YAG lasers produce a series of emission lines in the FT Raman range. The sample under study will either reflect or Rayleigh scatter these emissions, and spurious lines will be produced in the spectrum which could be easily mistaken for real Raman bands. For this reason it is essential to 'clean' the output from the laser by using either a laser premonochromator or a notch pass filter. Similar near infrared lines are also emitted by the He–Ne lasers used to frequency-calibrate FT Raman instruments. Unless a suitable filter is placed in front of the He–Ne laser massive spikes can appear in the FT Raman spectra, particularly if this reference laser is coaxial with the Raman light. Sharp emission lines will also produce the 'ringing' sidelobes discussed in Chapter 4, and the very sharp bands found in the FT Raman spectra of gases will inevitably cause the same problem.

All of the signal processing electronics in an FT Raman spectrometer are subject to electromagnetic interference. Thus, as well as the true optically generated signal, the interferogram measured may contain other cosinusoidal components resulting from this interference. These will be transformed and appear as spurious spikes in the spectrogram. For instance, on the Southampton prototype the 50 Hz oscillation of the mains electricity supply appears as a sharp line around 100 cm^{-1} absolute. However, spurious lines from this source can be identified by changing the scanning speed of the interferometer, as the frequency they appear at in the spectrogram will

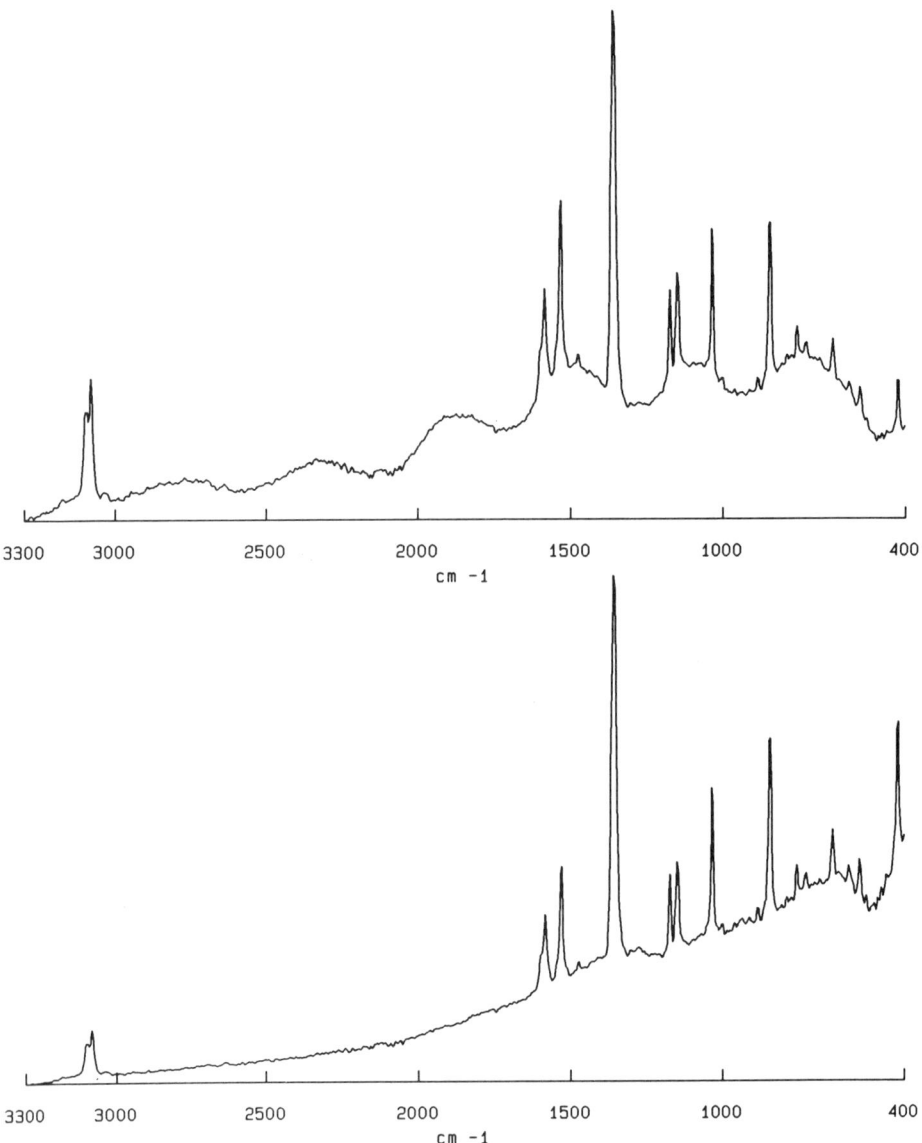

Fig. 5.13 — The raw and corrected FT Raman spectrum of a fluorescent impure sample of *ortho*-dinitrobenzene.

alter. It is desirable to screen the detector and amplifier electronics as much as possible to reduce the effects of electromagnetic interference. Similar problems may also result if the interferometer is subject to excessive vibration. If the interferometer experiences a sudden jolt, a sharp spurious spike appears in the interferogram and this transforms as a sinusoidal oscillation across the entire spectrogram.

Sec. 5.4] Sampling in FT Raman spectroscopy

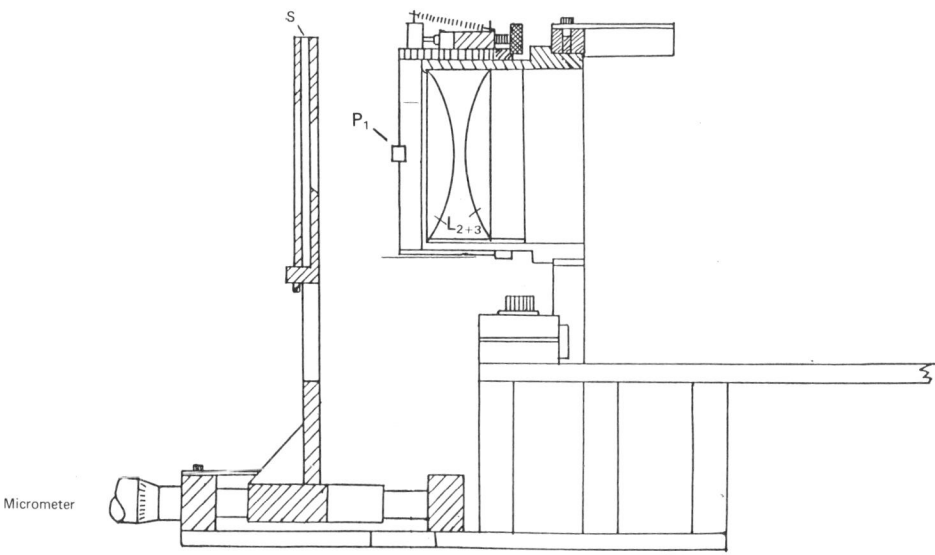

Fig. 5.14 — The sample area of the Southampton prototype FT Raman spectrometer. S, sample stand: P_1, laser reflection prism: L_{2+3}, collection lenses.

Other spurious lines can result from the method by which the interferogram is sampled. As mentioned in Chapter 4, the maximum frequency of light which can be measured on an FT instrument is governed by the sampling frequency. This defines the so called Nyquist limit. In an instrument using a He–Ne laser as a reference, and recording two data points per He–Ne wavelength, the Nyquist limit is 15 800 cm^{-1}. However, light at greater frequencies than this may still be passing through the interferometer and the detection system may pick up electromagnetic interference at a frequency greater than the sampling rate. If the Nyquist limit is NL, the interference pattern generated from light of frequency NL+X is indistinguishable from that generated by light of frequency NL−X. This effect is known as aliasing. Consequently, higher frequencies will 'fold back' round the Nyquist limit and produce spurious bands at lower frequencies which may well appear in the FT Raman range. FT Raman instruments incorporate electronic and software filters to try to eliminate this problem, but these will never work perfectly and small spurious bands are still observed. One final point concerns frequency shifts. In Chapter 4 it was explained that if an interferometer is not illuminated correctly, errors of several wavenumbers can occur in band frequencies. This will be apparent if spectral subtractions are attempted, when differential band shapes will be produced. These may be incorrectly interpreted as real band shifts and considerable care must be taken when performing studies of this sort.

5.4 SAMPLING IN FT RAMAN SPECTROSCOPY

The Southampton prototype FT Raman machine was developed for use as a routine instrument, so efforts were made to make sampling as easy as possible. The lens collection optics do not restrict the size of the sample and a variety of special cells have been developed. All sample cells are mounted on 3"×2" in standard infrared sample cards and these slot into a holder in the sample area (see Fig. 5.14). The He–Ne reference laser is split into three beams inside the interferometer, and these project out through the collection optics. The patch viewed by the spectrometer is marked by the coincidence of these three beams. The Nd^{3+}:YAG laser is prealigned so that it always illuminates the viewed patch. Thus, to align a sample in the spectrometer, all that is required is to slot the cell into the holder and move it backwards or forwards with the micrometer until the three He–Ne spots converge. This can usually be seen by eye from the side, but when bulky cells such as cryostats are being used, a small dentists' mirror is often useful. The whole procedure takes a few seconds. Any fine tuning of the sample alignment can then be accomplished by closing the sample lid, switching on the laser and monitoring the Raman signal while adjusting the micrometer. However, this generally makes very little difference as the focus at the viewed patch is very soft and the signal measured is very insensitive to sample position. The elegant simplicity of this sampling system stems from the use of a 180° backscattering geometry with a prealigned Nd^{3+}:YAG laser and for analytical purposes it is ideal. The following paragraphs discuss the cells devised for various samples.

5.4.1 Liquids

Generally, spectra from liquids are recorded by using silvered spherical glass cells or glass cuvettes, sometimes with the back surface silvered (see Fig. 5.15). The

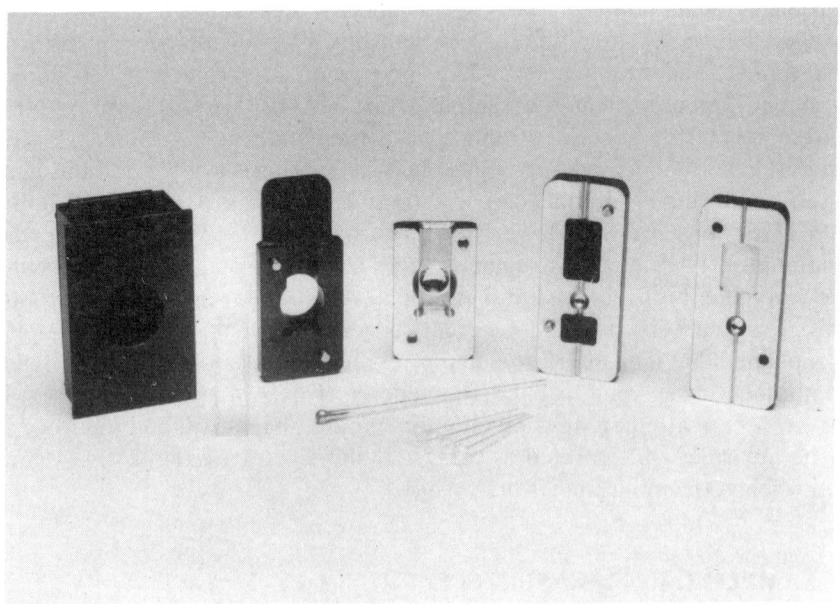

Fig. 5.15 — Typical FT Raman liquid cells. Courtesy Perkin Elmer Ltd.

spherical cells act as reflective integrating spheres and when their radius is optimized (about 7 mm) they are around five times better than an unsilvered cuvette [20]. However, they do not work so well with liquid samples that absorb Raman light, where it is important that the Raman light passes through only a short path-length of liquid. Very acceptable spectra can also be recorded by using half-silvered 8-mm sample bottles, which are cheap enough to be disposable.

5.4.2 Solids

Solid powders are usually compacted into cells consisting of a 3-mm hole drilled through a brass cylinder, into which a reflective metal rod is placed to compress the powder (see Fig. 5.16). The Raman intensity observed is related to the thickness of

Fig. 5.16 — Typical FT Raman solid cells. 2 cells to left courtesy of Perkin Elmer Ltd.

the sample, but does not increase once the sample depth exceeds 2 mm. Micro solid cells have been made with 0.7-mm diameter holes from which spectra of 0.006 mg of a strong Raman scatterer have been recorded. Where toxic or air-sensitive materials are studied a glass fronted cell can be used but there is an 11% drop in performance owing to reflection losses off the glass [8].

5.4.3 Gases

One of the earliest papers on FT Raman showed beautiful spectra of gases recorded at pressures of around 600 torr [5]. These experiments used visible excitation, and the data were collected for many hours with a highly specialized interferometer developed for astronomical research. Preliminary gas-phase experiments have been carried out at Southampton with near infrared excitation and a cell consisting of a narrow gold-coated glass tube (see Fig. 5.17). So far, only the Q branches of nitrogen and oxygen in air have been observed, but other cell designs and detectors are being devised and evaluated.†

5.4.4 Low temperatures

To date, no-one seems to have reported measurements at really low temperatures, but we have recorded several spectra, using cryostats containing liquid nitrogen as refrigerant. An example is shown in Fig. 5.18. Of the transfer-gas type, the cell holds liquid and solid samples at temperatures down to $-150°C$. We have also had some success at Southampton with Peltier-cooled devices. In these, the hot junction is water-cooled and the sample is held against the cold one. The whole Peltier cooler is isolated in a glass and metal evacuated cavity.

5.4.5 High temperatures

Blackbody emission by the sample restricts the temperature that can be used in near infrared excited instruments, but 200°C is just possible before the emission background becomes unacceptably intense. A cell designed at Southampton is shown in Fig. 5.19. It is incredibly simple, consisting of a machined-down miniature thermostatted soldering iron, held in our usual $3'' \times 2''$ in infrared card holder. The temperature at the cut-down copper solder tip is monitored with a type-K thermocouple. The tips are drilled with a 3 mm diameter, 2 mm deep hole to hold solids whilst liquid samples are contained in a 4 mm diameter partially silvered spherical glass cell.

5.4.6 Electrochemical measurements

The electrochemical cells designed in Southampton consist of a working electrode mounted in a KelF cylinder surrounded by a platinum-wire counter-electrode. A reference electrode with its attendant capillary enter from the side of the cell. The details of the cell are fairly self-evident and are described in Chapter 10, Fig. 10.4.

5.4.7 Adsorbed species

The Pyrex cells used for the study of adsorption processes are almost trivially simple, consisting of a horizontal tube ending with an optical flat 15 mm in diameter. The tube is fitted with a PTFE grease-free vacuum tap, to allow it to be connected to a vacuum system (see Fig. 10.16). The tap stopper is removable so that bulk catalyst can be placed in the tube. The complete cell will easily hold 1 g of sample and has a total mass of about 70 g. Adsorption processes are thus simple to follow gravimetrically with a conventional balance. Details of operation are again given in Chapter 10.

5.4.8 Microscopy

Conventional Raman spectrometers have often been fitted with a microscope attachment. This allows the Raman spectrum of a very small area of the sample to be collected. Not only does this permit the use of very small samples, but it also makes it possible to investigate variations in the structure of a sample with fine spatial resolution. By using a computer-controlled sample stage it is possible to produce an 'image' of the sample by plotting the frequency or intensity of a Raman band as a function of position [21]. Several papers have appeared in the literature showing

†A steadily increasing range of materials have now given vibrational spectra including CS_2, Cl_2+CH_3Br.

Fig. 5.17 — A very early FT Raman gas cell.

reasonably good FT Raman spectra recorded by using a microscope [22–25]. In none of these cases have the microscopes been fully optimized to work in the near infrared nor with an interferometer, and hence there is much room for improvement in the quality of the spectra produced. One commercial manufacturer, Bruker, has announced a microscope accessory.

5.4.9 Fibre optic coupling

Optical fibres developed to date perform extremely well in the visible and near-infrared regions of the spectrum. With this in mind, it has been suggested that fibre-optics could provide an excellent way of performing a Raman experiment remote from the spectrometer [26]. For instance, many fibre-optic probes could be placed in a chemical plant and interfaced to one spectrometer. The reduction in fluorescence obtained by using near-infrared excitation makes this a very real possibility. Williams and Mason [27, 28] have obtained excellent results by using a specially designed fibre-optic probe connected to an FT Raman spectrometer. The probe consists of a central fibre which carries the Nd^{3+}:YAG laser beam, surrounded by eighteen other fibres which collect the Raman light. By using this apparatus it has been possible to study remotely the kinetics of polymerization reactions.

5.4.10 Kinetic experiments

A prototype cell has been constructed at Southampton to perform reaction kinetics experiments using FT Raman spectroscopy (see Fig. 5.20). A small silicone oil bath is electrically thermostatted and stirred to maintain a constant temperature. This is fronted with a glass window behind which a small spherical liquid cell is held in the

Fig. 5.18 — Liquid nitrogen Dewar-based FT Raman cold cell.

oil. This slots into the spectrometer and the FT Raman spectrum of the thermostatted sample held in the liquid cell can be recorded. This apparatus has been used to study a nitration reaction for which the reaction order, rate constant and activation energy was derived from the spectra obtained.

5.5 LASER SAFETY

Unlike conventional Raman spectrometers, commercial FT Raman instruments have been designed to operate in a normal analytical laboratory alongside other pieces of equipment. In all cases, the sample area is enclosed, with a totally opaque lid fitted with more than one microswitch (or reedswitch) wired in series and interlocked to the laser power supply. The purpose of these is to switch off the laser if the sample lid is opened for any reason. It must be remembered, however, that the Nd^{3+}:YAG laser found in an FT Raman spectrometer produces a high-power

Fig. 5.19 — FT Raman high temperature cell.

invisible beam. Thus it is potentially very dangerous if proper safety precautions are not taken when it is operated. At Southampton, the instrument is prominently labelled to show that a powerful invisible laser is inside, and access to the room containing the instrument is restricted to those familiar with the hazards posed by lasers.

Occasionally it is necessary for the sample-lid interlock to be overridden; i.e. during installation, servicing† and repair, the instrument has to be operated in a potentially hazardous condition. On these occasions, only those workers who are working on the spectrometer are allowed in the room. Laser safety goggles that are completely opaque to radiation at the laser wavelength are worn. These are absolutely vital, particularly as the invisible nature of the beam means the user can be completely unaware of any stray laser reflections present in addition to the main beam. Similarly, care is taken to stop any stray laser beams from leaving the laboratory by using curtains.

Over-riding the sample lid interlocks requires the use of a key, and its use is recorded in a log book. As soon as the sample interlock is overridden other interlocks fitted to the laboratory door are automatically activated and warning lights above this door are illuminated. Whilst the sample-lid interlock is overridden, a warning buzzer should sound in the laboratory to remind workers that the laser is exposed. Should anyone inadvertently enter the laboratory while the laser is unsafe, the laser power supply automatically turns off.

---

† The Nd^{3+}:YAG laser used at Southampton is pumped by a discharge lamp, and this is replaced routinely every six months.

Fig. 5.20 — FT Raman kinetic cell. Courtesy Ventron Ltd.

5.6 PURCHASING AN FT RAMAN SYSTEM

At the time of writing several manufacturers have just launched FT Raman spectrometers. Many potential customers will have had little or no experience of Raman spectroscopy, and thus it seems appropriate to include a few guidelines on what to look for in good FT Raman spectrometer. The sensitivity of an FT Raman instrument is crucial since it decides which experiments are possible to perform. In theory, the quality of a spectrum can be improved by averaging more and more scans of the interferometer, and the signal-to-noise ratio will improve as a linear function of $\sqrt{x}$, where x is the number of co-added scans. However, this takes more time and consequently the cost per spectrum rises. A recent cost analysis comparing the throughput of samples for a number of analytical techniques has shown that FT Raman is very cost effective, on the assumption that a relatively small number of rapid scans would suffice in each case [29]. Alternatively, the signal-to-noise ratio can be improved by increasing the laser power, but since this increases the possibility of sample damage, it is not always an available option. Thus, it is always better to have a more sensitive spectrometer in the first place, because this allows the operator to use lower laser powers and shorter measurement times.

In principle, the signal-to-noise ratio should provide an absolute method of comparing the quality of two spectra. However, signal-to-noise ratios are rarely measured by a systematic method such as a standard computer program, but rather they tend to be estimated by eye. Consequently, values quoted are often highly subjective and this should be taken into account when making comparisons. Thus, in evaluation of the performance of an FT Raman spectrometer, rather than relying on the signal-to-noise ratios quoted in sales literature, it is a good idea to ask the manufacturer to record spectra from a few of your own samples, and compare the spectra obtained with your own eyes. For example, polytetrafluoroethylene (Teflon or ICI Fluon are typical and widely available) and sucrose (icing or castor sugar) are reasonably difficult samples from which to obtain good spectra. PTFE has low frequency bands, whilst sucrose is a relatively poor scatterer which has a very weak v_{OH} band near 3300 cm^{-1}. Requesting spectra run under the following conditions should enable you to make a fair comparison.

Laser Power: 400 mW (ask whether the laser was focused or unfocused — see Section 5.3).
Acquisition Time: 2 minutes (this should be approximately 25 scans, but request that the number of scans and their duration be stated).
Resolution: 4 cm^{-1} (this should be 1 data point every 2 cm^{-1}, see Section 4.3, but ask that this be confirmed).

In Fig. 9.16a you will notice that a very weak feature is present in sucrose near 3550 cm^{-1}. This is genuine as it is seen on visible laser excited instruments. Some F-T Ramm machines will find it.

It is also important to know what sort of detector was used when the spectra were recorded and whether this is fitted as standard. The very high frequency v_{OH} band may be lost if the detector is cooled to improve spectral quality (see Section 5.3). In order to be able to compare the spectra, they should all have been corrected for the instrumental sensitivity profile (see Section 5.3). In Fig. 5.21 we show fully corrected spectra of PTFE and sucrose recorded on our prototype machine. Since our battered old instrument is thoroughly worn out, manufacturers should certainly provide improved spectra. Do not be surprised if manufacturers offer widely different spectra or more likely insist on using more scans or more laser power than you request. You can make the obvious deductions.

Several commercial instruments incorporate 90° and even 0° excitation in addition to 180° back-scattering. Whilst these offer experimental flexibility, 180° back-scattering is much the most convenient if the instrument is to be used in a routine analytical environment. This system has the advantage that the Nd^{3+}:YAG laser can be prealigned and there is no restriction on the size of the sample. Unnecessary sample alignment will increase the cost per spectrum, and may have laser safety implications (see Section 5.5). The Southampton prototype is used as part of an undergraduate practical, and enables the unskilled students to produce spectra in minutes although they have never encountered the technique before. If such convenience and safety can be provided at Southampton, we feel that it should be on offer from all of the instrument manufacturers.

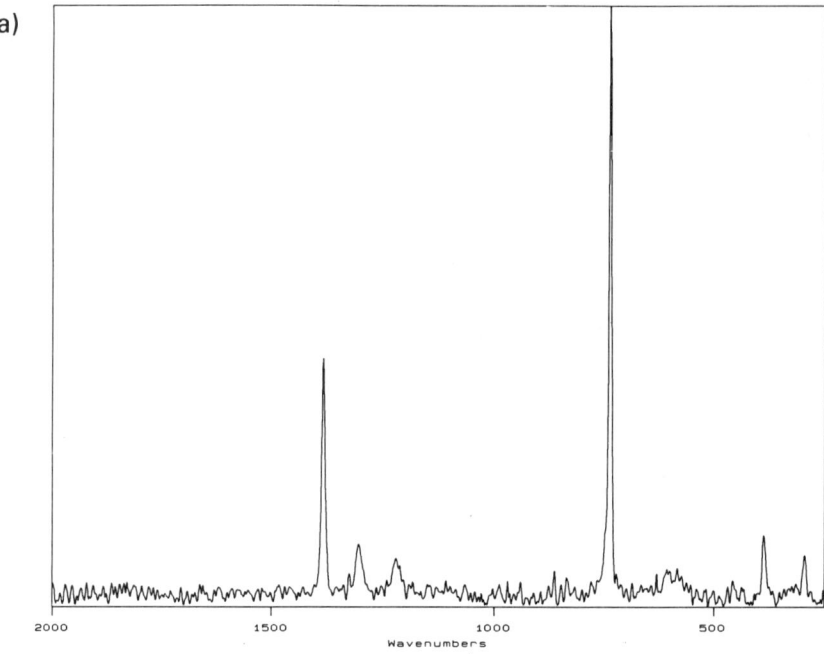

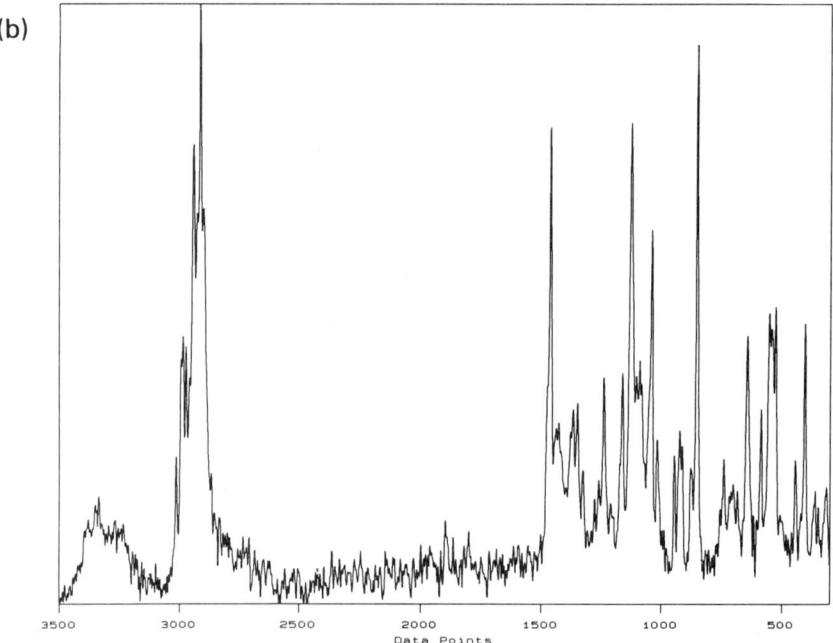

Fig. 5.21 — The FT Raman spectra of (a) PTFE, and (b) sucrose recorded on the Southampton Prototype at 4 cm^{-1} resolution with 400 mW of laser power taking 2 minutes to acquire.

Most of the commercial instruments so far announced have very small and restrictive sample areas — this is most odd and we suspect originates from those more concerned with product image. Our prototype may look ugly, but the sample area will accommodate cryostats, vacuum cells for adsorbed species, high-pressure cells and *in situ* electrochemical cells. One or two manufacturers have suggested that if large samples or cells are to be studied, the user can leave the lid open, but the laser safety consequences of this are completely unacceptable (see Section 5.5).

5.7 SURVEY OF CURRENTLY AVAILABLE COMMERCIAL INSTRUMENTS

By the Spring of 1991 five manufacturers (BioRad, Bomem, Bruker, Nicolet and Perkin Elmer) had announced products. Their exact specifications are still changing to include latest developments, but our survey will give readers a feel for the type of systems being marketed at the time of writing. Several manufacturers have quite rightly pointed out that by the time this book appears their specific components may well have altered. Each manufacturer was asked to provide a photograph, an optical diagram and to complete a pro-form designed to enable us to assemble a specification. The outcome is offered in the following pages in a very abbreviated form. Since we have not been able to try all of the instruments we can make no comments on the relative performance of the different manufacturer's products. After perusing all the information available the following points can be concluded:

(a) All of the machines use Nd^{3+}:YAG laser sources but their stability is very variable.
(b) All operate from $\approx$150 to 3500 cm^{-1} Stokes shift, and some below 150 cm^{-1} and the anti-Stokes region.
(c) All provide appropriate computer processing allowing infrared and Raman spectra to be displayed together.
(d) Considerable variation exists in the sample areas, one manufacturer being restricted to an on−axis mirror, two providing back-scattering lens systems and the remainder off-axis mirrors. As a result there is a considerable difference between each manufacturer's ability to study a wide range of samples. On-axis mirrors cannot be used to study anything other than small samples, and hence preclude the use of hot and cold cells, electrochemical cells, adsorbed species cells, etc. Adjustment of the sample in the beam and ease of use seem to be very variable. We do not have a complete set of data on the size of the sample areas, but our impression is that they are all miserably small. Certainly, a Dewar-based cold cell is impossible to mount in any of these instruments with the lids closed.
(e) All manufacturers seem to provide resolutions near the laser linewidth limit.

BIO-RAD
BIO-RAD
Digilab Division,
237, Putnam Avenue,
Cambridge,
Massachusetts 02139.
USA.

The Bio-Rad FT Raman accessory fits the FTS 60, FTS 60A and the FTS 40 series spectrometers to produce a combined near infrared/Raman bench. A fully combined FTIR/FTR system can be purchased, or the Raman accessory can be fitted to already existing spectrometers provided that they incorporate a quartz beam-splitter. The FTS 60A is capable of step-scan operation which may be useful in FT Raman experiments using pulsed lasers and time-resolved studies. A Nd^{3+}:YAG (1064 nm) laser manufactured by Spectron Laser Systems Ltd., Rugby, UK (model 301) is used as the excitation source providing up to 3 W at the sample (stability <0.1% RMS). Alternatively, a CVI laser (model: C-95cw) which has a maximum power at the sample of 750 mW (stability <1.0% RMS), or a diode-pumped Nd^{3+}:YAG laser are available as options. A laser power meter is built into the system.

The sample optics are based on a gold-coated off-axis mirror, and the operator can easily switch between 180° and 90° illumination. The focal length of the collection mirror is 1.2 in (30 mm). To collect a large solid angle of Raman scatter the mirror is deeply curved. This limits the size of the sample studied to about 2×2 in (50×50 mm) cross-section. The depth of the sample is limited to 6 in (150 mm) by the spectrometer case. Samples are held on an x, y, z, stage which allows movement of 20 mm in each direction. Sample holders for NMR tubes, capillaries and face-mounted solids are supplied with the machine. Additionally, Specac accessories can also be used. An additional He–Ne laser, colinear with the Nd^{3+}:YAG laser to facilitate sample alignment, is available as an option.

The standard dielectric filtering system allows shifts from 3500 to 120 cm^{-1} to be recorded. However, if the optional notch filter is purchased then Stokes bands from 2500 to 75 cm^{-1} and anti-Stokes bands from 100 to 1000 cm^{-1} can be observed. The detector generally employed is a thermoelectrically cooled InGaAs model. Alternatively, an extended-range germanium (North Coast Optical Systems Inc., Santa Rosa, California) detector can be used for high sensitivity applications. A 'white light' source is incorporated in the system to align the interferometer and to allow FT Raman spectra to be corrected for the varying instrument response. Data processing is achieved using the Bio-Rad workstation on which infrared and Raman spectra may be viewed, scrolled and plotted together. Resolutions of up to 0.1 cm^{-1} are available (at two data points per resolution element), but of course the best resolution obtainable in FT Raman mode is limited by the linewidth of the Nd^{3+}:YAG laser.

Sec. 5.7] Survey of currently available commercial instruments

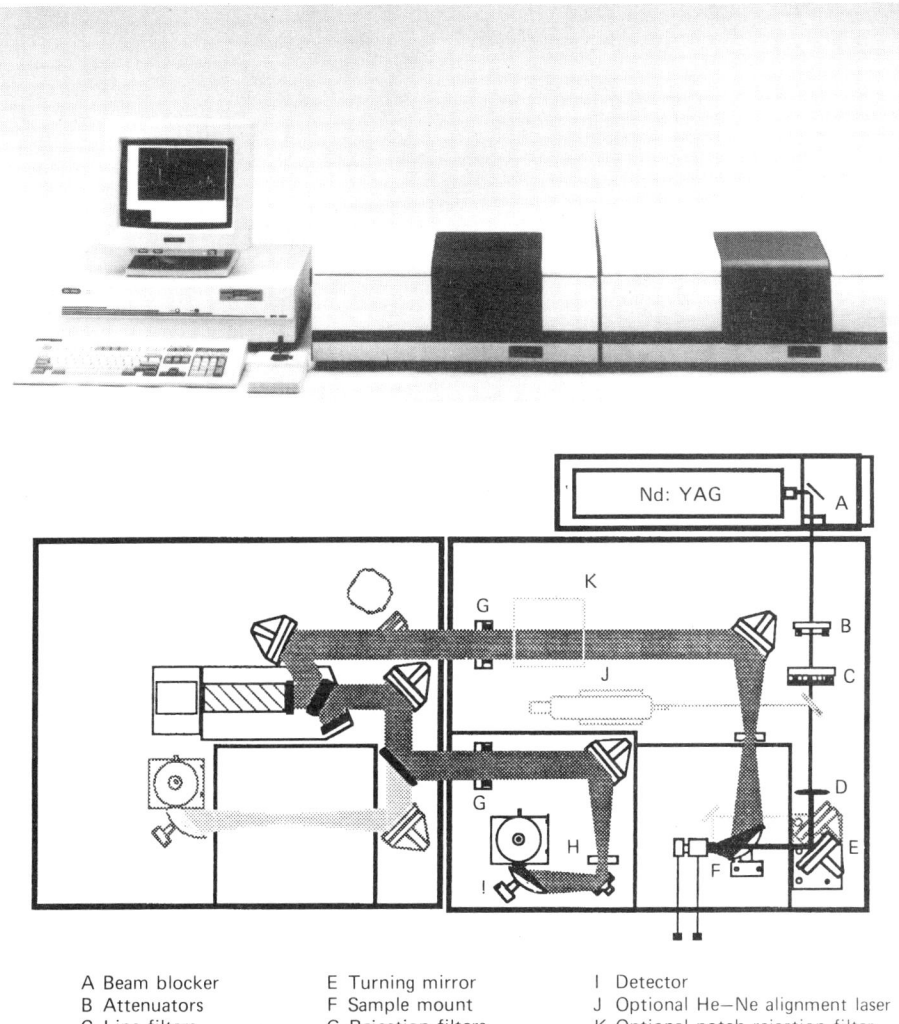

A Beam blocker
B Attenuators
C Line filters
D Focusing lens
E Turning mirror
F Sample mount
G Rejection filters
H Aperture
I Detector
J Optional He–Ne alignment laser
K Optional notch rejection filter

Fig. 5.22 — A photograph and optical diagram of the Bio-Rad FT Raman system.

Nicolet
Nicolet Analytical Instruments
5225-1 Verona Road
Madison
Wisconsin 53711
USA

The Nicolet FT Raman accessory is supplied as an option to the system 800 interferometer. In the infrared this provides very high performance with resolutions better than 0.08 cm^{-1} over a wide spectral range thanks to the use of exchangeable beam-splitters. For Raman purposes the CVI-C95 Nd^{3+}:YAG laser is used, providing up to 1 W at the sample with a stability of 1% RMS. When operating as a Raman instrument a silicon-coated CaF$_2$ or quartz beam-splitter is used with a cooled extended-range germanium detector. This provides a Stokes shift range of 3300–150 cm^{-1}, with an option to record Stokes shift down to 75 cm^{-1} and anti-Stokes shifts beyond 500 cm^{-1}. The highest resolution attainable in the FT Raman mode is 0.5 cm^{-1}. An on-axis $f/0.4$ mirror is used to collect the Raman scatter with the sample being mounted on an x, y, z adjuster. This collection system allows only small samples to be studied. Samples are held on a post near the focus of the mirror and a computer optimization routine is provided to assist alignment. Thus no 'knob twiddling' is required. Data processing is achieved on a Nicolet 680D DSP computer which enables infrared spectra to be displayed and plotted together.

Fig. 5.23 — A photograph and optical diagram of the Nicolet FT Raman system.

Sec. 5.7] Survey of currently available commercial instruments 119

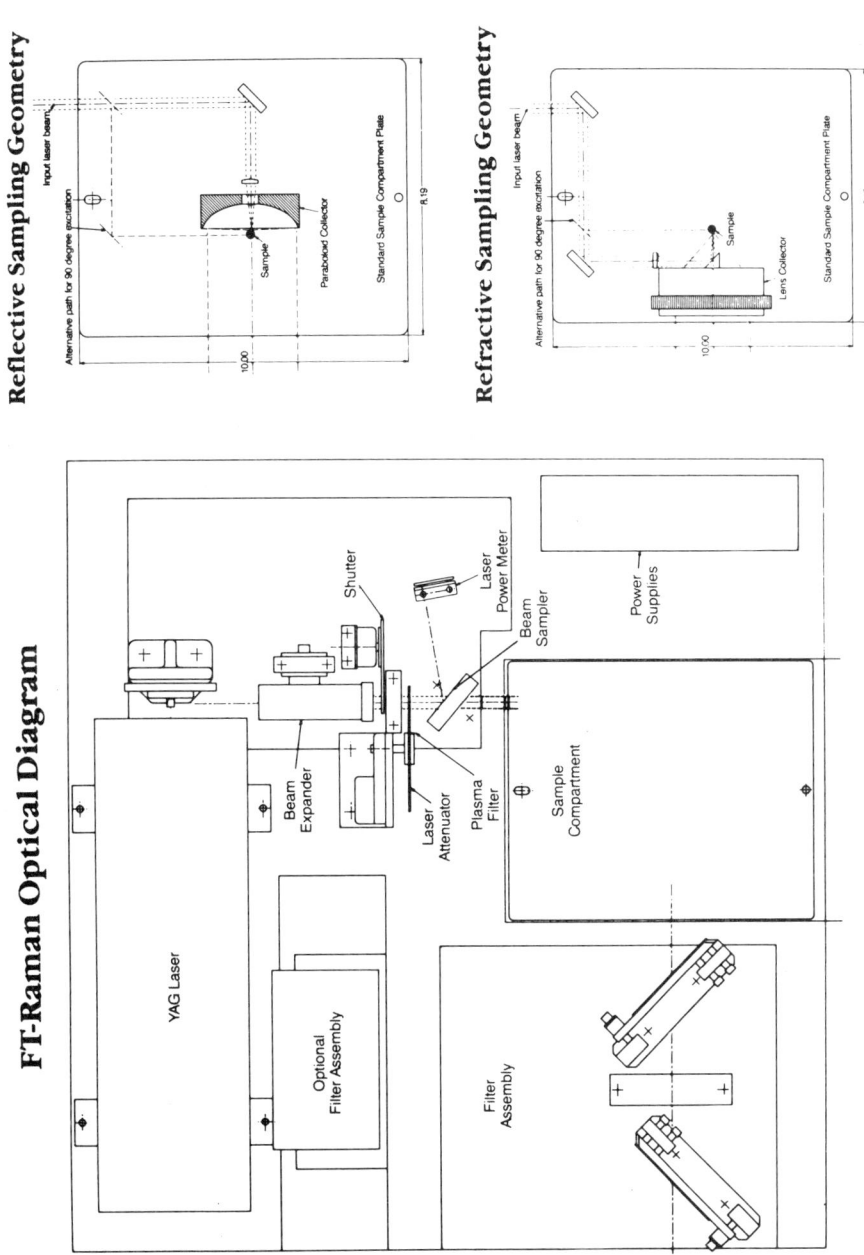

Fig. 5.23 — A photograph and optical diagram of the Nicolet FT Raman system.

Perkin Elmer

Perkin Elmer Ltd.,
Maxwell Road,
Beconsfield, Bucks.,
HP9 1QA, ENGLAND.

Perkin Elmer Corp.,
Analytical Instruments,
761 Main Ave.,
Norwalk, CT,
06859-0012 USA.

The 1700X Raman accessory can be supplied with new instruments which operate as a combined bench providing near or mid infrared coverage depending on the choice of beam-splitter. Alternatively the accessory can be fitted to existing model 1720 or 1760 FTIR's produced since January 1989. A specially stabilized laser manufactured by Spectron Laser Systems Ltd., Rugby, UK specifically for Perkin Elmer is used, providing between 2 mW and 3.5 W at the sample with a stability of 0.1% RMS. The laser is polarized as standard. The beam-splitters available are a long-range KBr which gives a working range of between 370 and 10000 cm^{-1}, quartz giving 2850 to 15000 cm^{-1} and CaF_2 with a range 15000 to 1200 cm^{-1}. Therefore, a combined mid infrared/Raman bench would be fitted with the long range KBr beam-splitter, whereas the near infrared/Raman bench would contain the quartz option which gives a two times improvement in the performance of the FT Raman system.

The sample optics employed are based on a 180° back-scattering lens system (f/0.6), although a 90° illumination arrangement with a half-wave plate is available as an optional extra. The Raman sample area has dimensions of (24×24×35 cm). The sample lid design will accommodate samples up to 13 cm high and 18 cm deep. All sample accessories are mounted on an x, y, z adjustable table and slot into a standard 3×2 in infrared sample slide. Sample accessories available are holders to study solids, gums, powders and one to hold 'odd-shaped' samples. Liquid containers available are capillaries, cuvettes and especially designed spherical cells which are held in reflective aluminized mounts to increase the Raman signal. Alignment of the samples in the laser beam is achieved by monitoring the Raman spectrum in real time and adjusting the sample position in the x, y and z directions.

The filtering system allows the range from 4000 to 150 cm^{-1} to be scanned routinely. The detector is an 'ultra' InGaAs model which can be operated either at room temperature for convenience, or at 77 K for improved signal-to-noise ratios but with a restricted spectral range. At room temperature the maximum shift recordable is 3550 cm^{-1}, whereas cooling the detector reduces this to 3000 cm^{-1}. Resolutions of 0.5, 1, 2, 4, 16, 32 and 64 cm^{-1} are available in both infrared and Raman mode, with two data points per resolution interval. Data manipulation is carried out on an IBM AT/PS2 or compatible computer, and infrared and Raman data may be displayed, scrolled and plotted together.

Sec. 5.7]   **Survey of currently available commercial instruments**

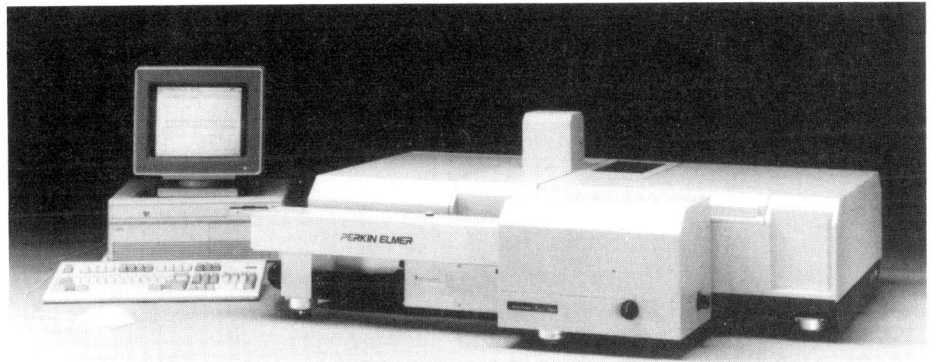

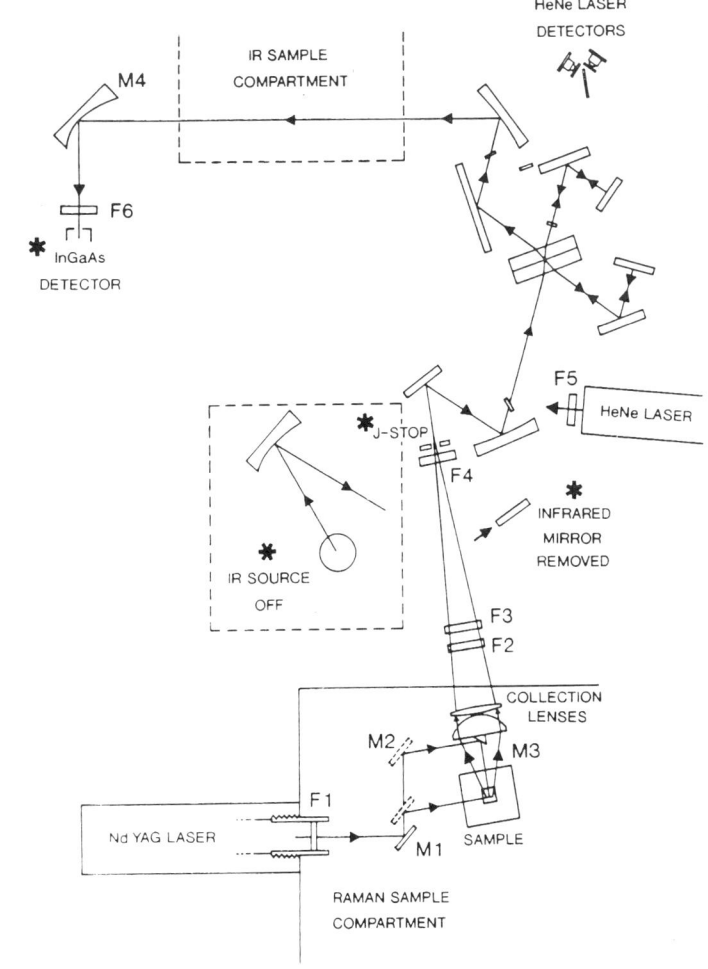

Fig. 5.24 — A photograph and optical diagram of the Perkin Elmer FT Raman system.

Bruker — FRA 106 Accessory

Bruker Instruments Inc.,
Manning Park,
Billerica,
MA 01821
USA

Bruker Analytische Messtechnik GMBH
Wikingerstrasse 13
D-7500 Karlsruhe 21
Germany

The Bruker FRA 106 FT Raman accessory is adaptable to Bruker IFS 66, 66V, 88, 113V and 120HR FT IR spectrometers. All of these instruments can operate in the near infrared and have easily exchangeable beam-splitters. For best performance in the near infrared, a silicon-coated calcium fluoride beam-splitter is used giving a range of 10000–1650 cm^{-1}. A germanium-coated potassium bromide beam-splitter with a range of 10000–450 cm^{-1} can also be used but with some loss in performance. Three lasers are available: (i) a standard c/w Nd^{3+}:YAG giving a maximum power of 2 W at the sample, (ii) a diode-pumped polarized Nd^{3+}:YAG producing 350 mW output and requiring no water cooling, and (iii) a more powerful 700 mW diode pumped Nd^{3+}:YAG device.

An $f/1$ lens collection optic is used and both 90° and 180° illumination is possible. Samples are held on an x, y, z Spindler and Hoyer stage and can be adjusted when the sample area cover is closed. The sample area cover limits samples to a maximum depth of 100 mm and a maximum height of 95 mm. A clamp holder for solids and liquids is supplied with the instrument. A microscope, low temperature cell, liquid cells, micro-sample holders and different focusing optics are available as accessories. The Bruker filtering system allows measurements to be made within 50 cm^{-1} of the laser on the Stokes and 100 cm^{-1} on the anti-Stokes side of the low line. An InGaAs detector is supplied as standard allowing a maximum Stokes shift of 2900 cm^{-1} when cooled to 77 K or 3500 cm^{-1} at room temperature. As an option a proprietary extended-range germanium detector can be used which allows Stokes shifts of 3500 cm^{-1} to be recorded when cooled to 77 K. Data processing is carried out on a Bruker work station or an IBM-compatible computer which both allow infrared and Raman spectra to be viewed, scrolled and plotted together. A computer correction routine is available to correct the spectra recorded for the instrumental response.

Sec. 5.7]   **Survey of currently available commercial instruments**   123

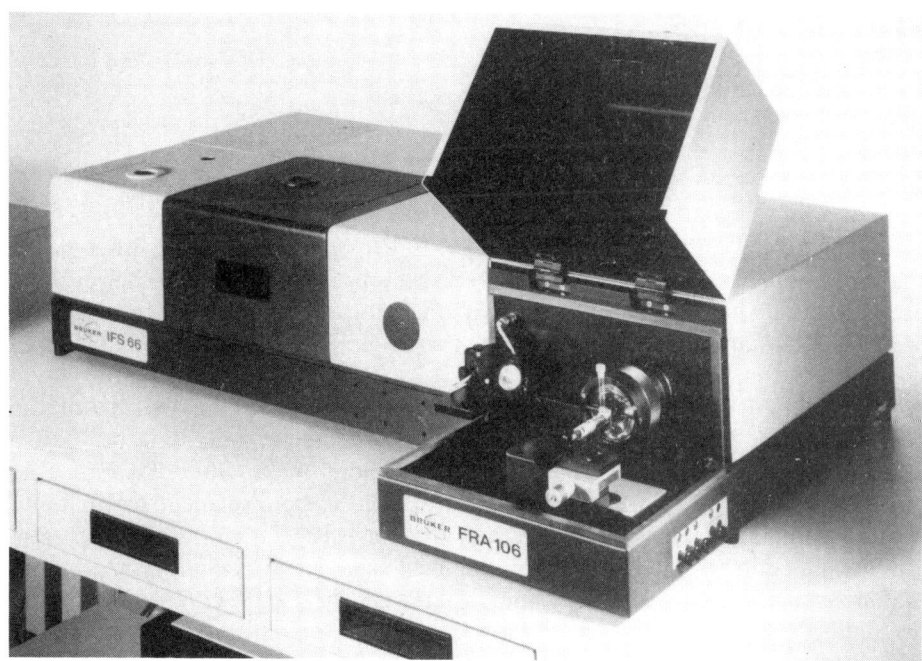

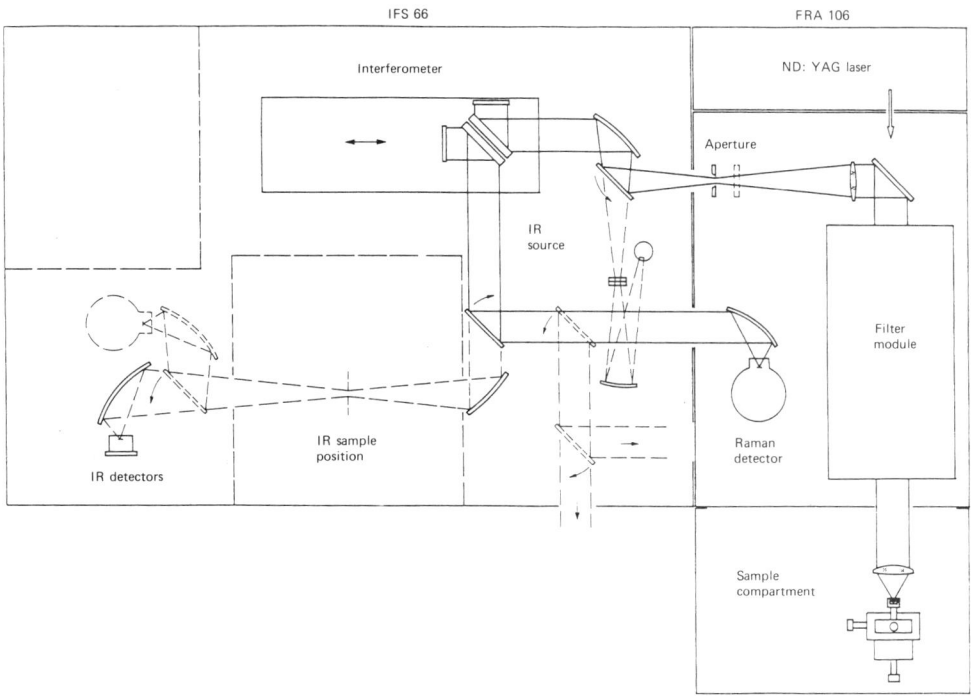

Fig. 5.25 — A photograph and optical diagram of the Bruker FRA 106 FT Raman system.

Bomem — Ramspec 152
Bomem Inc
459 Avenue St Jean Baptiste
Quebec
Quebec G2E 5S5
Canada

The Bomem Ramspec 152 is a combined FTR/FTIR spectrometer fitted with a long-range KCl beam-splitter. Mid infrared spectra can be recorded between 7800 and 450 cm^{-1}. The Nd^{3+}:YAG laser is manufactured by the Quantronix Corp. (model CW114) and provides a maximum power of 3 W at the sample (<4% RMS stability).
Stokes shift. The filters are changed by turning a dial on the outside of the spectrometer. A North Coast extended-range Ge detector operating at 77 K is supplied as standard. Data manipulation is carried out on an IBM AT-compatible computer which allows infrared and Raman spectra to be displayed, scrolled and plotted together. A 'white light' source is used to correct spectra for the instrumental response function, but it is not clear how well this source is characterized. The maximum resolution available in the mid infrared is 1 cm^{-1} and 2 cm^{-1} in the Raman (two data points per resolution interval).

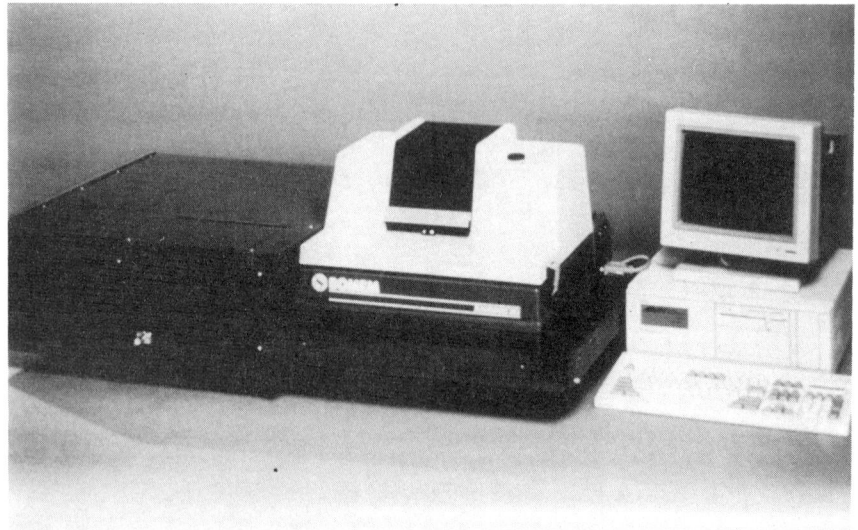

Fig. 5.26 — A photograph of the Bomem Ramspec 152 FT Raman system.

Bomem — DA8

The Bomem DA8 is a very high specification, research-grade spectrometer capable of scanning from the far infrared to the ultraviolet and hence can operate as a combined FTIR/FT Raman instrument. Two Nd^{3+}:YAG laser options are available — a Xe-lamp-pumped Quantronix Corp. model CW114 providing up to 6 W at the sample (stability 1% RMS), or a Melles Griot LDP2000M laser-diode-pumped device stable to 0.1% RMS and giving 1.5 W at the sample. Polarization options are available on request. Beam-splitters are interchangeable, KBr being used for the mid infrared and CaF_2, quartz or glass can be used for the near infrared and Raman. The YAG rejection filters allow Stokes shifts of 4000–40 cm^{-1} to be recorded. The spectrometer can be supplied with an extended-range germanium or an InGaAs detector. At 77 K these devices are sensitive over the Stokes shift ranges 3400–40 cm^{-1} and 3000–40 cm^{-1} respectively. The InGaAs detector can also be operated at room temperature and $-40°C$, allowing spectra over ranges 3300–40 cm^{-1} and 3200–40 cm^{-1} to be recorded. The laser line width limits the maximum resolution available in the Raman mode to 0.1 cm^{-1}, but resolutions as high as 0.0026 cm^{-1} (two data points per resolution interval) can be achieved in the infrared if the customer has sufficient funds!

Raman light is collected using an $f/1$ off-axis mirror and both 180° and 90° illumination is possible. Samples are mounted on a standard 2×3 in infrared sample holder with 20 mm of play in the x, y and z directions. The viewed patch on the sample can be identified by a white light source back-projected from the spectrometer. No sampling accessories are supplied as standard, but semi-silvered liquid bulbs, a powder press, capillary tube holder, film holder and cryostat are available as additional purchases. The sample area lid is 200 mm from the viewed patch. Data processing is accomplished on a VAX 3100 computer which allows infrared and Raman data to be displayed and plotted together. Spectra can be corrected for the instrumental response using a 'white light' source as in the Bomem Ramspec 152. The manufacturers claim that, with appropriate filtering and detection systems, the DA8 can record visible FT Raman spectra.

REFERENCES

[1] G. W. Chantry, H. A. Gebbie and C. Hilsum, *Nature*, 1964, **203**, 1052.
[2] D. B. Chase and T. Hirschfeld, *Appl. Spectrosc.*, 1986, **40**, 133.
[3] D. B. Chase, *J. Am. Chem. Soc.*, 1986, **108**, 7485.
[4] V. M. Hallmark, C. G. Zimba, J. D. Swalen and J. F. Rabolt, *Spectroscopy*, 1987, **2**, 40.
[5] D. E. Jennings, A. Weber and J. W. Brault, *Appl. Optics*, 1986, **25**, 284.
[6] S. F. Parker, K. P. J. Williams, P. J. Hendra and A. J. Turner, *Appl. Spectrosc.*, 1988, **42**, 796.
[7] P. H. Hendra and H. M. Mould, *Int. Lab.*, 1988, **18**, 34.
[8] G. Ellis, P. J. Hendra, C. M. Hodges, T. Jawhari, C. H. Jones, P. le Barazer, C. Passingham, I. A. M. Royaud, A. Sanchez-Blasquez and G. M. Warnes, *The Analyst*, 1989, **114**, 1061.
[9] G. Ellis, P. J. Hendra, C. H. Jones, K. D. O. Jackson and M. J. R. Loadman, *Kautschuk & Gummi Kunststoffe*, 1990, **43**, 118.
[10] P. J. Hendra, C. Passingham, G. M. Warnes, R. Burch and D. J. Rawlence, *Chem. Phys. Lett.*, 1989, **163**, 178.
[11] P. J. Hendra, P. Le Barazer and A. Crookell, *J. Raman Spectrosc.*, 1989, **20**, 35.
[12] M. Hanniet, T. Jawhari and I. A. M. Royaud, *J. Chim. Phys.*, 1989, **86**, 431.
[13] A. Sanchez-Blasquez, *Optica Pura Y Aplicada*, 1989, **22**, 25.

[14] P. J. Hendra, C. M. Hodges and T. Farley, *J. Raman Spectrosc.*, 1989, **20**, 745.
[15] A. Crookell, M. Fleischmann, M. Hanniet and P. J. Hendra, *Chem. Phys. Lett.*, 1988, **49**, 123.
[16] H. Sadeghi-Jorabchi, P. J. Hendra, R. H. Wilson and P. S. Belton, *J. Am. Oil Chem. Soc.*, 1990, **67**, 483.
[17] T. Jawhari and P. J. Hendra, *Spectrosc. World*, 1989, **1**, 6.
[18] E. N. Lewis, V. F. Kalasinsky and I. W. Levin, *Appl. Spectrosc.*, 1989, **43**, 156.
[19] C. J. Petty, G. M. Warnes, P. J. Hendra and M. Judkins; *Spectrochim. Acta*, 1991, **47A**, FT Raman special edition, in press.
[20] D. J. Cutler, *Spectrochim. Acta*, 1990, **46A**, 131.
[21] D. K. Veirs, J. W. Ager and G. M. Rosenblatt, *Proc. 12th Int. Conf. Raman Spectrosc.*, p. 898, Wiley, New York, 1990.
[22] R. G. Messerschmidt and D. B. Chase, *Appl. Spectrosc.*, 1989, **43**, 11.
[23] F. J. Bergin and H. F. Shurvell, *Appl. Spectrosc.*, 1989, **43**, 516.
[24] F. J. Bergin, *Spectrochim. Acta*, 1990, **46A**, 153.
[25] B. Schrader, A. Hoffmann, R. Podschadlowski and A. Simon, *SPIE 1145 Fourier Transform Spectroscopy*, 1989, 372.
[26] P. J. Hendra, G. Ellis and D. J. Cutler, *J. Raman Spectrosc.*, 1989, **19**, 413.
[27] K. P. J. Williams and S. M. Mason, *Spectrochim. Acta*, 1990, **46A**, 187.
[28] K. P. J. Williams, *J. Raman Spectrosc.*, 1990, **21**, 147.
[29] P. J. Hendra and M. Sweeney, *Spectroscopy World*, 1991, **3**, No. 2, 22.

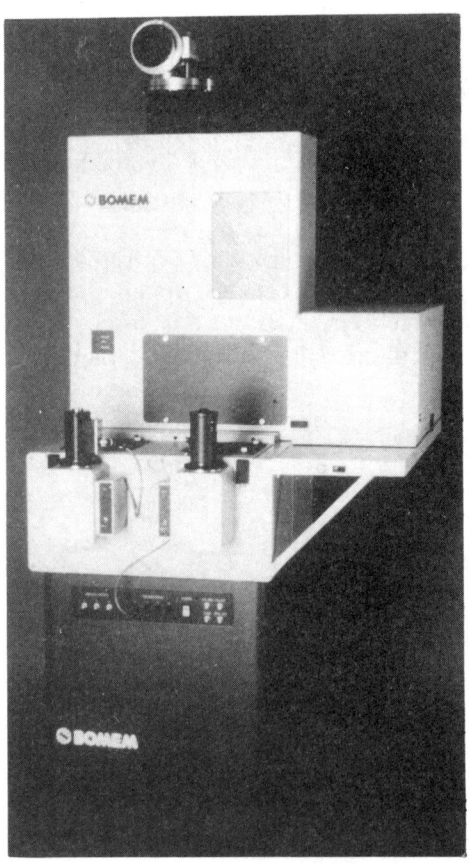
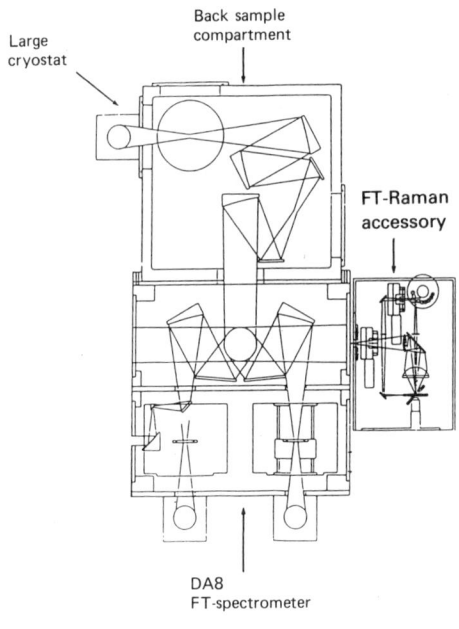

Fig. 5.27 — A photograph and optical diagram of the Bomem DA8 FT Raman system.

6
FT Raman intensities

6.1 INTRODUCTION

Bands in vibrational spectra have both frequency and intensity, and both of these quantities can convey information. Studying the intensities of spectral bands can enable us to make quantitative measurements and also compare the strength of the spectra given by various materials. This is true of any spectroscopic technique, but to date comparatively few quantitative studies have been undertaken using Raman spectroscopy. In this chapter we will discuss the basics of intensity measurements and highlight how in the future FT Raman spectroscopy is likely to be used as a routine quantitative tool. Most of this chapter is concerned with studies of liquids but extrapolation of these techniques to solids will be mentioned towards the end.

6.2 RAMAN INTENSITY FORMULAE

The theory relating to the intensity of Raman bands was devised by Placzek in 1934 [1]. He derived the following expressions:

Stokes scattering:

$$I_{v,ij} = \frac{2\pi^2 h}{c} \frac{(\bar{v}_0 - \bar{v})^4}{\bar{v}[1 - \exp(-h\bar{c}\bar{v}/kT)]} g_v \left(\frac{\delta\alpha_{ij}}{\delta Q_v}\right)_0^2 \qquad (6.1)$$

Anti-Stokes scattering:

$$I_{v,ij} = \frac{2\pi^2 h}{c} \frac{(\bar{v}_0 + \bar{v})^4}{\bar{v}[\exp(-h\bar{c}\bar{v}/kT) - 1]} g_v \left(\frac{\delta\alpha_{ij}}{\delta Q_v}\right)_0^2 \qquad (6.2)$$

where

$\bar{v}_0$ = absolute frequency of excitation radiation (cm^{-1})
$\bar{v}$ = frequency of band (Raman shift, cm^{-1})
T = absolute temperature of sample (K)
h = Planck's constant (J sec)
$\bar{c}$ = speed of light (cm sec^{-1})
k = Boltzmann constant (J K^{-1})
g_v = degeneracy of given vibrational mode

$\left(\dfrac{\delta \alpha_{ij}}{\delta Q_v}\right)_0$ = ij element of the polarizability derivative tensor for a given vibrational mode

These equations give the Raman intensity due to each element of the polarizability tensor for a particular vibrational mode per molecule. For an isotropic sample such as a liquid or a gas they can be re-written in terms of the mean polarizability, $\bar{\alpha}$, and the anisotropy, β (see Section 2.7). As in polarized Raman studies, the theoretical intensity measured depends on the experimental geometry.

Eqs. (6.1) and (6.2) are only needed for rigorous theoretical studies related to Raman intensities. For analytical experiments a much simpler model can be used and Raman intensities can be studied in a way similar to absorption spectroscopy. Infrared intensities are governed by the Beer–Lambert law:

$$\text{Log}_{10}\left(\dfrac{I_0}{I}\right) = ECL \tag{6.3}$$

$$= \text{Absorbance}$$

where

I_0 = incident light intensity
I = transmitted light intensity
E = proportionality constant called the extinction coefficient
C = sample concentration
L = sample path length

Intensities can be quantified by measuring the band height or band area. Band areas are statistically more significant as they use more than one data point in the spectrum and can be independent of resolution (see Section 3.4). Thus, in principle it should be possible to easily calculate the absorptivity of bands in IR absorption spectra and this value should give a true measure of band intensity. The absorptivity is a constant which reflects the value of parameters such as the transition moment for the vibration concerned. In practice, however, as the path length in infrared transmission spectroscopy is very small there may be considerable imprecision in its measurement, and absorptivities are not routinely calculated. However, they

Sec. 6.3] **Intensity measurements with a conventional Raman spectrometer**

are frequently used in UV/visible absorption spectroscopy. The situation is considerably more difficult for solids which have to be finely dispersed in a mulling agent or in a dispersant such as potassium bromide and have considerable particle size problems.

The intensity of Raman bands can be expressed in an equation analogous to the Beer–Lambert law:

$$I_{\text{Raman}} = KVCI_0 \tag{6.4}$$

where

I_{Raman} = intensity of Raman band
I_0 = intensity of exciting radiation
V = volume of sample illuminated by the source and viewed by spectrometer
C = concentration of the sample
K = constant for each band

At first sight this all seems very simple. However, the problem comes in the value of K, which in this expression is analogous to the absorptivity of the Beer–Lambert law. Whereas the frequencies and relative intensities of bands in a Raman spectrum are a function of the sample, the overall intensity of the whole recorded spectrum is determined by many instrumental factors. Different Raman spectrometers vary in their optical efficiency, method of illumination, wavelength response and particularly the skill of the operator. As a consequence estimates of absolute intensities vary wildly. This major difficulty stems from the fact that Raman is an emission process. Ideally one should collect all the light emitted, or know what fraction of the light is collected. This is very difficult to determine and thus it is impossible to compare values of K measured by different experimenters on different instruments. In infrared and UV/visible absorption spectroscopy all the instrumental factors are cancelled out by using a double-beam spectrometer or taking a background spectrum on an FT instrument. The next section discusses more fully the problem of making absolute intensity measurements using a conventional Raman spectrometer.

6.3 INTENSITY MEASUREMENTS WITH A CONVENTIONAL RAMAN SPECTROMETER

In Chapter 3 we discussed the difficulties in aligning a conventional Raman spectrometer. Both the sample and the laser beam have to be introduced into the tiny patch of space viewed by the spectrometer, and the intensity of the spectrum recorded is severely affected if any of these move by a few microns. This leads to very poor reproducibility in the intensity of Raman bands measured with a conventional Raman instrument. It is very difficult to determine how many molecules of the sample are being viewed by the spectrometer, how many of these are being irradiated by the laser and at what intensity. Various researchers have devised very complex apparatus in the hope of eliminating or controlling this problem [2]. However, the values of the absolute Raman cross-section (a formal measure of Raman intensity) reported in the literature for specific bands in specified samples disagree wildly and none of the error bars overlap! Even if the methods they devised could reliably measure Raman intensities they are too complex and time consuming to be employed in a routine analytical environment.

As well as problems of slight sample and/or laser misalignment, there is also a fundamental optical problem which causes errors in the measurement of Raman intensities with a conventional instrument. When a ray of light passes from one material to another with a higher refractive index, it is bent towards the normal of the interface between the two materials. This process is called refraction and is described by Snell's law (see Fig. 6.1):

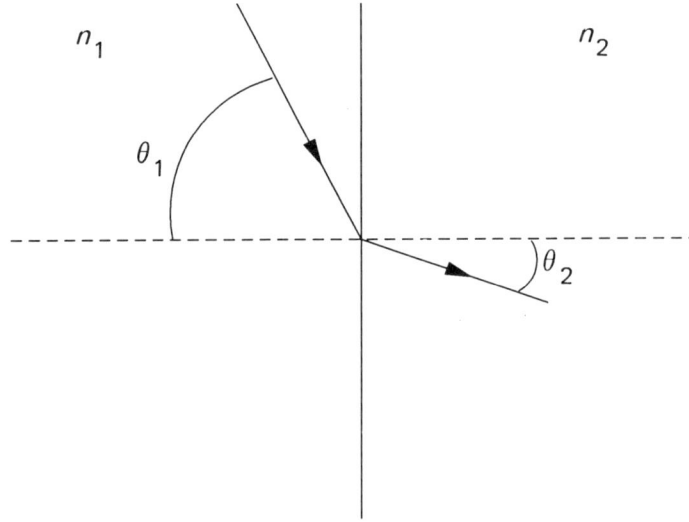

Fig. 6.1 — A ray being refracted as it moves between media. Here $n_2 > n_1$ so the ray is bent towards the normal.

$$\frac{\sin \theta_1}{\sin \theta_2} = \frac{n_2}{n_1} \tag{6.5}$$

where n_1 and n_2 are the refractive indices of the two materials, and θ_1 and θ_2 are the angles made with respect to the normal of the interface by the incident and refracted rays respectively. Consider a cuvette of liquid placed in the sample area of conventional spectrometer using 90° collection optics. The laser beam enters the sample parallel to the normal (i.e. orthogonal to the bottom of the cuvette) and is thus not refracted. However, the Raman light emitted is refracted as it passes the face of the cuvette. If the instrument is aligned correctly with one liquid sample which is then changed for another with a different refractive index the Raman light will follow a different path and may not illuminate the entrance slit correctly. The situation can also be analysed by considering the position of the patch in the liquid sample viewed by the spectrometer. If this is exactly superimposed with the laser beam, and then the sample changed, the instrument will collect light from a different patch, which may not by illuminated by the laser. This is shown in Fig. 6.2. This

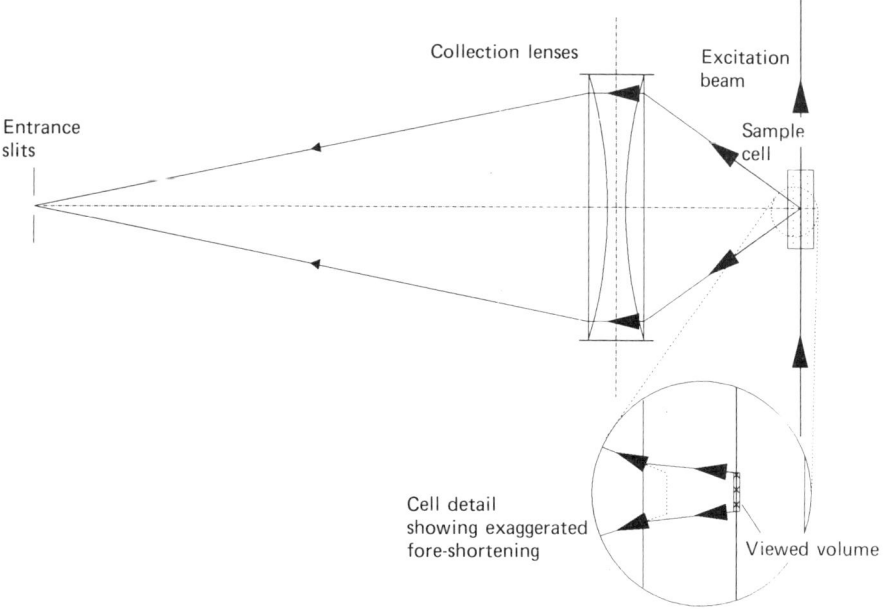

Fig. 6.2 — The front-end collection optics for a 90° scattering experiment using a conventional Raman spectrometer. The expanded view of the cuvette shows that the volume viewed by the spectrometer will change with the refractive index of the sample.

would require the sample to be realigned and there would be considerable imprecision in the intensity values obtained. The effect is called foreshortening. The problem could be rectified by using a 180° scattering arrangement but then the laser only illuminates a small round patch on the slit image and this would be optically wasteful.

6.4 METHODS OF QUANTITATIVE EXPERIMENTS

Although conventionally absolute measurements of Raman intensities are very difficult to make, it is still possible to perform quantitative experiments. These make use of either external or internal standards, the latter being far more reliable in conventional Raman spectroscopy. Band intensity ratios can also be very useful.

An example of the use of an external standard is as follows. Samples of carbon tetrachloride are to be examined which are contaminated with an unknown concentration of pyridine. Several external standards are prepared which consist of a known concentration of pyridine dissolved in carbon tetrachloride. The intensity of the 991-cm^{-1} pyridine band could be measured for each and a calibration curve drawn. The intensity of this band could be measured in unknown samples and the pyridine concentration determined. This method would require constant laser power and consistent sample alignment — something which is very difficult to ensure in conventional Raman spectroscopy. If the carbon tetrachloride was very heavily contaminated with pyridine, such that the ratio of the 991-cm^{-1} pyridine band to the 459-cm^{-1} band of carbon tetrachloride was not too large, plotting a calibration curve of this band ratio versus pyridine concentration would be very advantageous. This would allow measurements to be made irrespective of laser power and sample alignment. In this case the carbon tetrachloride is effectively being used as an internal standard. If the solvent has no useful Raman bands, then an known aliquot of an inert reference with a characteristic Raman band can be added to both the samples and the standard solutions (again this is a use of an internal standard).

Internal standards are far easier to use experimentally than external ones. However, their use does have some undesirable consequences, e.g. samples subject to quantitative analysis are often complex mixtures with many competing equilibria, and adding a further component for use as an internal standard may perturb this system. Also, the use of a solvent Raman band as an internal standard can be problematic. In an analysis of sulphur oxoacid anions in aqueous solutions, Milićev and Stergaršek [3] reported that the intensity of the 1650 cm^{-1} OH bending band in water changed significantly under the influence of different solutes. Thus, it would appear that it is highly desirable to use external standards in Raman experiments. This is the area in which an FT Raman spectrometer equipped with suitable collection optics is vastly superior to a conventional instrument.

6.5 MEASURING INTENSITIES WITH AN FT RAMAN SPECTROMETER

In Section 6.3 we discussed the problems of making absolute Raman intensity measurements. The problems associated with intensity measurements in conventional Raman spectroscopy can be eliminated by using a suitably designed FT Raman spectrometer. Much of the discussion is the result of experiments at Southampton [4], but other researchers have reported similar findings [5].

The Southampton prototype uses a discharge-lamp- pumped Nd^{3+}:YAG laser as an excitation source (Spectron Lasers, model 301). The output power from this laser stabilizes after it has been running for about 30 minutes. The laser beam is switched off by activating a flag inside the laser cavity which blocks the beam. Once this flag is opened the laser takes approximately two minutes to re-stabilize. If spectra are recorded after this two-minute period the band intensities measured are remarkably reproducible.

To demonstrate this fact ten spectra were recorded of benzene held in a 2-mm-path-length glass cuvette. Each time the cell was removed, emptied, refilled and returned to the spectrometer. After waiting for two minutes for the laser to stabilize the spectrum was recorded (50 scans, 12-cm^{-1} resolution, 300 mW). The standard deviation of the 992-cm^{-1} band intensity was found to be a remarkable 0.98%. If the benzene was exchanged *in situ* a standard deviation of 0.48% was obtained. Successive spectra recorded leaving both the cell and benzene in place gave a similar standard deviation indicating that exchanging the liquid alone had a negligible effect. The residual irreproducibility was attributed to noise in the spectra.

The extremely high reproducibility obtained using the Southampton prototype stems from the use of an interferometer and 180° back-scattering collection optics. Because the Jacquinot stop is large (8 mm), the ×6 collection optics mean that the instrument views a 1.3 mm diameter circular patch on the sample. This patch is very slightly overfilled by the laser beam whose position is fixed. Because the laser slightly overfills the viewed patch, the system is very optically efficient, unlike 180° scattering on a conventional instrument. The instrument also views a comparatively large sample volume, and this means that the measured Raman intensity is very insensitive to the position of the sample along the instrument's optical axis, i.e. its depth of focus is considerable. Figure 6.3 demonstrates this point by showing the effect of adjusting

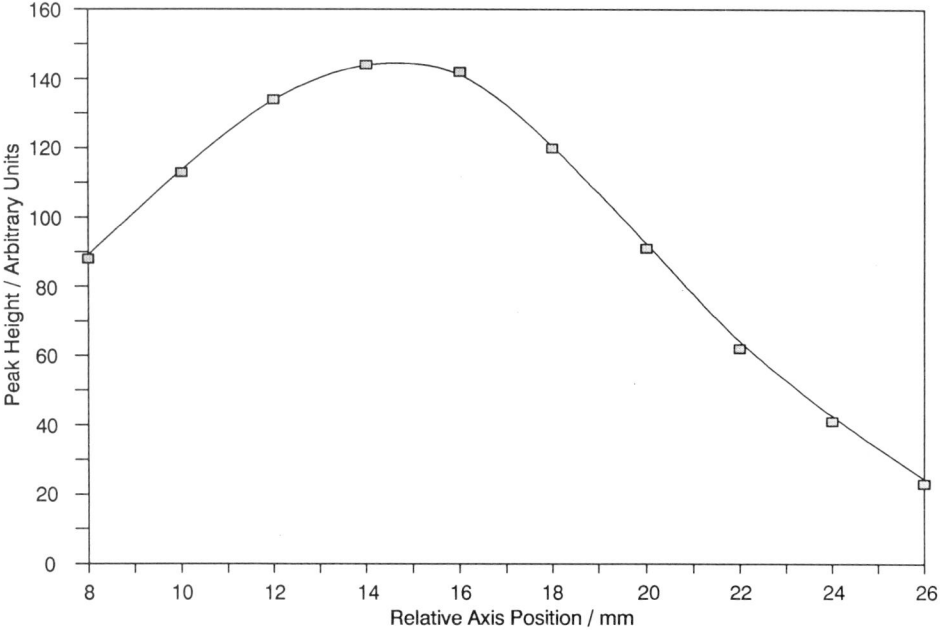

Fig. 6.3 — Graph of the intensity of the 992-cm^{-1} band of benzene held in a spherical cell as it is moved along the optical axis of the FT Raman spectrometer.

the position of a spherical cell containing benzene along the optical axis through the optimum or focus position. It is possible to position the sample more than a millimetre away from 'perfect' focus with a loss in intensity of only 2%. It is a simple matter to manufacture cell holders which are reproducible to the required degree, and hence quantitative experiments which involve exchanging sample cells present no problem. The 180° collection system and the depth of focus also ensures that any foreshortening caused by changes in sample refractive index is virtually eliminated.

All this information leads to a very appealing conclusion — it is now perfectly possible to perform quantitative Raman measurements using external standards without the need for scrupulous optical alignment. The experiments can be performed on a routine basis by comparatively unskilled staff and this must revolutionize quantitative Raman experiments. At Easter 1990, a 'Summer School' on FT Raman was held at Southampton University during which delegates were able to analyse the composition of *ortho-*, *meta-* and *para-*xylene mixtures by FT Raman spectroscopy using external standards.

A prerequisite for quantitative measurements using external standards is a laser source free from long-term drift, and current Nd^{3+}:YAG lasers are not perfect in this respect. Recently a highly stabilized discharge-lamp-pumped Nd^{3+}:YAG laser has been developed by Spectron Lasers which it is claimed is free from short-term instability and long-term drift. The gradual reduction in the costs of diode-pumped Nd^{3+}:YAG lasers also makes them an attractive source of stable laser output. The commercial FT Raman spectrometers on sale in January 1991 all used discharge-lamp-pumped lasers, but these are certain to be replaced by diode-pumped-devices in the next generation of instruments. The high reproducibility of FT Raman spectra has, however, only been studied in depth on the Southampton prototype. This uses a lens collection system, whereas some of the commercial instruments use mirror collection optics. Whether this reproducibility is available on all of these instruments is unknown to the authors. However, on a promising note, BioRad, whose instrument incorporates mirror collection optics, reported similarly high reproducibility at the 12th International Conference on Raman Spectroscopy in August 1990.

6.6 CALIBRATION OF FT RAMAN SPECTRA FOR INSTRUMENT RESPONSE

As mentioned in Section 5.3, FT Raman spectrometers have a varying spectral response owing to their component filters, beam-splitters etc. The unfortunate consequence of this is that spectra recorded on FT Raman instruments will differ from instrument to instrument. This disagreement is likely to be worst between different manufacturers, but may also be present on otherwise identical instruments as a result of variations in filter and detector characteristics from batch to batch. This situation is clearly very undesirable and could only too easily result in spurious correlations and hence incorrect conclusions being derived from spectra published in the chemical literature. Thus we contend that it is thoroughly irresponsible to publish FT Raman spectra without correcting them for instrumental response. This situation has not arisen in conventional Raman spectroscopy as in general the instruments used to date have used monochromators to reject the Rayleigh line rather than interference filters. This gives conventional Raman spectrometers a gently sloping, featureless instrumental response curve primarily due to varying photomultiplier detectivity†.

† Early pioneers in the use of lasers as sources for Raman spectroscopy were frequently blissfully unaware that bands at high shifts from the He–Ne source at 632.8 nm were grossly attenuated. Many spectra were published in the late 1960s which showed v_{CH} bands near 3000 cm^{-1} as weak features when they were in fact often the strongest bands in the spectrum!

Sec. 6.6] Calibration of FT Raman spectra for instrument response

The response function of an FT Raman spectrometer, however, goes up and down like a roller coaster.

A method has been refined at Southampton to correct recorded FT Raman spectra [6], and similar techniques are being used by *some* of the commercial instrument manufacturers. This method is not new and has been described previously [7,8,9]. To calibrate the instrument an emission source is required whose characteristics are known exactly. This source must have a broad, featureless spectrum across the range of FT Raman instruments. One possible choice would be a highly fluorescent sample with no Raman bands, but the exact profile of this fluorescence is extremely difficult to quantify. The best emission source available at the moment is a blackbody emitter whose temperature is known accurately [10,11]. The theoretical blackbody spectrum can then be calculated using the Planck distribution (Eq. (6.6)).

$$L^B(\lambda, T) \, d\lambda = \frac{2hc^2}{\lambda^5} \frac{1}{\exp(hc/\lambda kT) - 1} \, d\lambda \; W \, m^{-2} \, sr^{-1} \quad (6.6)$$

where
h = Planck's constant (J sec)
c = speed of light (m sec^{-1})
λ = wavelength (m)
k = Boltzmann constant (J K^{-1})
T = absolute temperature (K).

Eq. (6.6) has to be slightly rewritten for our purposes. It gives the radiation emitted over all space per unit wavelength interval (i.e. $d\lambda = 1$). An FT Raman spectrometer records a spectrum in wavenumbers, a unit of frequency. It is therefore necessary to change the variables to express the energy emitted in terms of unit frequency.

$$\nu = \frac{c}{\lambda} \quad (6.7)$$

$$d\nu = \frac{c}{\lambda^2} \, d\lambda \;, \quad d\lambda = \frac{\lambda^2}{c} \, d\nu \quad (6.8)$$

This is accomplished by substituting ν for λ and $d\nu$ for $d\lambda$ using Eqs. (6.7) and (6.8). This gives Eq. (6.9).

$$L^B(\nu, T) \, d\nu = \frac{2h\nu^3}{c^2} \frac{1}{\exp(h\nu/kT) - 1} \, d\nu \; W \, m^{-2} \, sr^{-1} \quad (6.9)$$

where
ν = frequency (Hz)

This is then written in terms of wavenumbers by substituting $v = 100c\omega$ and $dv = 100c\,d\omega$ to give Eq. (6.10).

$$L^B(\omega, T)\,d\omega = \frac{2 \times 10^8 \omega^3 hc^2}{\exp(100hc\omega/kT) - 1}\,d\omega \text{ W m}^{-2} \text{ sr}^{-1} \qquad (6.10)$$

where
ω = frequency (absolute cm^{-1})

Eq. (6.10) gives the energy emitted over all space in watts per wavenumber interval. Traditionally Raman spectra have been presented in photon counts per second rather than watts†. In the near infrared range in which FT Raman spectrometers currently operate, the absolute frequency in wavenumbers of the Raman scattered radiation lies roughly in the range 9400–5900 cm^{-1}. The photon energy in this region increases by approximately 50% from 5900 cm^{-1} to 9400 cm^{-1}. To plot 'correct' FT Raman spectra on a counts per second ordinate scale Eq. (6.10) must be divided by $100\,hc\omega$ to give Eq. (6.11).

$$L^B(\omega, T)\,d\omega = \frac{2 \times 10^6\,c\omega^2}{\exp(100hc\omega/kT) - 1}\,d\omega \text{ counts s}^{-1} \text{ m}^{-2} \text{ sr}^{-1} \qquad (6.11)$$

Using Eq. (6.11) (counts per second) or Eq. (6.10) (watts) a theoretical blackbody spectrum from a blackbody of absolute temperature T can be calculated for the FT Raman spectral range on a computer. Figure 6.4 shows theoretical blackbody spectra calculated at a variety of temperatures. Next the blackbody spectrum of a blackbody source is recorded on the FT Raman instrument. This could be from a standardized lamp or a small furnace. Although it is experimentally easier to use a lamp, this does present problems in measuring the temperature of the emitting filament since the latter is sealed in a vacuum! Filament temperatures can be measured using an optical pyrometer, but measurements from these devices are often highly subjective. The temperature of a furnace (ideally above 1500 K) can easily be measured fairly precisely using a high temperature thermocouple.

FT Raman instruments are very sensitive and thus light from any sensibly sized blackbody emitter will have to be severely attenuated to avoid detector saturation. However, any attenuation will have to be uniform at every frequency, otherwise the blackbody spectrum will be perturbed by the near infrared spectrum of the attenuator. This problem is encountered if pieces of paper are used as attenuators. Similarly, any diffusing device such as a frosted quartz window may preferentially scatter light at higher frequencies if too coarsely ground. The collected emission must pass the complete optics of the instrument (including the collection optics) and fill the interferometer uniformly. It is best that the spectrometer views a defocused image of

---

† Raman spectra recorded in the 1970s were often measured using picoammeter or phase-sensitive detection systems. These tended to integrate the pulse data output of the photomultipliers rather than count the pulse rate. Although imprecise, there is some correlation between spectra recorded by pulse and analogue techniques.

Sec. 6.6]   **Calibration of FT Raman spectra for instrument response**

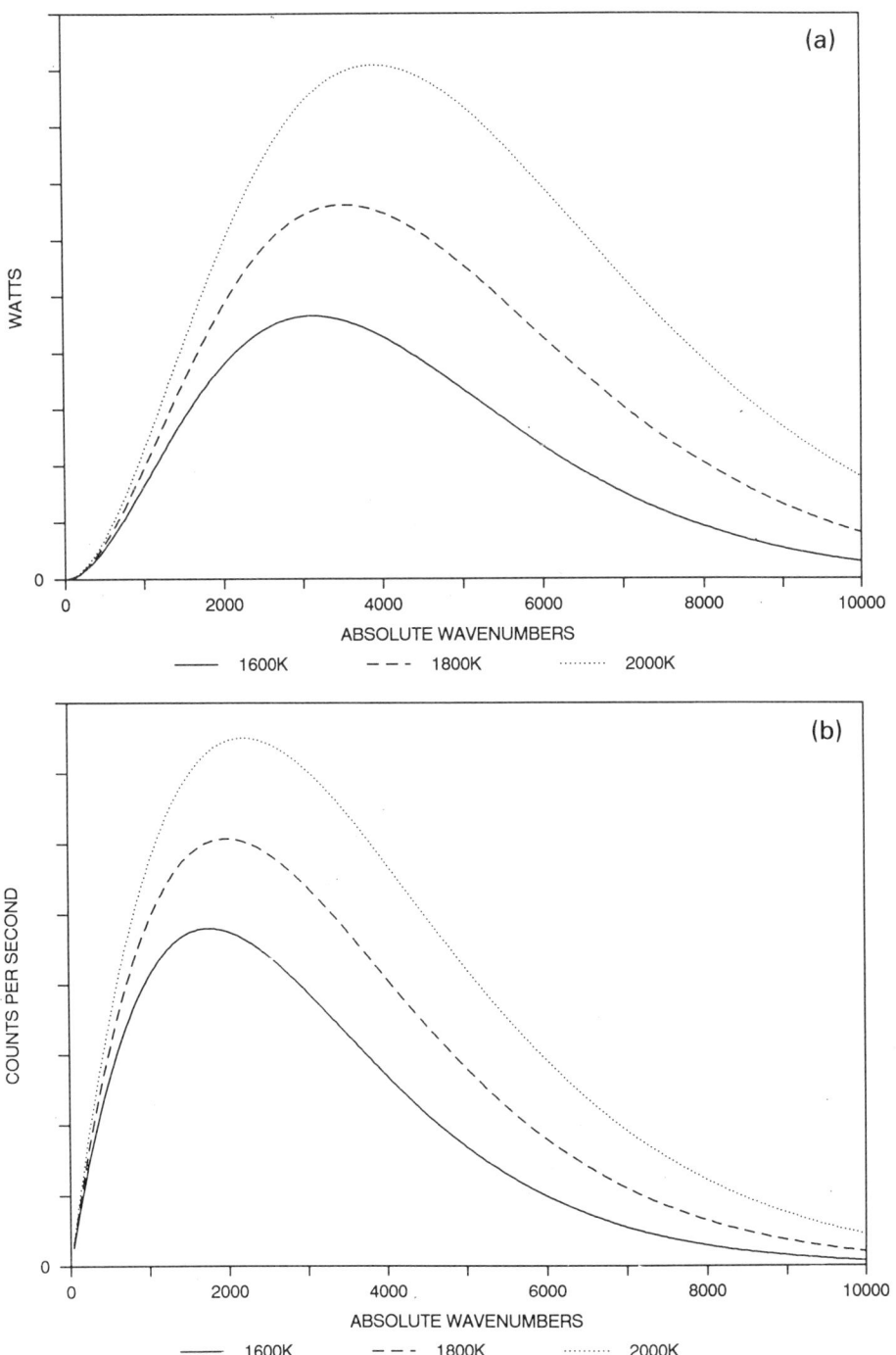

Fig 6.4 — Blackbody curves for a variety of temperatures on an absolute wavenumber abscissa scale, (a) in watts, (b) in counts per second. As the temperature is increased the overall intensity increases and the maximum of the curve shifts to higher frequency. At these temperatures the high frequency tail of the curve is in the range of a FT Raman spectrometer equipped with a Nd^{3+}:YAG laser.

the blackbody emitter in case one part of the source is at a different temperature to another. The hottest regions will radiate most intensely according to the Stefan–Boltzmann relationship (Eq. (6.12)).

$$M = \sigma T^4 \text{ W m}^{-2} \tag{6.12}$$

where
T = absolute temperature (K)
σ = Stefan–Boltzmann constant (W m^{-2} K^{-4})

At Southampton, a tube furnace filled with small pieces of fire brick is used and this is placed on the optical axis of the spectrometer about 1.5 m from the collection optics. A virtually closed iris is placed at the viewed patch of the spectrometer and several attenuators consisting of small holes drilled in thin metal sheet are placed between this and the furnace. The area around the collection lens is draped with heavy black cloth. Blackbody spectra are recorded at the various instrumental settings generally used (e.g. resolution, filter tilt angle, apodization function, etc.). It is important to obtain a very high signal-to-noise ratio in these spectra so as not to contaminate with noise any spectra which will be multiplied by the resulting correction function. However, the integrated area under the spectrum must not be too great to avoid any non-linearities in the detection system†. The temperature of the furnace is recorded, and this is used to calculate the theoretical blackbody spectrum. The real and calculated spectra are then scaled to give the same maximum intensity (see Fig. 6.5). A set of correction functions is then obtained by dividing the calculated blackbody spectrum by the experimentally recorded ones (see Fig. 6.6). Subsequent FT Raman spectra can then be corrected for the instrumental response by multiplying them by the appropriate correction function (recorded at the same resolution, filter angle, etc. as the spectrum).

Multiplying recorded spectra by the correction curve has a marked effect. Figure 6.7 shows how the oscillations on the fluorescent background of a spectrum of PES/PESES copolymer are removed. Similarly, Fig. 6.8 shows how the relative band intensities change in the spectrum of methyl iodide. Because the correction curve varies across the spectral range, the noise level in different regions of corrected spectra will not be the same. For cosmetic purposes it is worth making zero regions of the correction curve where the filters are opaque or the detector is not sensitive. This avoids excessive noise at the edges of corrected spectra. One other feature which is encountered is the near infrared absorption of airborne water vapour. Absorption bands due to the P and R vibration rotation branches of water appear around 7200 cm^{-1} absolute (2200 cm^{-1} shift from 1064-nm Nd^{3+}:YAG) in both the recorded blackbody spectrum and any spectrum with a background. Obviously, when the Raman spectrum appears on a flat background with no Raman bands in this region this will not be a problem. On spectra which do show a background the

---

† In a Raman-derived interferogram the centre burst is very weak. In a blackbody emission spectrum the total collected radiation is high, hence the centre burst is intense and may well cause problems with non-linearity in the detector and preamplifier circuits.

Sec. 6.6] Calibration of FT Raman spectra for instrument response

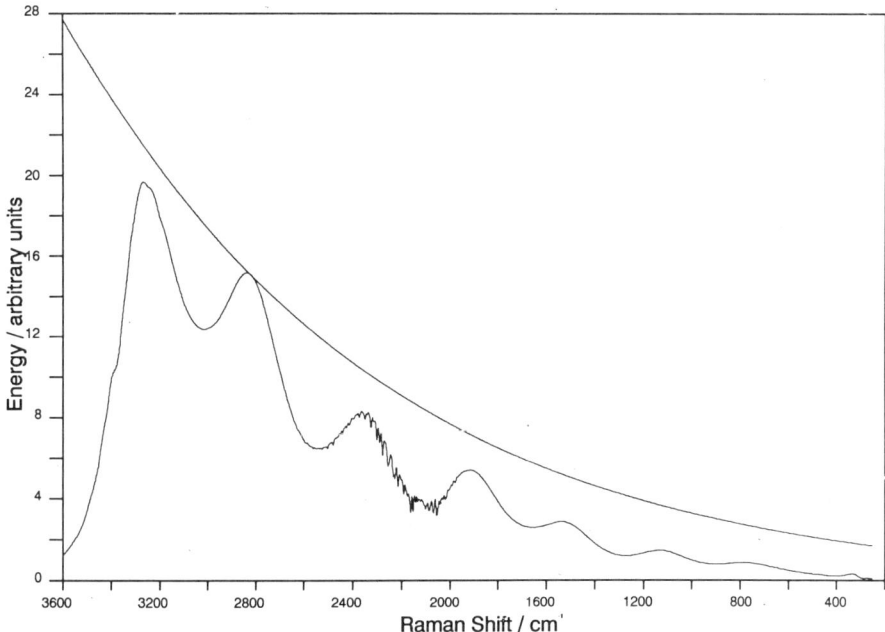

Fig. 6.5 — Calculated and recorded blackbody spectra superimposed. The abscissa axis is labelled in shift from the 1.064-μm line of a Nd^{3+}:YAG laser. The calculated spectrum is in counts per second. The recorded blackbody curve shows the filter characteristics and absorptions due to atmospheric water vapour.

correction curve may over-compensate for these sharp lines and invert them. This is because the spectrum of a distant blackbody source will suffer from water vapour absorption to a greater extent than a fluorescent sample in the FT Raman spectrometer. If this causes problems, the correction curve may be smoothed, but over-smoothing will perturb any other sharp features on the correction curve such as the filter cutoff (see Fig. 6.9). The only true solution is to always purge the instrument with dry air and the optical path between the furnace and the spectrometer.

The correction curves produced will only be correct for one filter/detector/beamsplitter combination. If any of these components is upgraded a new set of correction curves will have to be recorded. If this is likely to occur on a regular basis a furnace is a rather inconvenient source to use. It is, therefore, a good idea to use a secondary standard such as a large-area filament lamp or the fluorescence profile from a well characterized sample. The spectra of these could be corrected using the furnace correction curves to obtain their true spectra. If any subsequent modifications are then made to the spectrometer it is possible to re-record these spectra under the new conditions and use them in conjunction with the previously corrected true spectra to prepare a new correction curve. In this method it is only necessary to use the furnace once.

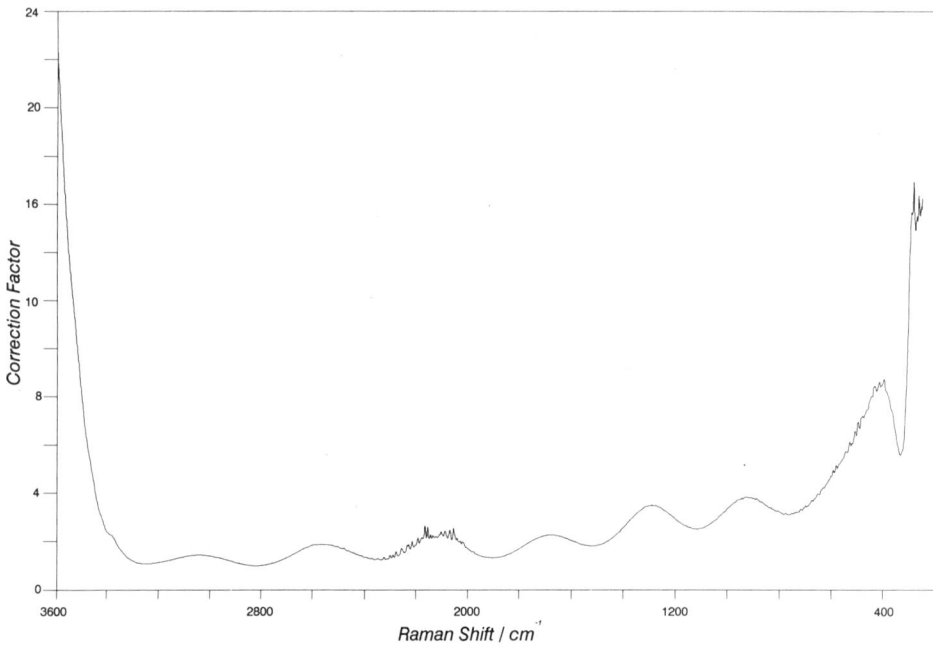

Fig. 6.6 — The correction curve obtained by dividing the theoretical blackbody curve by the recorded one shown in Fig. 6.5. Where the detector is not sensitive (greater than 3500 cm^{-1} shift) and the filters are not transmissive (less than 300 cm^{-1} shift) the correction factor is very high and these regions of the corrected spectra should be ignored.

An ideal blackbody is a theoretical concept and real thermal emitters may have slightly different emission characteristics. The Planck distribution describes the maximum amount of energy that can be emitted by a radiator of unit size at a given temperature. Real radiators emit less radiation than this, the ratio of the intensity of the theoretical and real emission being known as the emissivity. Thus an ideal blackbody has an emissivity of one, and real emitters have an emissivity less than one. Emissivity can be a function of wavelength and temperature. A source for which emissivity is independent of wavelength is known as a greybody. If an accurate relative correction curve for an FT Raman instrument is to be measured, a greybody emitter should ideally be used. In this respect a simple furnace may not be the best source available. Instrument manufacturers have experience of developing very well characterized sources for infrared spectrometers. As such they should be able to produce a suitable source for calibrating FT Raman spectrometers whose exact emission profile could be described by a polynomial curve.

Eqs. (6.1) and (6.2) show that the intensity of spontaneous Raman scattering increases as the fourth power of the frequency of the light being scattered. Thus Raman scattering is more efficient in the visible than in the near infrared. The relative intensity of bands in a Raman spectrum is thus dependent on the v^4 relationship, and the exact form of the spectrum will vary with the excitation wavelength. Figure 6.10(a) shows a plot of $(v_0 - v_{vib})^4$ for argon ion (514.5 nm) and

Sec. 6.6] Calibration of FT Raman spectra for instrument response

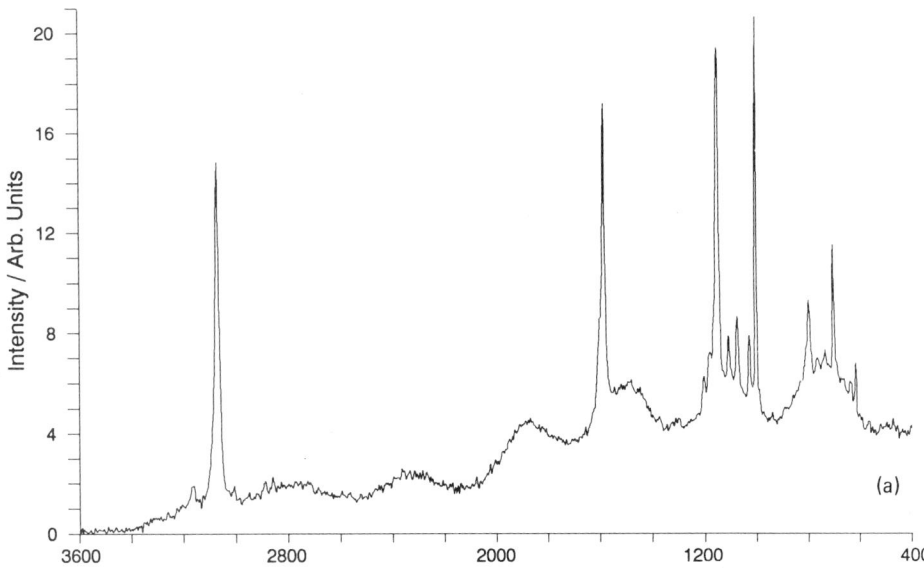

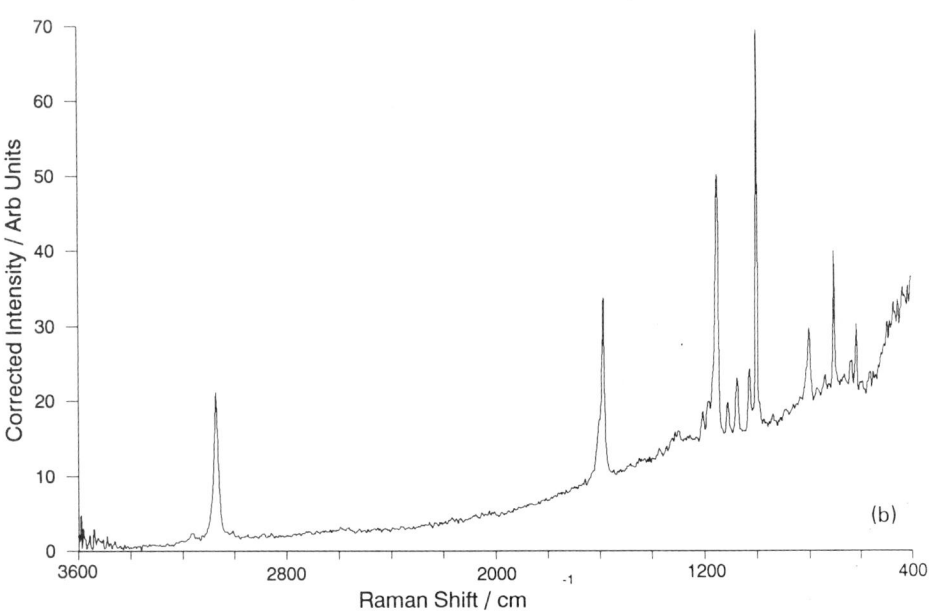

Fig. 6.7 — (a) A raw uncorrected FT Raman spectrum of a PES/PESES copolymer showing oscillations due to filter characteristics. (b) The same spectrum after multiplication by the correction curve. The artificial oscillations are removed revealing the true gently sloping background. Courtesy of C. J. Petty.

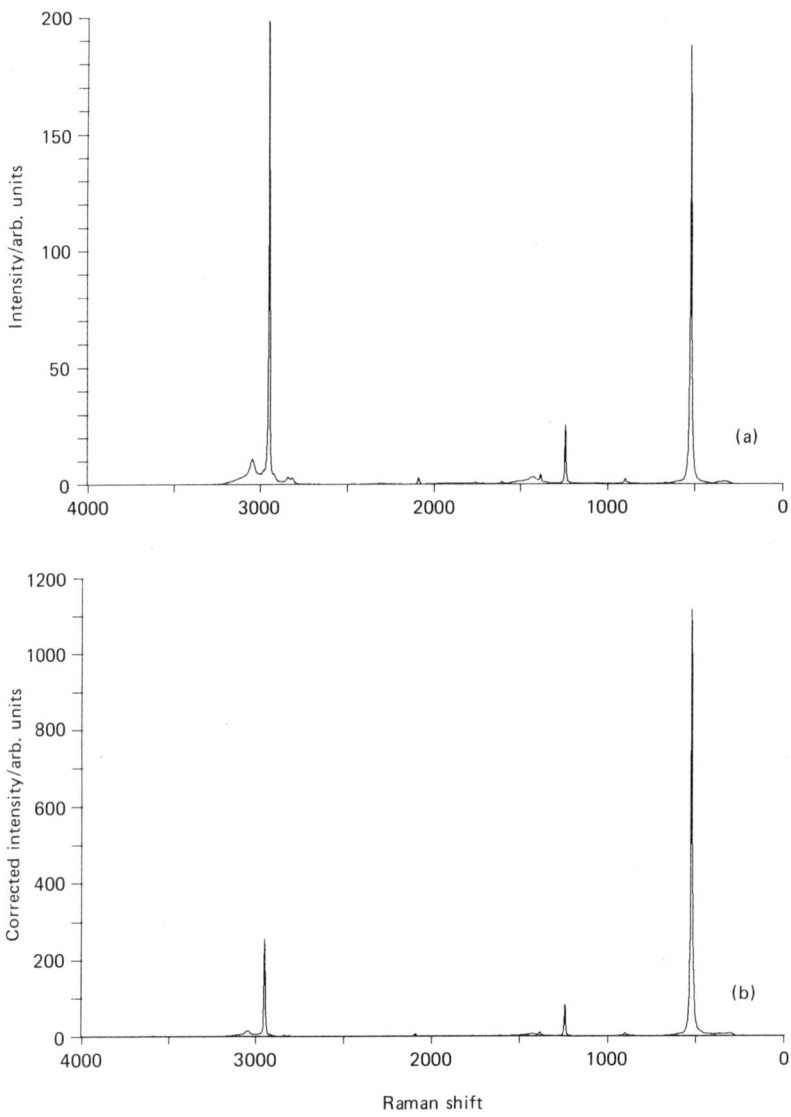

Fig. 6.8 — FT Raman spectrum of methyl iodide (a) before and (b) after multiplication by the correction curve, showing how the relative band intensities change. The corrected spectrum is in counts per second. Courtesy of C. J. Petty.

Nd^{3+}:YAG (1064 nm) laser sources. These curves have been normalized so that they are equal at $\nu_{vib} = 0$ cm^{-1}. The scattering efficiency for Nd^{3+}:YAG-excited spectra falls off much more sharply as the Stokes Raman shift increases. Most of the Raman spectra published to date have been recorded in the visible with a range of fairly similar excitation wavelengths (488.0, 514.5, 632.8, 647.1 nm) so the ν^4 effect

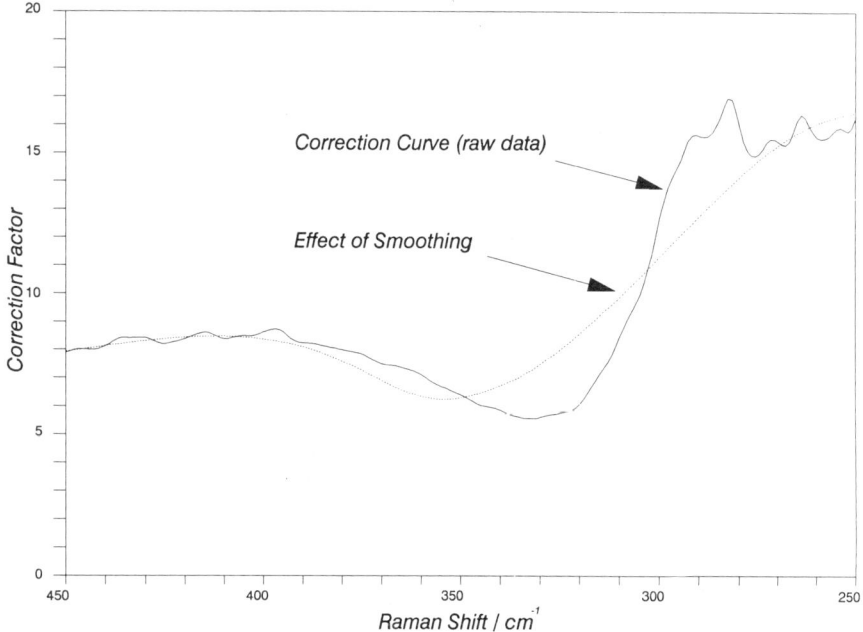

Fig. 6.9 — The correction curve at the filter cutoff, unsmoothed and smoothed, showing how sharp features in the correction curve are degraded by over-smoothing. Courtesy of C. J. Petty.

on relative band ratios was very slight. Figure 6.10(b) shows the true ratio of the two curves in Fig. 6.10(a), showing how the relative band intensities will be very different for visible and near infrared excited Raman spectra.

To date virtually none of the Raman spectra published in the literature have been corrected for instrumental sensitivity. As the instrumental response curve for a conventional instrument is relatively flat there was comparatively little need to do so. By chance, uncorrected FT Raman spectra actually look very like visible Raman spectra, even though none of them are the true spectrum of light emitted by the sample! Although the scattering efficiency curve for a visible laser source falls less steeply than that for a near infrared laser (Fig. 6.10(a)), the efficiency of a photomultiplier tube falls towards the high Stokes shift end of the spectrum. The two effects tend to cancel each other out. This is by no means an argument for not correcting FT Raman spectra, but serves to illustrate how one can be lulled into a false sense of security by being unaware of variations in instrumental sensitivity. Considerable care should be taken when attempting to use band ratio data measured in the visible to analyse Raman spectra recorded in the near infrared.

6.7 STANDARDIZATION OF FT RAMAN SPECTRA

There are various ways in which the intensity of vibrational bands can be expressed. The simplest method is to compare the relative intensities of bands in the same

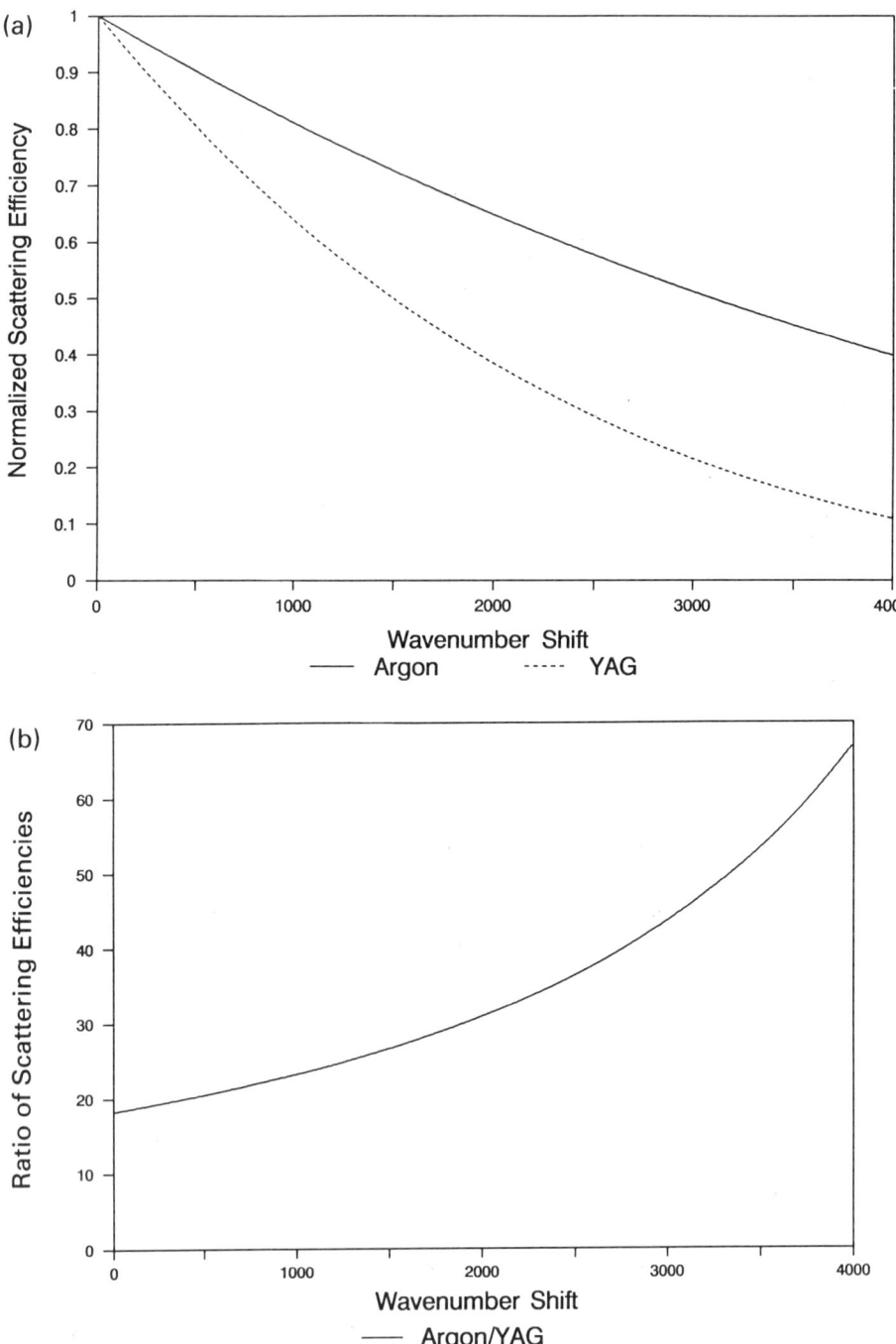

Fig. 6.10 — (a) Plot of $(v_0 - v_{vib})^4$ for laser lines at 514.5 nm (argon ion) and 1064 nm (Nd^{3+}:YAG); (b) the ratio of the two curves in (a). The two curves in (a) have been scaled so that they superimpose at 0 cm^{-1} shift. In reality the scattering efficiency from the Nd^{3+}:YAG laser at 0 cm^{-1} shift is an 18th that for the argon ion laser, and this value falls to a 65th by 4000 cm^{-1} shift as shown in (b).

spectrum. This is usually done by assigning bands as strong, medium or weak, and thousands of spectra have been published annotated in this way. The method does, however, have several drawbacks. Firstly, most of the assignments are made by using the band heights, and ignoring the band widths, rather than their area, the true measure of band intensity. Secondly, as the intensities of the spectral bands are only compared relative to one another within one spectrum, there is no way of telling how the intensity of a band labelled as strong in one spectrum compares with that of a band in the spectrum of another material. What 'strong' actually means in this case is 'an intense band in this spectrum but bearing no relation at all to bands in any other spectrum'. This is particularly problematic in Raman spectroscopy because there is wide variation in the intensity of Raman scatter from different materials. Finally, there is no way of making quantitative measurements using this system of classifying band intensity.

The high reproducibility of and ability to use external standards in FT Raman spectroscopy offers a straightforward solution to this problem. Work at Southampton has been carried out to develop a standard intensity scale on which the FT Raman spectra of liquids can be published [12]. This centres around using an organic liquid which can be used as an internationally agreed external standard, so the intensity of bands in subsequent spectra can be expressed in terms of the intensity of a band in this standard. In this way the varying performance and ordinate scales of different instruments would be discounted. Since the intensity of spontaneous Raman scattering is a linear function of the excitation intensity (Eq. (6.4)), if the standard and sample need to be run at different powers a simple correction can be made.

The chemical chosen as the external standard must have a number of key properties:

1. Ideally it must be non-toxic, or if it is at all hazardous it must be kept in a sealed cell.
2. It must be a good Raman scatterer so that spectra of high signal-to-noise ratio can be obtained quickly.
3. It must possess a broad Raman band so that its peak height will be comparatively unaffected by resolution and apodization changes.
4. It must be readily available.
5. It must not absorb in the range 1.064–1.7 μm.

After searching through atlases of Raman spectra it has been proposed that the 1170 cm^{-1} band of hexachlorobutadiene (HB) be used (see Fig. 6.11). This strong band is sufficiently broad that its height is unaffected by changes in resolution down to 4 cm^{-1}, and only reduces by 6% on moving to 8 cm^{-1} resolution. It has been used for many years as a mulling agent in infrared spectroscopy. Unfortunately, it has recently been listed as a suspected carcinogen. However, the external standard can be sealed permanently in a vessel identical to the sample cell to be used, and there is no need for the user to handle it directly.

Most important in the standardization of FT Raman spectra is the correction of the sample and HB spectra for the instrumental response function. After this has been done the external standard can be used in two ways. Firstly, if the area of the

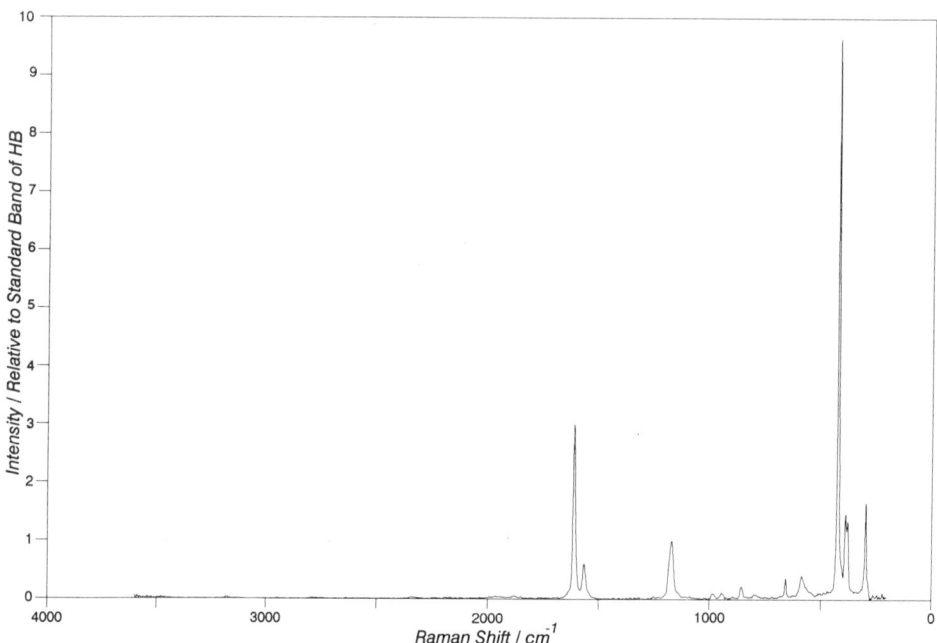

Fig. 6.11 — The FT Raman spectrum of hexachlorobutadiene.

HB band is measured, the area of bands in the sample can be expressed as a percentage of this area (with appropriate corrections for differing laser powers). These can then be tabulated and a large mass of data on the relative Raman cross-sections for various bands in many liquids would hence be rapidly accumulated. This would provide a source of information which could be used to find relationships between sample structure and Raman intensity. It would also be possible to perform quantitative experiments without the need to prepare a set of standard solutions by using the relative Raman cross-section of the analyte on the HB scale.

The second method proposed uses band heights and allows FT Raman spectra to be presented on a meaningful ordinate scale. Spectra could be presented with percentage of the HB 1170 cm^{-1} band height on the ordinate axis, stating the resolution used. Naturally relative band heights will change with resolution, but the scale would provide a method of assessing the intensity of Raman bands at a glance. To give an indication of how this works we present in Fig. 6.12 two acetone spectra run at different laser powers, resolutions, scan speeds, apodization functions, etc. Both of these have been corrected and appear on an HB ordinate scale. This shows how well two spectra run on different FT Raman instruments but corrected and presented on this scale agree. Large libraries of FT Raman spectra could be easily collated on this scale and the computer searching techniques commonly used in other forms of spectroscopy would be available for Raman.

There is, unfortunately, one problem which complicates the recording of Raman spectra: self-absorption of the Raman light. In a transparent liquid Raman light is

Sec. 6.8]   **Examples of the use of FT Raman in quantitative analysis**   147

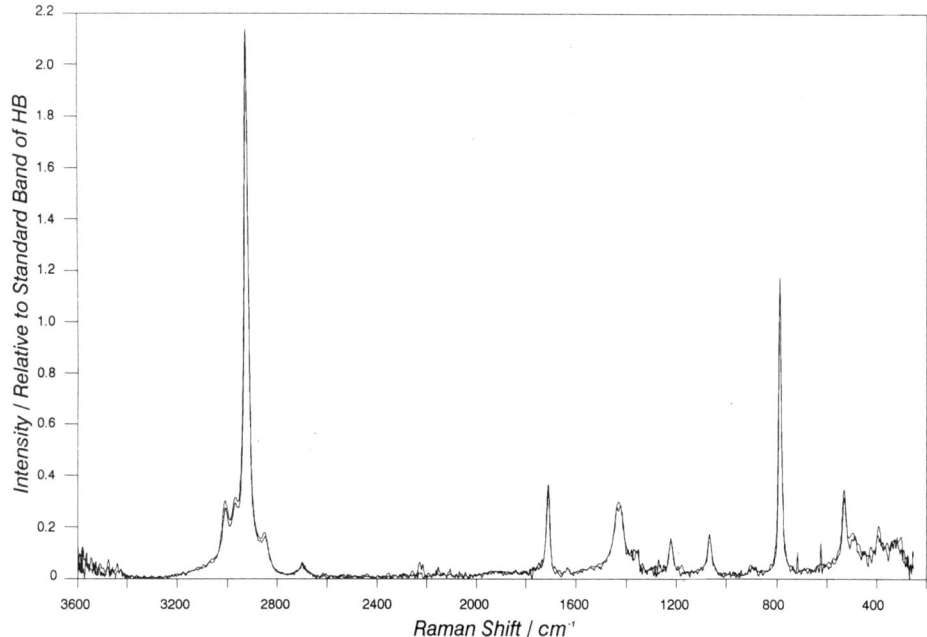

Fig. 6.12 — Two spectra of acetone run under different conditions presented on the HB ordinate scale. The two spectra superimpose virtually perfectly. Any differences in band height are due to using different resolutions; the band areas are the same.

scattered from within the sample volume. As it passes through the sample towards the spectrometer it may be re-absorbed and the spectrum recorded is a convolution of the Raman spectrum and near infrared absorption spectrum. In most cases the effect is slight but it presents severe difficulties when recording the spectra of aqueous solutions. Water has a very strong near infrared absorption which cuts in at 7400 cm^{-1} (2000 cm^{-1} Raman shift from an Nd^{3+}:YAG laser). The effect can be diminished by using D$_2$O as a solvent instead of H$_2$O (see Fig. 6.13). The FT Raman spectrum of an aqueous solution of phenol obtained with both solvents is shown in Fig. 6.14. Unfortunately, it is often not possible to substitute D$_2$O for H$_2$O and this problem is likely to preclude many biological studies if an Nd^{3+}:YAG laser is used as an excitation source.

6.8 EXAMPLES OF THE USE OF FT RAMAN IN QUANTITATIVE ANALYSIS

To date several papers have been published specifically on making quantitative measurements with an FT Raman spectrometer [4,5,13–17]. Many other published papers also contain examples of extracting quantitative data from FT Raman spectra. Some of these use external standards; others use simple band ratios.

The degree of unsaturation in oils and margarine has long been a topic of applied research, particularly in the light of possible links with heart disease. The standard

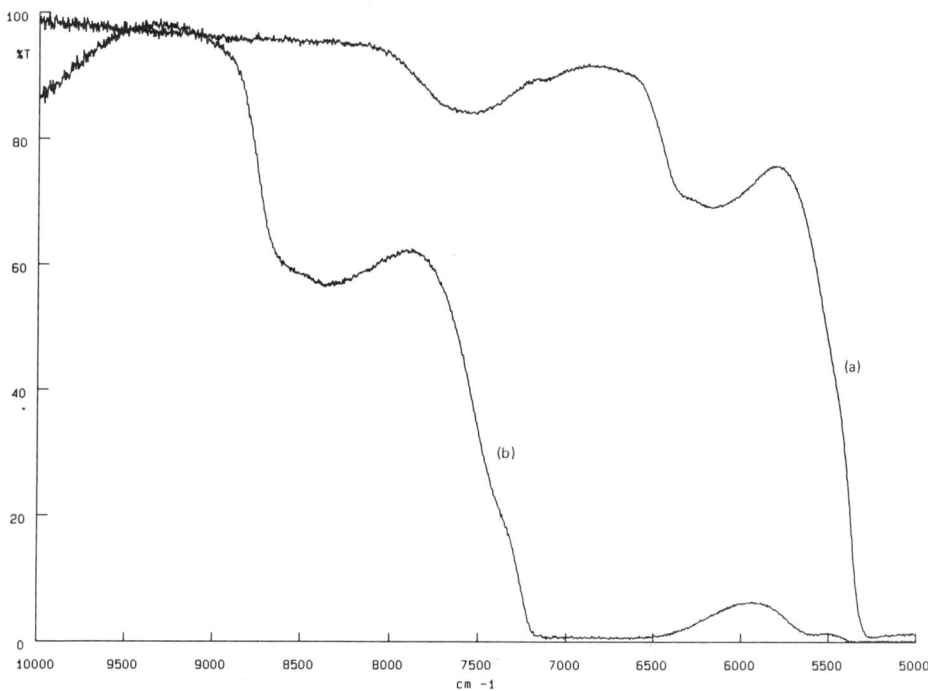

Fig. 6.13 — The near infrared spectrum of (a) D_2O and (b) H_2O in a 5-mm path length quartz cell. H_2O absorbs strongly in the current near infrared FT Raman range (approx 9500–5500 cm^{-1}), but the absorptions of D_2O are much less intense making it a superior solvent for FT Raman studies.

chemical test for unsaturation is to iodinate the double bonds in a quantitative experiment. The results of this test are expressed in terms of standard iodine values [18]. Infrared spectroscopy has been used as an alternative in the past but it suffers from two main problems. Firstly, the $\nu_{C=C}$ band (1660 cm^{-1}) is weak and can be obscured by the much stronger $\nu_{C=O}$ band (1745 cm^{-1}) of fatty acids. Secondly, margarines are normally fats emulsified with water, which gives a strong OH bending vibration also around 1660 cm^{-1}. Wilson et al. [15] have, however, measured the degree of unsaturation in oils and margarines by FT Raman spectroscopy. FT Raman is an ideal tool to investigate this problem as both the $\nu_{C=O}$ and ν_{OH} vibrations are only weakly Raman active, whereas the $\nu_{C=C}$ vibration gives rise to a strong Raman band. Margarines and oils give high quality, fluorescence-free Raman spectra when excited in the near infrared. The ratio of the $\nu_{C=C}$ band (1601 cm^{-1}) to the δCH_2 scissoring vibration (1444 cm^{-1}) was calculated and plotted as a function of iodine value. Figure 6.15 shows the plot obtained for five commercial oils using band heights, and Fig. 6.16 shows a similar plot for seven margarines obtained using band areas. Better results are obtained using band areas, as margarines contain both *cis*- and *trans*-isomers giving rise to two overlapping $\nu_{C=C}$ bands. The calibration curves

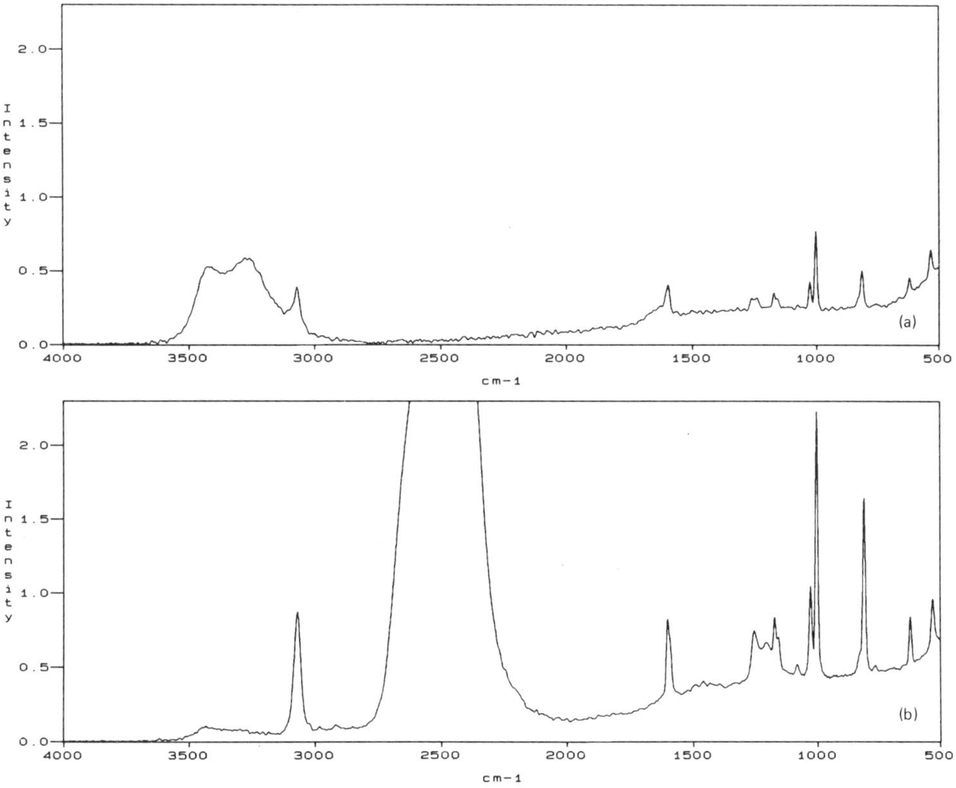

Sec. 6.8]   Examples of the use of FT Raman in quantitative analysis   149

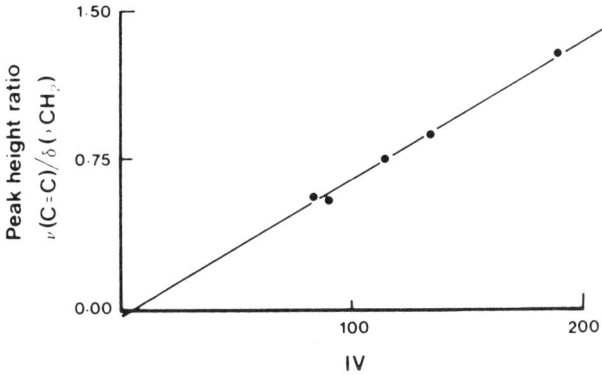

Fig. 6.14 — The FT Raman spectrum of phenol dissolved in (a) H_2O and (b) D_2O recorded using 1 W of laser power. The spectrum from the D_2O sample is approximately three times as intense as that from the H_2O sample. H_2O gives bands at 3400 and 1650 cm^{-1}, whilst D_2O gives bands at 2500 cm^{-1} and 1200 cm^{-1}. Uncorrected data.

Fig. 6.15 — Band height ratio $v_{C=C}/\delta(CH_2)$ versus iodine value for five commercial oils. Reproduced with thanks from ref. [15].

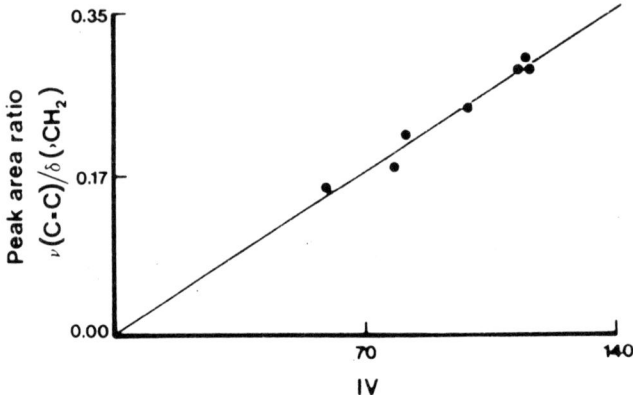

Fig. 6.16 — Band area ratio $v_{C=C}/\delta(CH_2)$ versus iodine value for seven commercial margarines. Reproduced with thanks from ref. [15].

shown in Figs. 6.15 and 6.16 could be used to measure the degree of unsaturation of a sample in a few minutes.

External standards have been used in FT Raman experiments to demonstrate the feasibility of the technique to study intermolecular interactions [4]. Acetone and chloroform, a far from ideal mixture, was investigated. Pure acetone was used as an external standard, with the intensity of acetone bands in acetone/chloroform mixtures being expressed in terms of their intensity in pure acetone. Figure 6.17

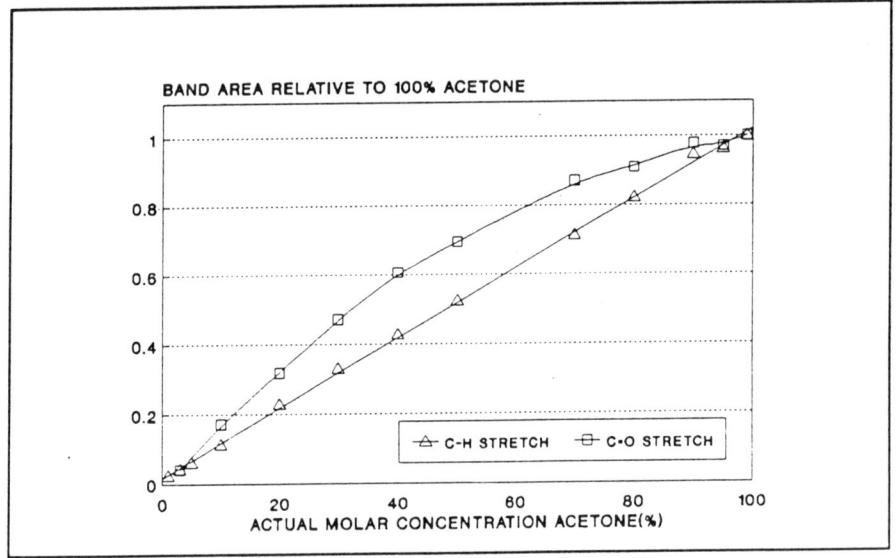

Fig. 6.17 — The behaviour of the CH stretching and CO stretching bands of acetone in acetone/chloroform mixtures.

shows the results obtained from the CH stretching and CO stretching bands. The CH stretching intensity of acetone follows a simple linear plot, whereas the CO band shows non-linear behaviour as a result of hydrogen bonding.

The analysis of mixtures of hydrocarbon fuel mixtures is very important in the petroleum industry. Both the infrared and Raman spectra of hydrocarbons of the general formula C_nH_{2n+2} show little variation as n changes. It is therefore not possible to make quantitative measurements based on simple band ratios from the spectra of mixtures of these compounds, as all of the bands overlap severely. The growth of chemometrics has spawned a number of numerical methods which can be used very successfully to obtain quantitative data from the spectra of mixtures. These methods are being used increasingly as the cost of sufficiently powerful computers falls and instrument manufacturers offer the appropriate software packages as additional purchases. The techniques currently used are based on multiple linear regression (MLR), principal components regression (PCR) and partial least squares (PLS) [19]. PCR and PLS use a step called principal components analysis (PCA), which is also known as abstract factor analysis [20].

FT Raman spectroscopy has considerable potential in the analysis of liquid fuel mixtures. Hydrocarbons produce strong Raman spectra which can be recorded quickly. The use of fibre-optic coupling may make it possible to record spectra remotely in a chemical plant and use subsequent analysis of these spectra to control a manufacturing process. Lorber *et al.* [16] assessed the effectiveness of PLS, PCR and MLR techniques in analysing mixtures of unleaded petroleum, super unleaded petroleum and diesel fuels by recording their FT Raman spectra. In each case a series of calibration samples was prepared by mixing these three components in known proportions. The predictive ability of each of the techniques was assessed by comparing the root mean square errors between the actual composition of the calibration samples and that predicted by the computer model. PCR and PLS were found to be more accurate than MLR, notably because they can partially accommodate non-linearities in the spectral data arising from intermolecular interactions in non-ideal mixtures. PLS incorporates more information from the calibration data into the model and can be more accurate than PCR if the spectra analysed contain noisy, information-poor regions.

Williams *et al.* [17] have used FT Raman spectroscopy combined with CIRCOM, a commercial factor analysis package, to measure the cetane number and cetane index of gas oils. These are measures of the ignition quality of a fuel in an internal combustion engine. Conventionally the cetane number of a fuel is measured in a 'standard' engine under controlled conditions. These measurements are costly, time consuming and require a large sample of gas oil. The cetane index is a secondary standard which relates the density and boiling range of gas oils to the cetane number. The FT Raman spectra of 18 different samples of known cetane number and cetane index were recorded. The CIRCOM package was then used to identify five principal component spectra which when added with suitable coefficients could reproduce the original spectra. These coefficients were analysed by multiple linear regression to correlate them with the cetane number and cetane index values. This produces a regression equation which directly relates the coefficients to cetane number and cetane index. The CIRCOM package is then able to measure the cetane number and index of subsequent samples by resolving their spectra into a combination of the principal components and substituting the coefficients into the regression equation. The correlation coefficients for cetane index and cetane number were 0.93 and 0.77 respectively. Clearly the correlation with cetane number is not good, but there could be significant errors in the cetane numbers of the calibration samples owing to difficulties in comparing results from 'standard' engines. However, the beauty of factor analysis packages such as CIRCOM is that it may be possible to correlate spectral changes to any property of the sample likely to affect its vibrational behaviour. Once a calibration model has been built, it may be possible to measure many properties of a sample by solely recording its spectrum [21]. The application of these techniques to FT Raman is ripe for development.

6.9 QUANTITATIVE STUDIES ON SOLIDS

So far all of this chapter has centred on isotropic liquids. The quantitative analysis of solids presents significant experimental difficulties. Raman intensity from powdered samples falls as the particle size decreases (see Fig. 6.18). Also, the number of

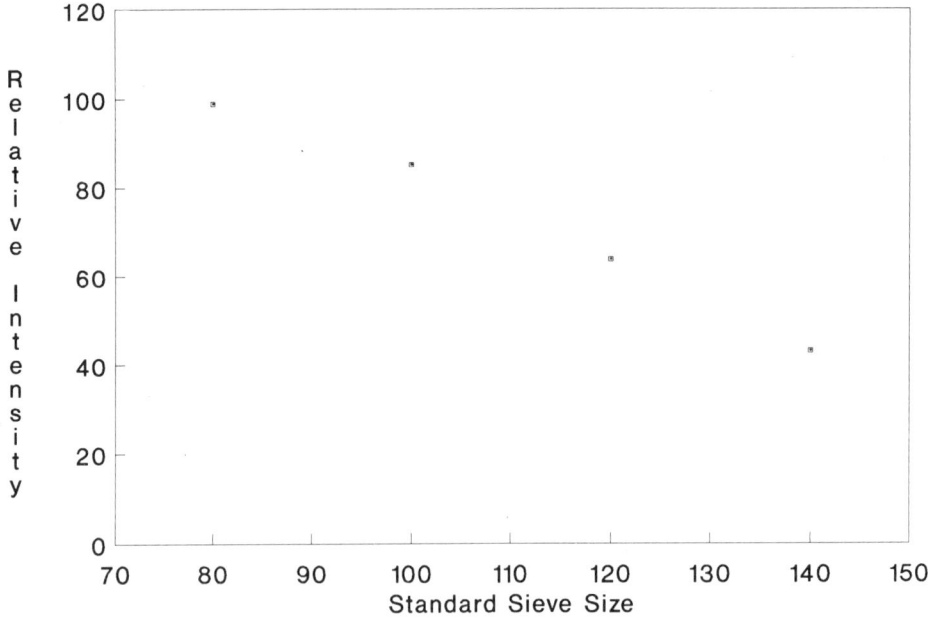

Fig. 6.18 — Raman intensity as a function of particle size for potassium chromate powder. The x axis labels are the mesh sizes of the standard sieves through which the sample powder passed measured in wires per inch.

sample molecules in the laser beam depends on how densely the particles are packed. Unfortunately, these problems hinder the use of external standards for solid samples.

However, relative band ratios can still be used to extract quantitative data from the spectra of solid samples. Ellis et al. [22,23] have investigated the FT Raman spectra of poly(aryl ether sulphone) (PES) and poly(aryl ether ether sulphone) (PEES) and amorphous copolymers of the two. The repeat units for these two polymers are shown in Fig. 6.19. The authors found clear correlations between the

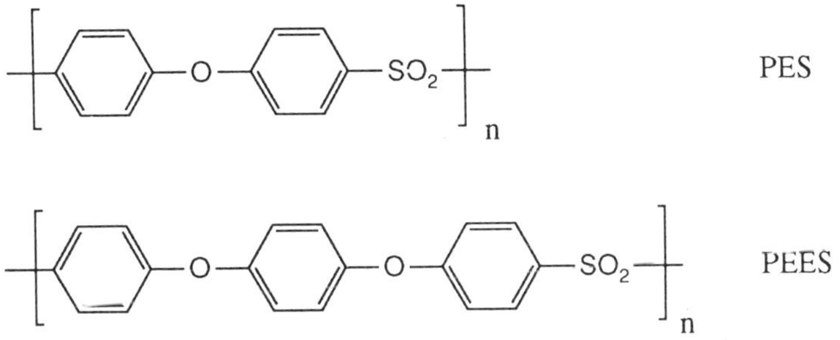

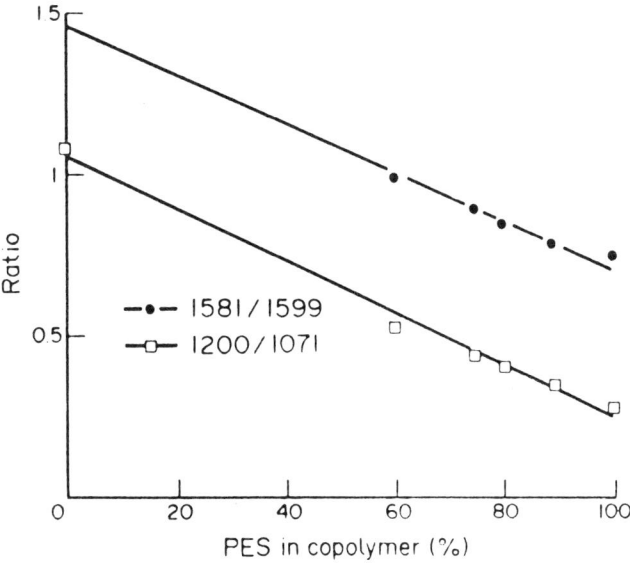

Fig. 6.19 — Repeat units for (a) PES and (b) PEES polymers.

relative band ratios and copolymer composition (as measured by ^{1}H NMR). This is shown in Fig. 6.20. The exact reason for this close correlation is unclear, but the

Fig. 6.20 — Band intensity ratios 1581 cm^{-1}/1599 cm^{-1} and 1200 cm^{-1}/1071 cm^{-1} for PES/PEES copolymers as a function of %PES content. Reproduced with permission from *Spectrochimica Acta*, **46A**, J. K. Agbenyeya et al., copyright 1990, Pergamon Press PLC.

observation could be used to characterize these copolymers in a manufacturing plant.

Hendra et al. [24] have studied single number nylons by FT Raman spectroscopy. These materials have the general formula $[NHCO(CH_2)_n]_m$, where the nylon number is given by $n + 1$ (i.e. nylon 3 is $[NHCO(CH_2)_2]^m$). The alkyl chain has a similar vibrational behaviour to a long-chain saturated hydrocarbon. By studying different nylon samples for which the value of n increases, the ratio of the CH_2 bending band to the CO stretching band (amide I) was found to increase linearly with n (see Fig. 6.21). Thus FT Raman spectroscopy can be used to determine the nylon number of an unknown nylon sample.

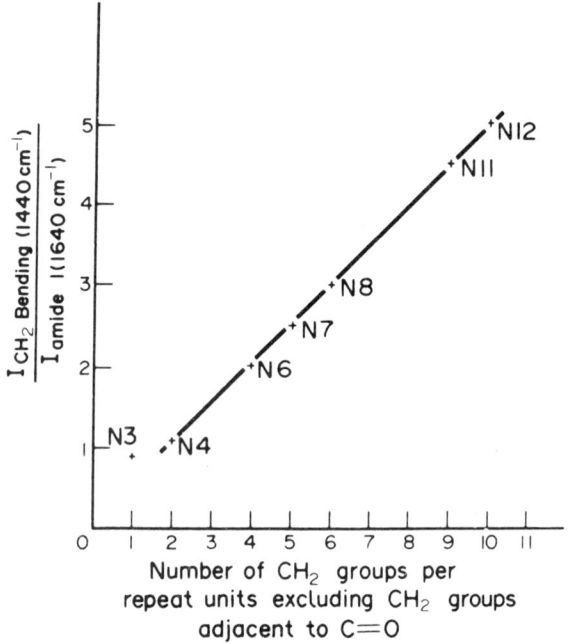

Fig. 6.21 — Graph of the ratio of the CH_2 bending mode to the CO stretching band for nylons as a function of the number of CH_2 groups per repeat unit excluding those adjacent to C=O. The points are labelled with the corresponding nylon number. Reproduced with permission from *Spectrochimica Acta*, **46A**, P. J. Hendra et al., copyright 1990, Pergamon Press PLC.

REFERENCES

[1] G. Placzek, *Handbuch der Radiologie*, Ed. Erich Marx, Akademische Verlagsgesellschaft VI, **2**, 209–374 (1934). Translation: *The Rayleigh and Raman Scattering*, University of California Radiation Laboratory (UCRL) Trans 526(L) (1962).
[2] *Raman Spectroscopy of Gases and Liquids*, Ed. A. Weber, Springer-Verlag, Berlin, 1979.
[3] S. Milićev, A. Stergaršek, *Spectrochim. Acta*, 1989, **45A**, 225.

References

[4] C. J. Petty and G. M. Warnes, *12th International Conference on Raman Spectroscopy*, p. 798 (1990). Ed. J. R. Durig and J. F. Sullivan, John Wiley & Sons, 1990.
[5] Bio-Rad Inc., Poster at 12th International Conference on Raman Spectroscopy, Columbia, South Carolina, August 1990.
[6] C. J. Petty, G. M. Warnes, P. J. Hendra and M. Judkins, *Spectrochim. Acta*, 1991, **47A**.
[7] D. P. Strommen and K. Nakamoto, *Laboratory Raman Spectroscopy*, Wiley & Sons, 1984.
[8] H. H. Perkampus, K. Kortum and H. Bruns, *Appl. Spectrosc.*, 1969, **23**, 105.
[9] M. D'Orazio and B. Schrader, *J. Raman. Spectrosc.*, 1974, **2**, 585.
[10] F. Grum and R. J. Becherer, *Radiometry*, volume one of 'Optical Radiation Measurements', Academic Press, New York, 1979.
[11] A. P. Thorne, *Spectrophysics*, 2nd Ed, Chapman & Hall, 1988.
[12] C. J. Petty, P. J. Hendra and T. Jawhari, *Spectrochim. Acta*, 1991, **47A**.
[13] T. Jawhari, P. J. Hendra, H. A. Willis and M. Judkins, *Spectrochim. Acta.*, 1990, **46A**, 161.
[14] T. Jawhari and P. J. Hendra, *Spectroscopy World*, 1989, **1**, No 1, 6.
[15] H. Sadeghi-Jorabchi, P. J. Hendra, R. H. Wilson and P. S. Belton, *J. Am. Oil. Chem. Soc.*, 1990, **67**, 483.
[16] M. B. Seasholtz, D. D. Archibald, A. Lorbe and B. R. Kowalski, *Appl. Spectrosc.*, 1989, **43**, 1067.
[17] K. P. J. Williams, R. E. Aries, D. J. Cutler and D. P. Lidiard, *Anal. Chem.*, 1990, **62**, 2553.
[18] C. Pagrout, *Standard Methods for the Analysis of Oils, Fats and Derivatives*, Pergamon Press, Oxford, England, 1979.
[19] K. R. Beebe and B. R. Kowalski, *Anal. Chem.*, 1987, **59**, 1007A.
[20] E. R. Malinowski and D. G. Howery, *Factor Analysis in Chemistry*, John Wiley & Sons, New York, 1980.
[21] P. M. Fredericks, J. B. Lee, P. R. Osborn and D. A. J. Swinkels, *Appl. Spectrosc.*, 1985, **2**, 303 and 311.
[22] G. Ellis, A. Sanchez, P. J. Hendra, J. M. Chalmers, J. G. Eaves, W. F. Gaskin and K. N. Kruger, *J. Mol. Struct.* in press.
[23] J. K. Agbenyega, G. Ellis, P. J. Hendra, W. F. Maddams, C. Passingham and H. A. Willis, *Spectrochim. Acta*, 1990, **46A**, 197.
[24] P. J. Hendra, W. F. Maddams, I. A. M. Royaud, H. A. Willis and V. Zichy, *Spectrochim. Acta*, 1990, **46A**, 747.

7

The application of FT Raman spectroscopy in organic chemistry

7.1 GROUP FREQUENCIES

In its early days Raman spectroscopy was one of very few physical techniques which could provide information on organic compounds. Hibben, in 1939, collected a vast amount of data on Raman spectra of organic materials and attempted to define group frequency correlations [1]. However, it has rarely been possible to record spectra from a full range of related compounds because of problems of fluorescence, and sadly many correlations have been made on such a limited number of compounds that they are of suspect reliability. More recent publications have appeared which have attempted to bring the situation up to date. They include a book on infrared and Raman spectroscopy by Colthrup, Daley and Wiberley [2] and books solely on Raman by Baranska, Labudzinska and Terpinski [3], and Dollish, Fately and Bentley [4]. Very recently an atlas containing the infrared and Raman spectra of over one thousand organic compounds has been published [5]. In this chapter we have attempted to distil some of the more reliable correlations and hope that they will be extended in the near future using the more complete data available from FT methods.

7.1.1 The X–H stretching region

A considerable number of species produce characteristic bands in the X–H stretching region (see Fig. 7.1). In general ν_{NH} and ν_{OH} vibrations are weaker than CH features, whilst aliphatic CH bonds give rise to stronger bands than aromatic ν_{CH} groups. In fact, the ν_{CH} modes often give rise to the strongest bands in the spectrum of aliphatic species, e.g. low density polyethylene (see Fig. 8.2). However, in aromatic compounds the ν_{CH} modes can be considerably weaker then the skeletal vibrations. This can be seen in Fig. 7.4 which shows the FT Raman spectrum of anthracene. In infrared spectra hydrogen-bonded OH and NH species produce

Sec. 7.1] Group frequencies

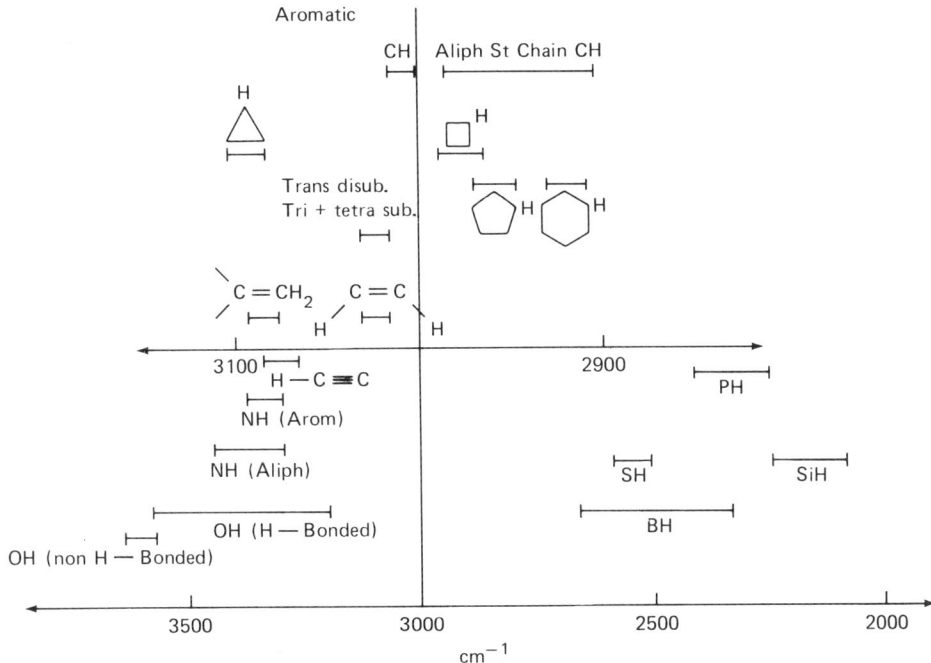

Fig. 7.1 — The frequencies of Raman bands due to X–H species.

intense but very broad absorptions. In Raman spectroscopy, these groups give rise to much weaker bands, but they can yield considerable information. ν_{XH} bands where X=B, S, Si P, etc. appear in sparsely populated regions of the spectrum and the analytical value of these correlations can be valuable in appropriate circumstances.

7.1.2 Saturated rings

The breathing motions of carbon rings give quite useful correlations, as shown in Fig. 7.2. As a carbon ring becomes more strained, it becomes more rigid and the

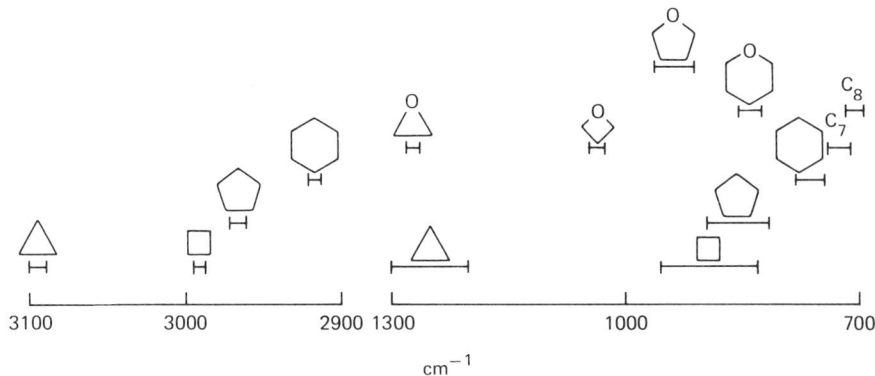

Fig. 7.2 — The frequencies of Raman bands due to saturated rings.

breathing frequency rises. In six-membered rings, inclusion of up to two nitrogen or oxygen atoms has relatively little effect on the breathing frequency, but three substituents or the inclusion of a sulphur atom reduces the intensity of the breathing mode and can make its assignment difficult. Many other correlations between frequency and structure have been proposed, and a more comprehensive discussion is given in ref. [4].

7.1.3 Unsaturated species ($\Delta v = 2300-1600$ cm^{-1})

Perhaps the most valuable correlation in Raman spectroscopy is the band due to stretching of the C=C group which appears near 1650 cm^{-1}. It is almost invariably one of the most intense bands in the Raman spectrum whether the olefinic group is in a chain or in a ring. Typical frequencies of a non-conjugated C=C group within a chain are shown in Fig. 7.3. Conjugation and substitution of the olefinic group with

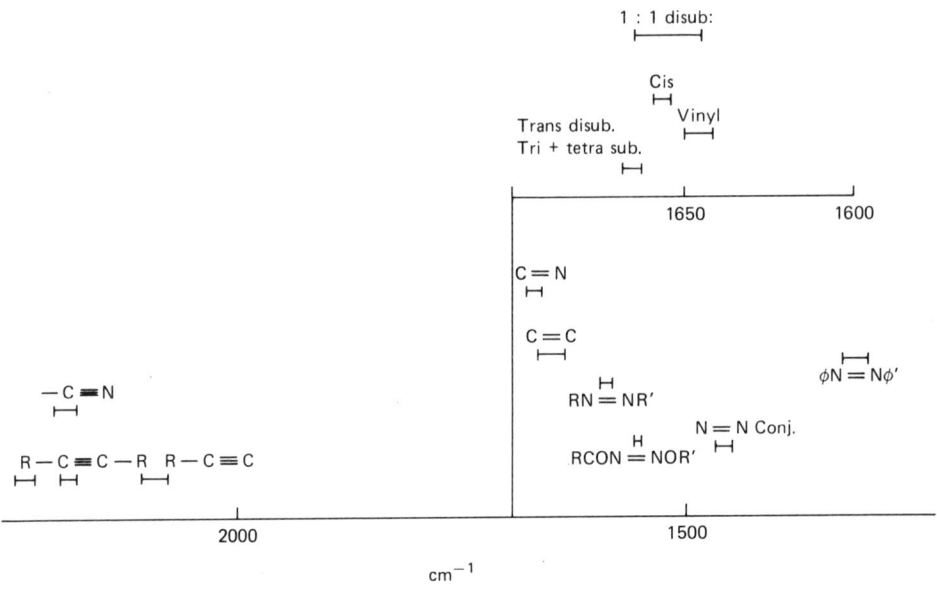

Fig. 7.3 — Frequencies of Raman bands due to unsaturated groups.

halogen atoms invariably lowers the frequency of $v_{C=C}$. The inclusion of the group in a ring usually results in its frequency lying outside the limits given in Fig. 7.3, but its intensity is normally maintained. Symmetry requires that where the C=C (or C≡C group) lies at the centre of mass of a molecule (e.g. in *trans*-XCH=CHX species) the stretching vibration centred on the C=C group will only occur in the Raman spectrum. This is indeed the case, but a very weak band can occur in the infrared spectrum if the C=C or C≡C group lies near the centre of mass. Thus, the correlation is hopelessly unreliable in the infrared, and Raman spectroscopy is far more suited to the analysis of unsaturated materials.

Alkyne (C≡C) and cyanide (C≡N) groups give rise to very strong Raman bands near 2200 cm^{-1} shift. Disubstituted acetylenes normally show a doublet (2225 and 2300 cm^{-1}) where of course only one band is expected. The origin of this second band is the subject of some conjecture [4]. The frequency of the $\nu_{C≡C}$ mode is normally lowered by conjugation. The same effect is seen in the nitriles where the cyanide vibration is at a lower frequency in aromatic and other conjugated compounds. Thus, the frequency ranges in these cases are:

> Alkyl–C≡N   2260–2240 cm^{-1}
> Aryl–C≡N   2240–2230 cm^{-1}
> C=C–C≡N   2235–2215 cm^{-1}

The C=N and N=N species behave very similarly to the C=C system, their frequency being slightly lowered by conjugation or the presence of aromatic groups. For example,

> Alkyl–CH=N–Alkyl   1675–1665 cm^{-1}
> Aryl–CH=N–Alkyl   1655–1645 cm^{-1}
> Aryl–CH=N–Aryl   1640–1630 cm^{-1}

The N=N frequency in ureas, RCO–N=N–COR', appears around 1555 cm^{-1}.

7.1.4 Aromatics

A search of available sources shows that correlations here are poorly developed probably because previous investigators have found their efforts hindered by fluorescence. Several points do, however, arise quite clearly. Mono-, di-, tri- and tetra- substituted benzene rings all give a sharp intense band near 1600 cm^{-1} and a ν_{CH} vibration near 3050 cm^{-1}. The CH stretching band system decreases in complexity as substitution is increased, i.e. as the number of hydrogens on the ring is reduced. The precise frequency of the 1600 cm^{-1} band appears generally to vary little with substitution. However, inclusion of chlorine, bromine or iodine reduces the frequency to below 1600 cm^{-1}. Other correlations have been suggested which rely on the ability to make polarized measurements [4], but at present it is unclear whether analysts are going to use polarization in a routine manner when using FT Raman instruments.

Perhaps anthracene needs special attention. Albeit not the most important compound known to organic chemists, it has been almost universally adopted as a 'standard' for checking the performance of FT Raman instruments on a routine basis. Although it is not really significant it is satisfying to know the origin of the bands so frequently observed. In Fig. 7.4 we show the FT Raman spectrum of anthracene with some of the more prominent bands identified [6,7].

7.1.5 Oxygen containing species

Carbonyl groups produce bands which are much weaker in the Raman than in the infrared; however, some useful correlations do exist and these are shown in Fig. 7.5.

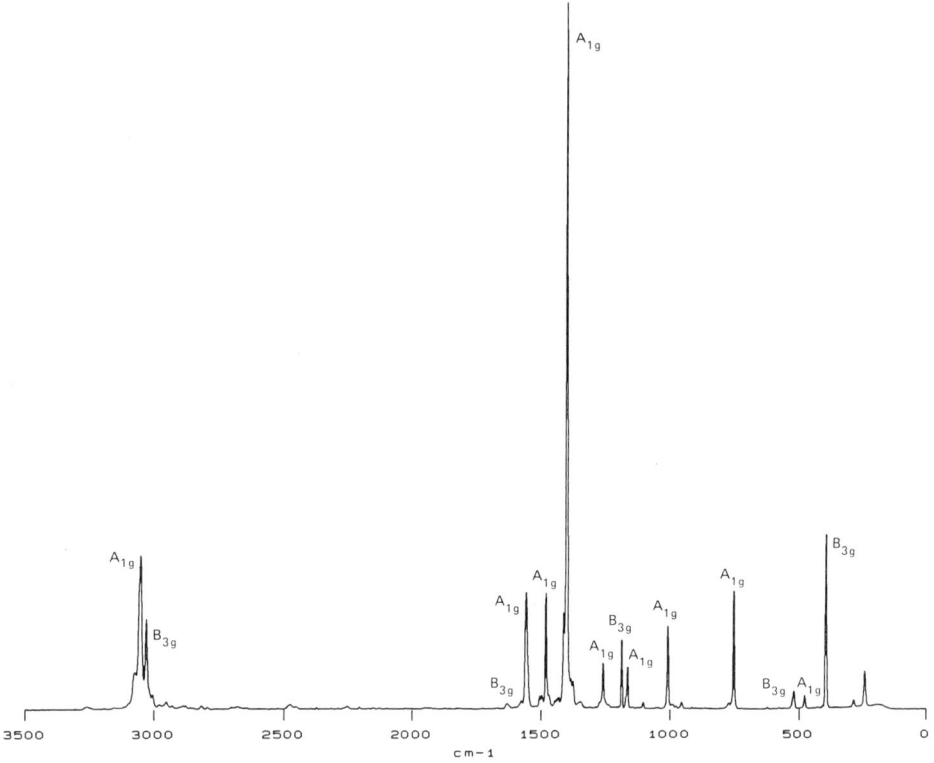

Fig. 7.4 — The FT Raman spectrum of anthracene (D_{2h}) showing the symmetry labels of the principal bands.

The nitro group also gives useful Raman correlations particularly since the symmetric NO_2 stretch is normally very prominent, and its Raman frequency appears in the range 1370–1330 cm^{-1} for aromatic nitros, and 1390–1355 cm^{-1} for aliphatic nitro-containing molecules. Raman bands can also be observed due to nitroso and sulphone groupings, and the frequencies of all of these are shown in Fig. 7.5.

7.1.6 Sulphur-, phosphorus-, silicon- and halogen-containing groups
Chlorides, bromides and iodides show strong Raman features, whereas fluoride bands are comparatively weak. Unfortunately, these molecules vibrate at frequencies similar to those of other groups and their vibrations are often coupled with other atoms in the system, and this limits their analytical usefulness. Species containing sulphur, phosphorus and silicon also give rise to strong Raman bands. However, their frequencies are very variable and although correlations based on small ranges of compounds have appeared their value in a general analytical sense is suspect. Thus, Baranska *et al.* [3] mention several correlations in open structures, whilst Dollish *et al.* [4] prefer to confine themselves to a more detailed study of some heterocycles. The C–S bond vibrates over the extensive range 705–570 cm^{-1},

Sec. 7.1] **Group frequencies** 161

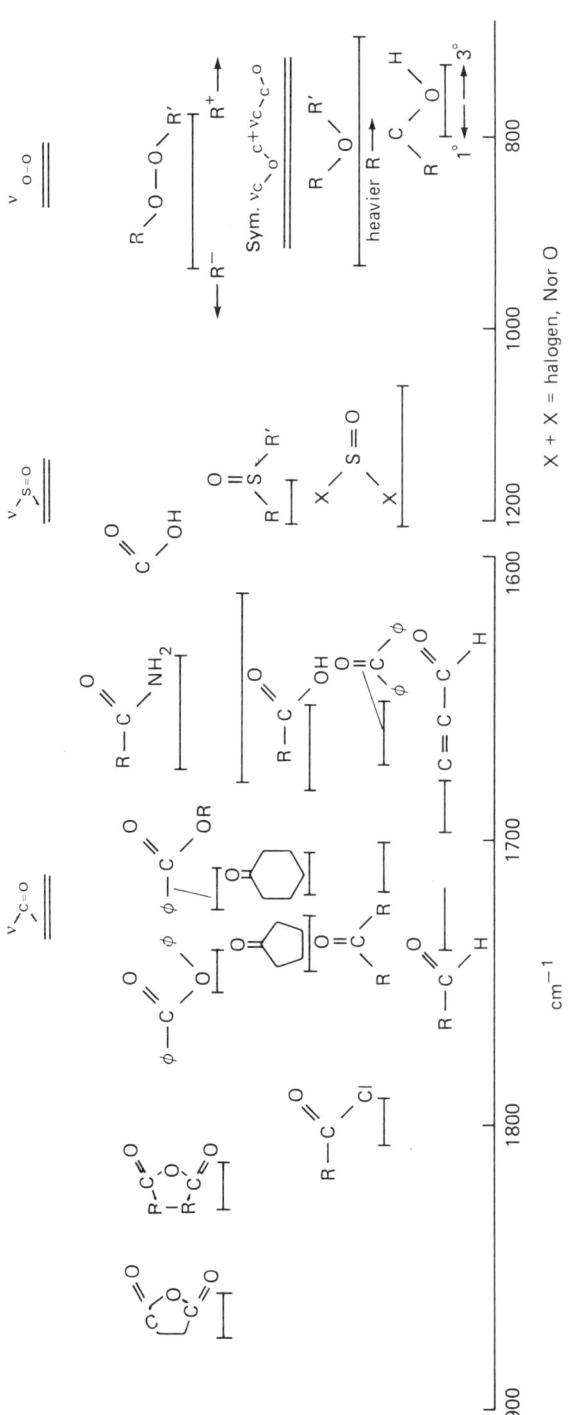

Fig. 7.5 — The frequencies of Raman bands due to oxygen-containing groups.

whereas the S–S group usually gives rise to a band near 500 cm^{-1}. In general, polysulphides show more intense ν_{S-S} vibrations than disulphides and do so at frequencies below 500 cm^{-1}. The frequencies of Raman bands from all of these groups are collected in Fig. 7.6.

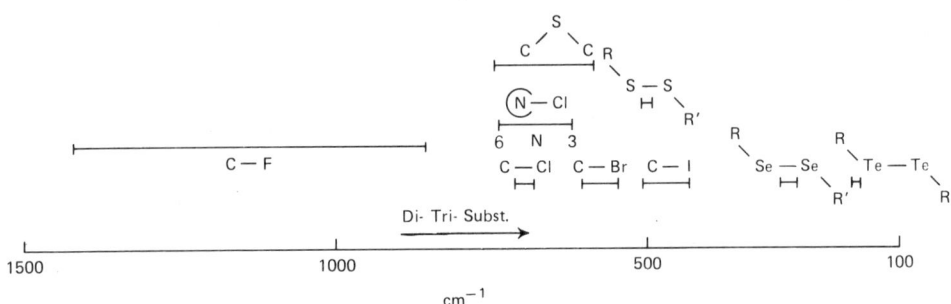

Fig. 7.6 — The frequencies of Raman bands due to groups containing sulphur, phosphorus, silicon and the halogens.

Clearly as diagnostic tools the assignments given in the previous pages are rather incomplete and a more comprehensive set of correlations would be very desirable. The considerable reduction in fluorescence obtained when using an Nd^{3+}:YAG laser as an excitation source will herald the way to recording spectra from long series of related compounds, and subsequent compilation of reliable group frequencies. There is clearly much work to be done. The rest of this chapter describes some preliminary applications developed at Southampton University and elsewhere where FT Raman has already proved its considerable potential in the analysis of organic chemicals.

7.2 SPECIFIC APPLICATIONS

7.2.1 Alkaloids

FT Raman methods have allowed us to produce fluorescence-free spectra from a series of addictive drugs such as heroin, codeine and morphine (see Fig. 7.7). It can be seen that the bands due to the phenyl group change dramatically in both relative intensity and frequency as the phenyl group in these complex materials is substituted. This is the exact inverse of the infrared spectra of these chemicals, which are very similar. Initial results on the morphine analogues led us to investigate a number of other addictive drugs in collaboration with the local police force. Even those 'street samples' which had been mixed with a range of highly fluorescent fillers often gave useful spectra [8]. Figure 7.8 shows the FT Raman spectrum of pure amphetamine sulphate ('speed') compared with a street sample which had been mixed with sorbitol. The spectrum of the sorbitol can be subtracted, leaving that due to the narcotics. Similar analyses can be carried out rapidly on other drug samples and this

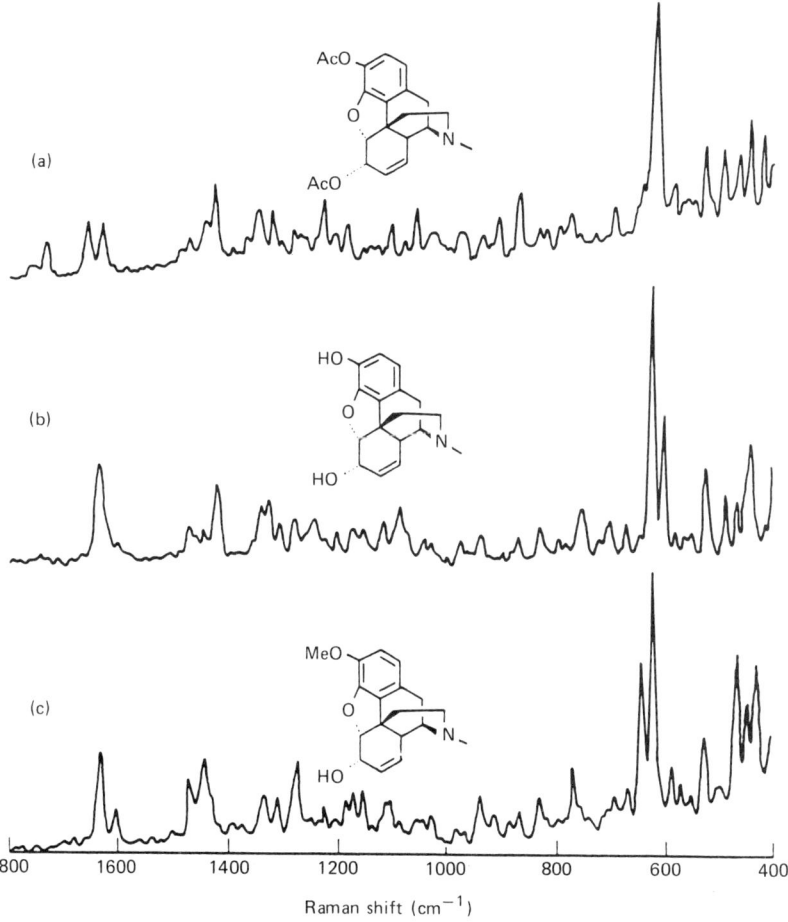

Fig. 7.7 — The FT Raman spectra of (a) heroin, (b) morphine and (c) codeine. Courtesy of Dr C. M. Hodges.

procedure may prove a very effective initial screening process in the processing of forensic samples. If the presence of an alkaloid was indicated by the FT Raman spectra, a full quantitative analysis could then be carried out using chromatographic methods.

7.2.2 Explosives

Raman spectroscopy is particularly suited to the analysis of explosives as nitro groups give rise to intense Raman bands, as mentioned in Section 7.1.5. However, many explosives are coloured aromatic materials which fluoresce under visible excitation. Initial studies on explosives at Southampton using FT Raman have proved very successful, with fluorescence proving to be far less troublesome than expected. Explosives are intense scatterers and give excellent quality spectra, as witnessed by

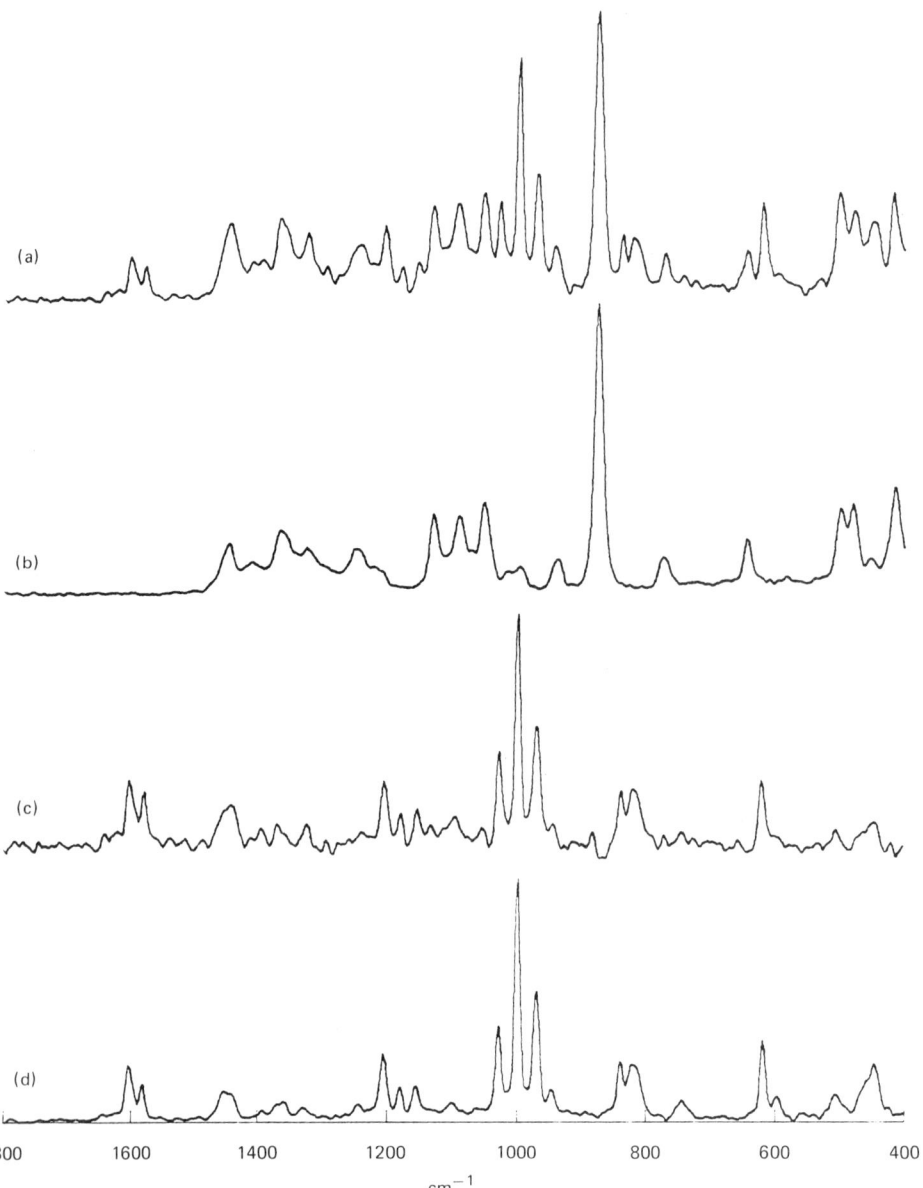

Fig. 7.8 — The FT Ramman spectrum of (a) 'street sample', 23% amphetamine sulphate in sorbitol and (b) pure sorbitol. A computer subtraction gives (c) the predicted spectrum of amphetamine sulphate, whilst (d) shows the actual spectrum recorded from a sample of pure amphetamine sulphate. Reproduced with permission from *Spectrochmicia Acta*, **46A**, C-M. Hodges and J. Akhavan, copyright 1990, Pergamon Press PLC.

Fig. 7.9. Absorption of the laser radiation can still be a problem where samples are deeply coloured, since it can cause burning and destruction of the sample. On one occasion this problem was encountered with a sample of explosive with the obvious

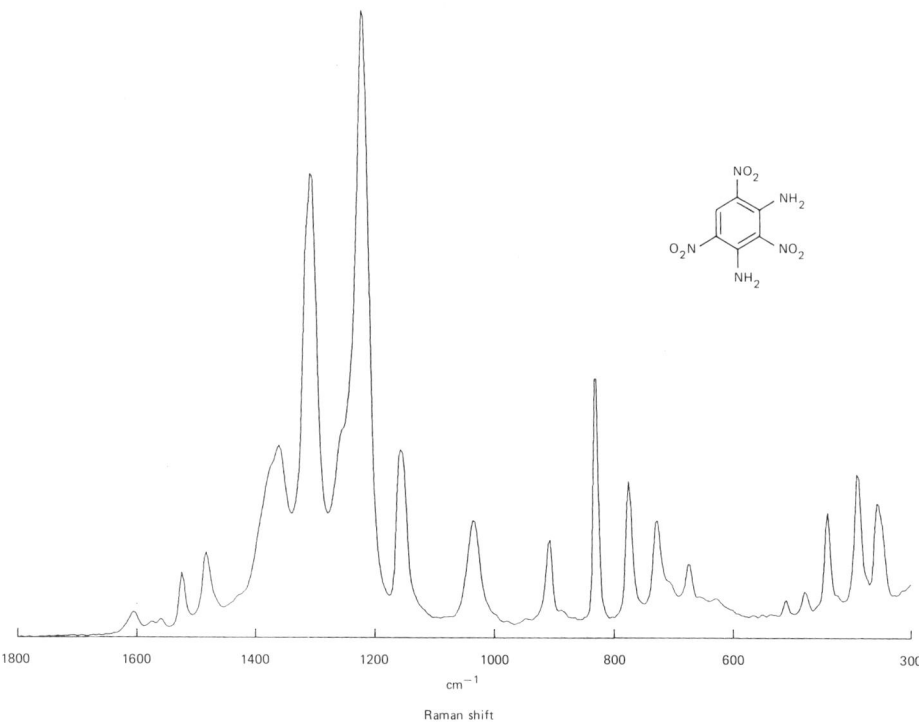

Fig. 7.9 — The FT Raman spectrum of the explosive diaminotrinitrobenzene (DATB).

consequence. However, because the sample size was small, about 20 mg, no damage was done to the equipment, to the infinite relief of the postgraduate researcher concerned.

Nitration reactions involved in the commercial production of explosives often involves the use of very strong nitric acid and/or nitric acid/N_2O_5 mixtures. These reactions are hard to monitor because of the hazards involved and the need to precisely control the reactant temperature. NMR is attractive, in principle, but failure of the sample tubes would have very expensive consequences. Infrared is not suitable because of the presence of water and the lack of adequate corrosion-resistant windows. FT Raman spectroscopy has, however, demonstrated its value and useful kinetic data have been obtained [9]. Heating by the laser seems to be quite manageable whilst fluorescence or colour from products is no problem. The highly intense Raman scatter from the nitrated species makes it possible to record kinetic points early in the reaction co-ordinate. To maintain temperature control a sample cell has been designed which is suspended in circulating thermostat fluid (see Fig. 5.20).

An unexpected observation from the study of explosives was that nitro compounds apparently give intense spectra regardless of the skeleton, i.e. not only the bands associated with the nitro groups but the whole spectrum is very strong.

Aliphatic and aromatic polynitro compounds are intense scatterers even though their non-nitrated substructures alone might be weak scatterers. The reproducible nature of FT Raman spectra has been demonstrated in recent work, covered in Chapter 6. All spectra of the polynitro compounds were recorded under similar instrumental conditions (sample form and laser power) and therefore their intensities are comparable. The origin of the spectral intensity has been investigated but there appears to be no systematic relationship between the number of nitro-containing species and the intensity of the corresponding Raman band. No correlation was observed in similar studies on substituted aromatics and polyhalo species. However, in some systems there does appear to be direct correlation between the band intensity of a particular group and the number of those groups in the molecule. For example, the phosphate band intensity relative to other bands in the spectrum increases linearly in adenosine mono-, di- and tri-phosphate. One possible explanation for the non-systematic intensities observed with nitro compounds is resonance enhancement, a non-linear process. Several tests exist for this phenomenon (e.g. the variations in overall intensity and variations between bands with excitation wavelength and correlation between intensity and the absorption spectrum) but no evidence for enhancement could be found in these cases.

The quality and intensity of the spectra obtained from explosives led us to believe that FT Raman might have forensic applications in this area. Frequently only minute samples of explosives are available, thus we have explored the minimum size of sample required to produce identifiable spectra. Using a solid sample holder of only 0.6 mm in diameter and a depth of sample of about 1 mm, the sample mass is restricted to less than 10 µg yet if carefully placed at the viewed point in the instrument excellent spectra can be recorded with ease. Forensic scientists are interested in the detection of explosives as films on the surfaces of post-explosion fragments (wire, cloth, paper, etc.) but because of the small mass of the sample within the scattering volume no useful spectra have been recorded. It may well be that methods can be developed to improve on this situation, e.g. waveguide spectroscopy, solution and use of surface-enhanced Raman spectroscopy (SERS), or for very tiny specimens, microscopy, but none could be regarded as routine.

However, when larger samples are available FT Raman is a potentially very useful forensic method for identifying the composition and hence possibly the source of impounded explosives. A case in point is the by now infamous plastic explosive Semtex, whose composition varies significantly from batch to batch. Recently Akhavan et al. [10] have shown that samples of Semtex contain varying amounts of the explosives RDX and PETN contained in a filler. Figure 7.10 shows the very dissimilar spectra from three different samples of Semtex. Conventional identification of explosives and propellants is normally carried out by solution techniques and followed by physical methods of analysis. The first step is normally a separation technique, the separated materials then being examined chromatographically and/or spectroscopically. FT Raman and infrared studies of the unseparated material could prove useful and would be relatively cheap to perform. Raman spectroscopy could provide information about the explosive, the binders are weak scatterers and make little contribution to the spectrum. Information about the binders, typically hydrogen-bonded polyols, would be available from the infrared.

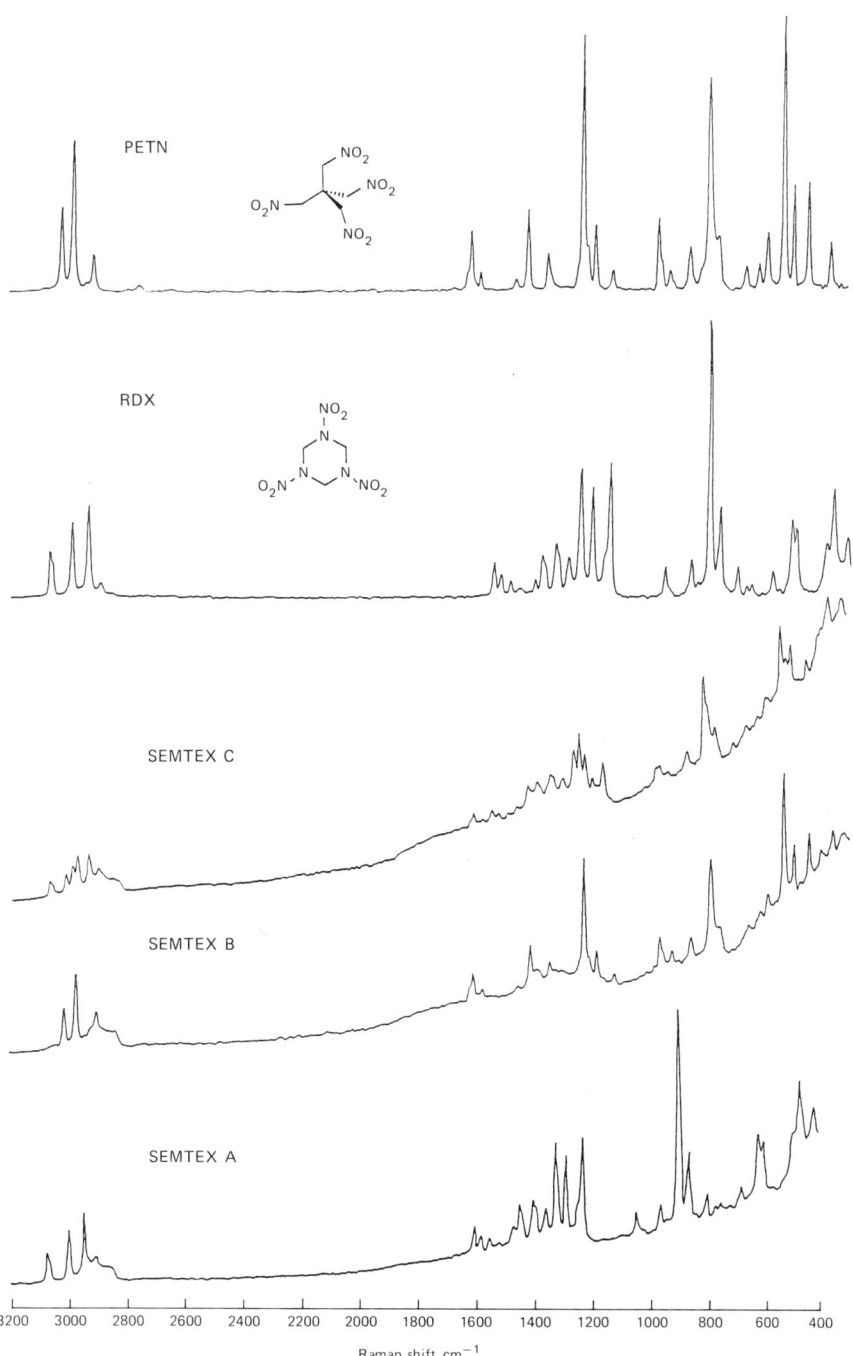

Fig. 7.10 — FT Raman spectra of different samples of Semtex and the explosives RDX and PETN. Courtesy of Dr J. Akhavan.

7.2.3 Dyestuffs

Another exploitation of the wide difference in Raman scattering intensity between compound types comes from a recent study of dyed fibres. Dyestuffs very frequently give intense Raman spectra. When dyed fibres are studied the fibre is often a weak scatterer and hence even a few per cent of a dyestuff is enough for a spectra to be readily recorded. The Raman spectrum of the dyed fibre is primarily the spectrum of the dye, whereas the infrared spectrum is predominantly that of the fibre. Fig. 7.11

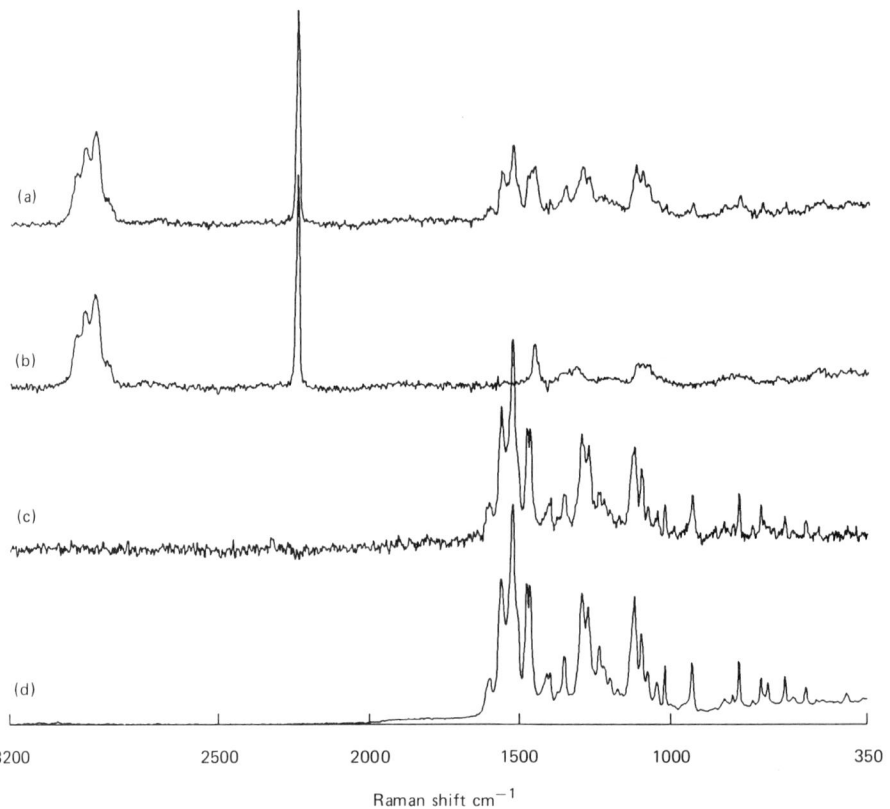

Fig. 7.11 — The FT Raman spectrum of (a) red Courtelle fibre, (b) undyed Courtelle fibre, (c) computer subtraction showing the spectrum of the dye adsorbed to the fibre and (d) the spectrum of the pure dye.

shows the spectrum of 'Courtelle', a commercial polyacrylonitrile polymer used in the textile industry, before and after dyeing. It can be seen that the subtracted spectrum of the dyestuff is identical to that of the solid dye, suggesting that the dye-fibre interaction in this case is weak with only physisorption occurring. Several authors have shown the spectrum of dyes such as Rhodamine 6G or sodium

Sec. 7.2] **Specific applications** 169

fluorescein to illustrate their control of fluorescence, but only recently have these spectra been analysed. As part of an SERS study, Majoube *et al.* [11] compared the spectra of solid Rhodamine 6G and DCM (4-dicyanomethylene 2-methylene 6-(*p*)dimethylaminostyrl-4*H*-pyran) with the materials absorbed to SERS active electrode and sol surfaces. Although the SERS spectra are readily recorded in the visible, the solids can only be successfully studied if near infrared sources are used. As mentioned in Chapter 5, it is not essential to use the 1.064μ source; Shimidzu has reported very nice spectra of Rhodamine 6G recorded with a Kr$^+$ 752.5-nm line [12].

7.2.4 Pharmaceutical materials
Pharmaceutical compounds give excellent FT Raman spectra, even when diluted with carriers. Spectra of indomethacin, an anti-inflammatory drug, and promethazine, an anti-histaminic, are shown in Figs. 7.12 and 7.13. Commercial drugs are

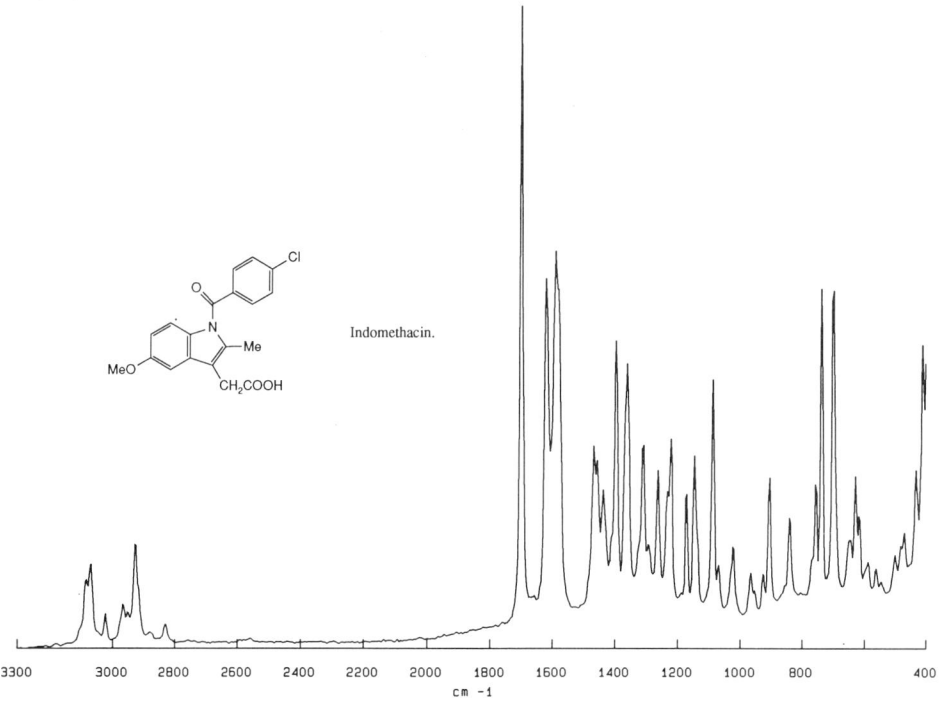

Fig. 7.12 — The FT Raman spectrum of indomethacin.

often used in very small doses and are compounded in an inert matrix to make them into tablet form, or to ensure that they are released slowly in the body. The spectrum of the 'filler' can be subtracted from that of the commercial drug to obtain a spectrum of the pure pharmaceutical. This is demonstrated in Fig. 7.14, where we show the

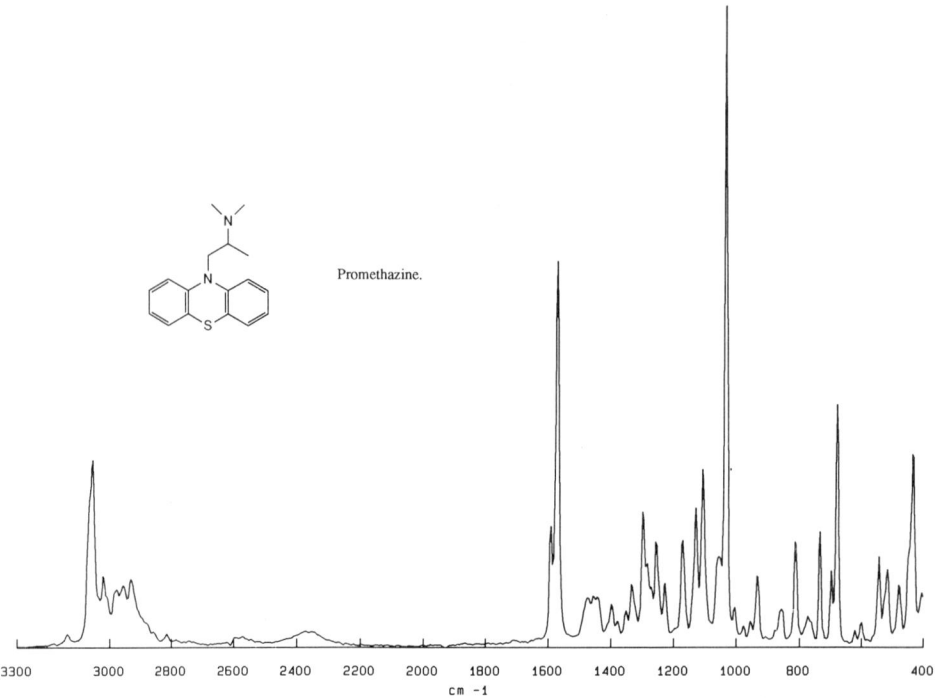

Fig. 7.13 — The FT Raman spectrum of promethazine.

spectrum of sodium diclofenac compounded with sodium alginate, a delivery system, and the results of the spectral subtraction. Useful spectra can still be recorded from compounded drugs even when the actual pharmaceutical is present in quite low concentrations. This stems from the fact that the pharmaceuticals are strong Raman scatterers, whereas the fillers produce only weak spectra. This is a peculiarity of Raman spectroscopy where relative intensities from different materials can vary by as much as 100:1. This variation is unknown in infrared spectroscopy, all compounds producing similar overall absorption. The literature already contains a number of reports of FT Raman spectra of pharmaceutical compounds including the antidepressant valium or diazepam and the related materials nordazepam, pinazepam and halazepam. Their structures are shown in Fig. 7.15. Neville and Shurvell give a detailed description of their infrared and Raman spectra, identifying all the bands [13]. More recently Nielson et al. have reported FT Raman spectra of ellipticines (anti-cancer agents) and partially analysed them [14].

When drugs are taken orally it is often desirable to control the rate at which they are ingested into the body. Tudor has shown that it is feasible to follow these processes using FT Raman and has studied drugs containing cimetidine as the active ingredient [15]. A pill was exposed to simulated body fluids and then sliced in half. FT Raman spectra were recorded across the diameter of the cut face and used to determine the concentration of active pharmaceutical as a function of position. The absorption of water which is inevitably involved makes similar experiments by

Sec. 7.2] Specific applications 171

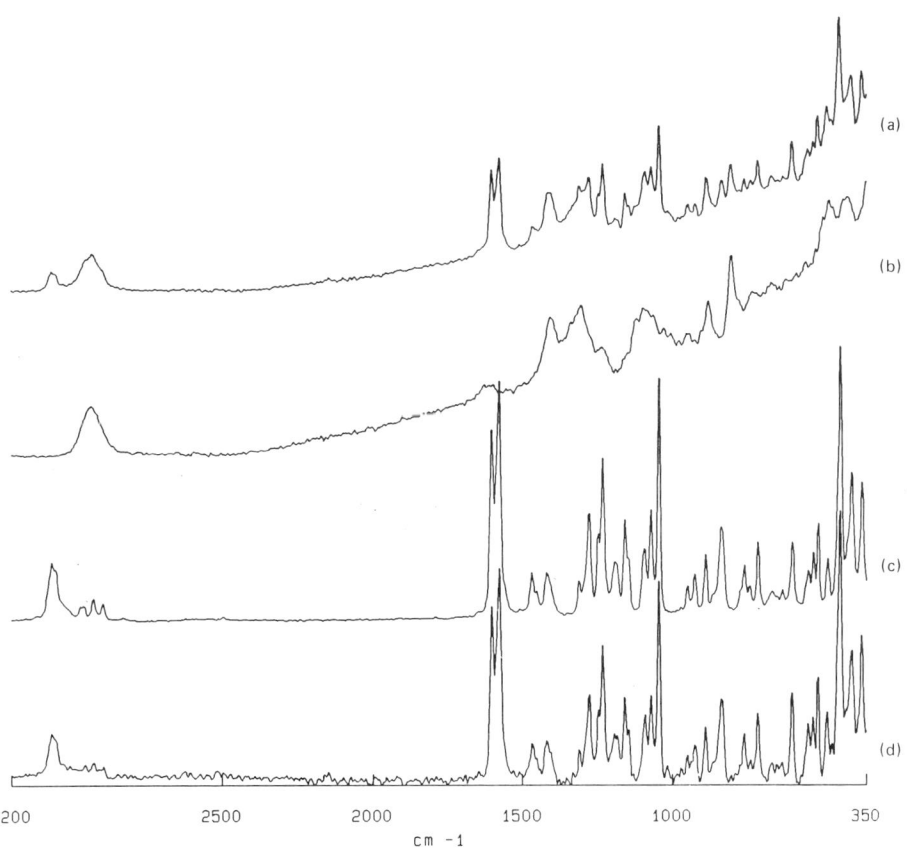

Fig. 7.14 — (a) The FT Raman spectrum of sodium diclofenac in the filler sodium alginate, (b) sodium alginate, (c) sodium diclofenac and (d) a computer subtraction showing the sodium diclofenac spectrum.

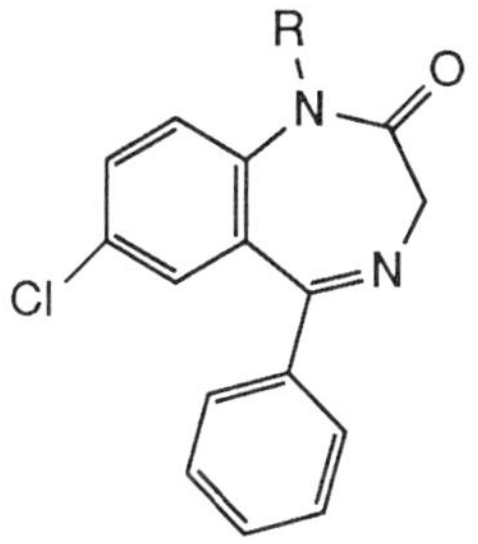

Halazepam   R = CH$_2$CH$_3$.
Diazepam   R = CH$_3$.
Nordazepam   R = H.
Pinazepam   R = CH$_2$CCH.

Fig. 7.15 — The structures of diazepam, nordazepam, pinazepam and halazepam.

infrared spectroscopy, whether by diffuse reflection or photoacoustic methods, unattractive, but Raman remains convenient and precise. The analysis is complicated by the fact that cimetidine exists in several different polymorphs and these produce slightly different Raman spectra [16]. This polymorphism is very important medicinally as the different crystal forms are absorbed at different rates into the body. FT Raman spectroscopy offers a way to analyse mixtures of cimetidine polymorphs. To achieve this, the most relevant approaches are deconvolution of the overlapping bands followed by direct correlation of the area under the bands with the proportion of the polymorphs present or analysis of the whole spectrum using 'correlation factor analysis. Several of these procedures have been used by infrared spectroscopists for some time, e.g. the Perkin Elmer CIRCOM routine, but their extension to Raman methods is only just beginning [17]. This extrapolation may not be trivial as Raman bands tend to be narrower than the infrared counterparts for which the routines were written. Further, Raman spectra frequently occur over a fluorescence background and of course do not obey the Beer–Lambert law. The assumptions made in writing the software in the first place will need careful appraisal, but the widespread application of chemometric techniques to Raman spectroscopy is certain.

7.2.5 Structural studies

Several structural analyses based on FT Raman methods have appeared recently in the literature [18,19] to join the enormous number of examples derived from conventional Raman spectra. In order to follow the polymerization of a high temperature thermosetting composite suitable for reinforcement with glass or Kevlar, Williams *et al.* performed a structural analysis of the precursors and polymerization intermediates using FT Raman and infrared methods [19]. Spectra were obtained of *N*-phenyl maleimide, its deuterated analogue and methylene dianiline bismaleimide. Depolarization ratios were measured for solutions of these compounds in methylene chloride and were used to assist band assignments. *N*-phenyl maleimide spectra are shown in Fig. 7.16.

7.2.6 Polymorphism

The polymorphic investigation of cimetidine discussed in Section 7.2.4 is not unique. Recently a Raman study of napthazarin has been reported [20]. This molecule has three possible static point symmetries — C_{2v}, C_{2h} and D_{2h}. The C_{2h} structure is thought not to exist owing to stability and energy considerations. Napthazarin also forms three polymorphs — A, B and C. By comparing FT Raman and infrared spectra the C_{2v} configuration was found to be the most likely at room temperature. In this paper, vibrations of the OH groups were considered in detail and were found to be very atypical. Normally v_{OH} is strongly infrared active and weak in the Raman spectrum, occurs near 3300 cm^{-1} and is broad. In both the infrared and Raman spectra of napthazarin no bands definitely due to v_{OH} could be discovered. This most unexpected result may be due to the extremely strong form of hydrogen bond present. Perhaps significantly, a similar situation prevails in spectra of nickel dimethylglyoximate where again the hydrogen bond is very strong.

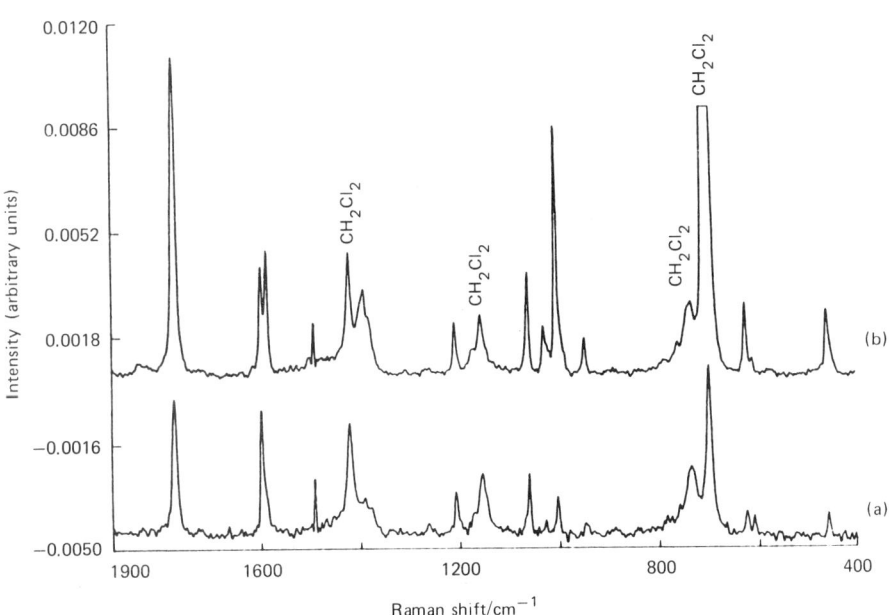

Fig. 7.16 — Polarized FT Raman spectra of a solution of N-phenylmaleimide in methylene chloride (a) perpendicular polarization and (b) parallel polarization. Uncorrected data. Reproduced with permission from *Spectrochmica Acta*, **46A**, J. F. Parker *et al.*, Copyright (1991) Pergamon Press PLC.

7.3 CONCLUSION

This perhaps is the place to consider some of the points made earlier in this chapter regarding Raman spectra in general. An infrared spectrum tends to be dominated by bands due to vibrations of certain highly polar groups, e.g. the amide bonds, $v_{C=O}$, v_{OH} or v_{CCl}. In these cases the features tend to broaden out, particularly where hydrogen bonding is possible, and obscure considerable ranges of the fingerprint region. In Raman spectroscopy this seems not to occur. Since the bands are almost invariably sharp, very little overlap occurs and the spectra recorded are far more detailed. When series of closely related molecules have been studied, infrared spectroscopists have been used to observing similar spectra with subtle differences indicating the variation along the series. This is normally apparent in the infrared because the vibrations of the structure common to the series almost invariably dominates the spectrum. In the Raman this may not always be the case. It is perfectly possible for the variants in structure to contribute strongly to the spectra and hence to generate very different results through a series. The morphine alkaloid spectra presented in Fig. 7.7, another example where very similar compounds give rise to very different spectra, can be found in Fig. 9.14, which shows the FT Raman spectra of α and β-D-Glucose. Clearly, intramolecular force fields and intermolecular packing affect the vibrational spectra as a whole (as would be expected), but the variations appear predominantly in the Raman rather than the infrared spectra. These types of differences could easily be exploited in future for routine analyses such as quality control monitoring or production control.

REFERENCES

[1] J. H. Hibben, *The Raman Effect and its Chemical Applications* Reinhold Publishing Corp, New York, 1939.
[2] N. B. Colthrup, L. H. Daley and S. E. Wiberly, *Introduction to Infrared and Raman Spectroscopy*, Academic Press, New York, 1975.
[3] H. Baranska, A. Labudzinska and J. Terpinski, *Laser Raman Spectrometry: Analytical Applications*, Ellis Horwood, Chichester, UK, 1987.
[4] F. R. Dollish, W. G. Fately and F. F. Bentley, *Characteristic Raman Frequencies of Organic Compounds*, Wiley, New York, 1974.
[5] Bernhard Schrader, *Raman/Infrared Atlas of Organic Compounds*, VCH Verlagsgesellschaft, Weinheim, Germany, 1989.
[6] D. J. Evans and D. B. Scully, *Spectrochim. Acta*, 1964, **20**, 891.
[7] N. Neto, M. Scrocco and S. Califano, *Spectrochim. Acta*, 1966, **22**, 1981.
[8] P. J. Hendra, C. M. Hodges and T. Farley, *J. Raman Spectrosc.*, 1989, **20**, 74.
[9] D. Viehoff, University of Southampton, unpublished work.
[10] J. Akhavan, *Spectrochim. Acta,* 1991, in press.
[11] M. Majoube and M. Henry, *Proceedings of the Twelfth International Conference of Raman Spectroscopy, Columbia, South Carolina*, Ed. J. R. Durig and J. F. Sullivan, Wiley, New York, 1990, p296.
[12] R. Shimidzu, *Proceedings of the Twelfth International Conference of Raman Spectroscopy, Columbia, South Carolina*, Ed. J. R. Durig and J. F. Sullivan, Wiley, New York, 1990, p892.
[13] G. A. Neville and H. F. Shurvell, *J. Raman Spectrosc.*, 1990, **21**, 9.
[14] J. M. Espinsosa, D. H. Christensen, G. O. Sorenson, O. Faurskov Nielsen, J. Aubard, M. Schwaller and G. Dodin, *Spectrochim Acta*, 1991 in press.
[15] A. M. Tudor, C. D. Melia, P. J. Hendra, D. Lee, R. Mitchell and M. C. Davies, *Spectrochim. Acta*, 1991, **47A**, in press.
[16] M. C. Davies, J. S. Binns, C. D. Melia and D. Bourgeois, *Spectrochim. Acta*, 1990, **46A**, 277.
[17] K. P. J. Williams, R. E. Aries, D. J. Cutler, and D. P. Lidiard, *Anal. Chem.*, 1990, **62**, 2553.
[18] C. D. Dyer, J. D. Kilburn, W. F. Maddams and P. A. Walker, *Spectrochim. Acta*, 1991, **47A**, in press.
[19] J. F. Parker, S. M. Mason and K. P. J. Williams, *Spectrochim. Acta*, 1990, **46A**, 315.
[20] S. O. Paul, C. J. H. Schutte and P. J. Hendra, *Spectrochim. Acta*, 1990, **46A**, 323.

8

The analysis of polymeric materials using FT Raman spectroscopy

8.1 INTRODUCTION

Since the first Raman spectrum of a polymer, polystyrene, was published by Signer and Weiler in 1932 [1], there have been several reviews which have summarized the application of classical Raman spectroscopy to the analysis of polymers [2,3,4].

Raman spectroscopy is particularly suited to the study of organic polymeric materials because of the sensitivity of the technique to the structure of molecules containing non-polar species such as C–C and C=C which commonly make up the backbone of synthetic polymers. This in theory allows subtle changes in conformation, crystallinity and stereoregularity of the backbone to be monitored. Similar studies on inorganic polymers, such as silicones, can also yield useful information.

It was not, however, until the advent of laser Raman spectroscopy [5,6] and the coincident improvements in instrumentation in the late 1960s that the analysis of polymers became a practical subject. The review by Koenig [2] in 1971 documents the Raman spectra of over 20 different polymers which illustrate the information that can be obtained on the structure as well as composition of the material. A later review in 1986 by Gerrard and Maddams records the expansion of the field and includes accounts of low frequency work and orientation in polymers [3].

Although the development of laser Raman spectroscopy promised to revolutionize the study of polymers, Raman has remained in a poor second place in the industrial laboratory to infrared spectroscopy, primarily because of the problems of sample fluorescence [7]. The problems caused by sample fluorescence have been discussed in Chapter 3. By their very nature, polymers are impure and in their applications they are usually compounded with anti-oxidants and other additives which tend to exacerbate the fluorescence problem. Techniques capable of reducing this fluorescence (extraction, recrystallization and burning out the fluorescence by exposure to the exciting beam) are time consuming and therefore tend to make the technique uneconomic. In addition, many of the additives and pigments used in polymer technology are deeply coloured and may absorb the visible laser excitation, causing sample heating which may well obliterate any Raman signal and even destroy the sample.

Infrared absorption spectroscopy has remained the preferred technique for polymer analysis, despite the frequent problems associated with sample preparation. Reflectance techniques require suitably shaped samples, whereas transmission methods demand a sample which is sufficiently transparent to allow the penetration of infrared radiation. Many polymers fail to meet these experimental criteria without careful preparation. The techniques employed to form a transmitting sample include casting thin films from solvent, compression moulding or the manufacture of KBr pellets. These preparations often alter the nature of the sample, including the very properties which one is attempting to monitor. Polymers are also frequently infusible and insoluble, further restricting sampling options, e.g. crosslinked systems cannot be dissolved without degradation. Raman, being a scattering technique, suffers from none of the sampling limitations.

Many polymers are filled with reinforcing agents, such as silica, carbonates and clays, when in applications. These materials tend to restrict study in the IR because of the broad intense absorptions due to polar groups (Si–O and C=O) which mask the weaker ones from the polymer. These species frequently give rise to very weak bands in Raman spectra, as illustrated in Fig. 8.1. This characteristic is very useful in the study of composites, and several applications are described later in this chapter.

The recent advent of near infrared excited FT Raman spectroscopy [8,9,10] has allowed the re-evaluation of the Raman technique for the routine analysis of polymeric systems. The lower photon energy used does not excite the same levels of fluorescence it does in visible Raman spectroscopy. This has allowed a massive expansion of the number of samples which are accessible. The availability of analytical-grade FT spectrometers also allows the rapid acquisition of data. Thus commercial polymers, copolymers and blends can be examined, allowing properties such as crystallinity or composition to be monitored at different stages in a manufacturing process, even when the polymer is contaminated with reinforcing agents or dyes [11]. Figure 8.2 shows the FT Raman spectrum of a yellow plastic duck. Figure 8.3 shows the sampling arrangement used. The spectrum of the duck indicates that it is composed of low-density polyethylene. FT Raman could also be used to study polymeric components 'in service'.

The following descriptions of some of the applications of FT Raman spectroscopy serve not to review the total field but to illustrate the usefulness of the technique to the analyst.

8.2 GENERAL ANALYSIS

FT Raman spectroscopy can provide both qualitative and quantitative information on the composition of polymers, copolymers and blends. The $\nu_{(C=C)}$ feature which is weak in infrared absorption is dominant in the Raman scattering. Its frequency is also sensitive to its chemical environment and this has, for example, permitted the

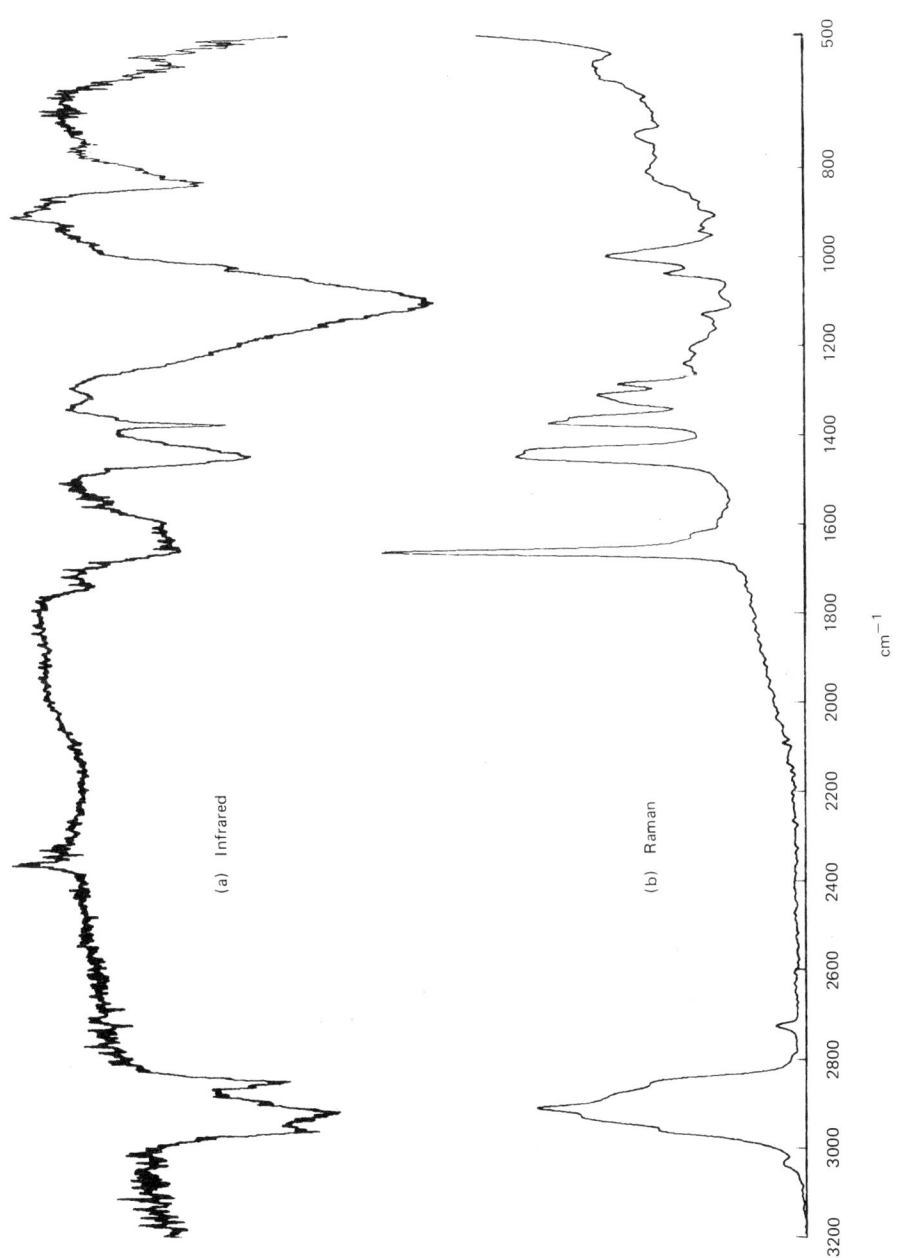

Fig. 8.1 — The ATR FTIR and FT Raman spectrum of vulcanized natural rubber filled with silica.

178   **The analysis of polymeric materials using FT Raman spectroscopy**   [Ch. 8

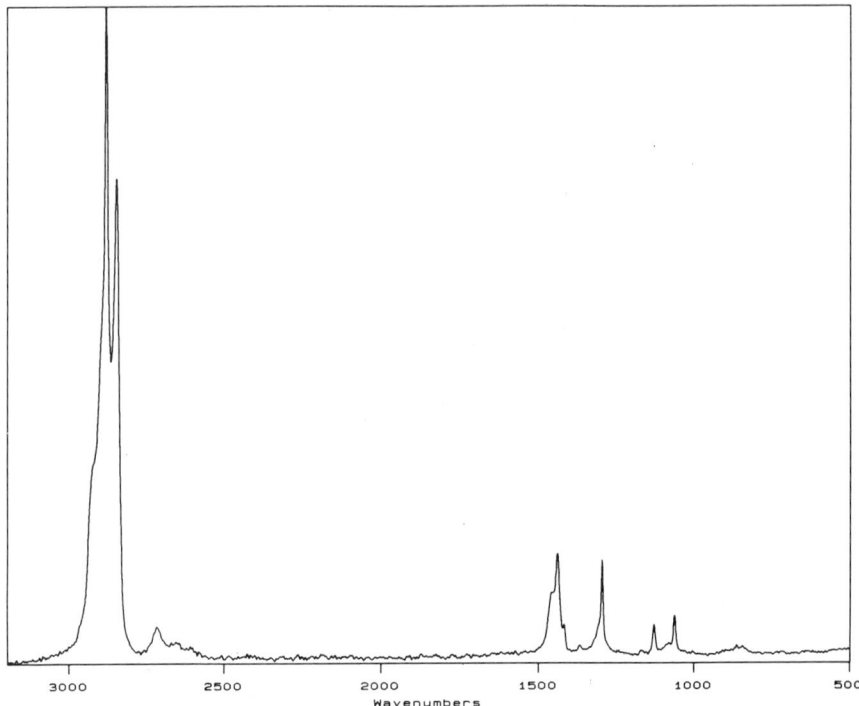

Fig. 8.2 — The FT Raman spectrum of a moulded plastic duck. Uncorrected data.

different isomers of polybutadiene to be identified and quantified [12]. Figure 8.4 shows the FT Raman spectra of several polybutadiene rubbers. The bands due to the *cis*, *trans* and pendant vinyl groups are readily identified. The *cis:trans:*vinyl ratio for the three rubbers, derived from the relative band areas, compare very well with those values determined using NMR. It must also be emphasized that this type of analysis can also be carried out on vulcanized systems. Copolymers of butadiene and styrene can be readily studied. The Raman method is particularly useful for the quantitative determination of the vinyl isomer in styrene/butadiene rubber because the standard technique, NMR, is unreliable in this case [12]. The analytical process can be extended to cyanide-containing species, which show a band near $2340\,cm^{-1}$ due to the $v_{(C\equiv N)}$ vibration, such as nitrile or acrylobutadienestyrene (ABS) rubber. Analyses of poly(aromatic sulphone) copolymers and poly(aromatic ketone) composites have been described [13] (see Section 8.6). Preliminary work by Williams and Mason on the quantitative analysis of commercial polyurea/polyurethane foams has also been reported [11]. Agbenyega *et al.* [13] also describe work on several copolymers, including ethylene oxide/vinyl chloride and tetrafluoroethylene/hexafluoropropylene systems. The latter case is of interest, since the spectra of copolymers containing up to 10% hexafluoropropylene do not show any significant differences from that of the homopolymer. However, because spectra of excellent signal-to-noise ratio had been recorded, difference spectroscopy was practical and

Fig. 8.3 — A photograph demonstrating the ease of sampling, showing the position of the duck in the sample compartment before the spectrum shown in Fig. 8.2 was recorded.

was used to analyse the copolymers. The spectrum of pure poly(tetrafluoroethylene) was subtracted from those of the copolymers. Care must always be taken in this type of analysis because very often properties of the copolymer, such as its crystallinity, change as a function of composition and new spectral bands may appear on this account.

An FT Raman study on the identification of single-number nylons has been published [14]. Fluorescence-free spectra were obtained from nylon 3 to nylon 12 inclusive, a polymeric homologous series varying only in the length of the CH_2 chain adjacent to the secondary amine group. The strong bands in the Raman spectra of nylons arise mainly from the backbone methylene sequences, with the main differences between the Raman spectra arising in the region between 1000 and 1500 cm^{-1}. This region of the spectrum contains bands characteristic of CH_2 chain twisting and wagging motions and C–C skeletal modes. In addition, it includes very weak bands called 'band progressions', which are also observed in short-chain paraffins and can be used to identify the different nylons. In contrast the infrared spectra of the various nylons tend to be very similar since they are dominated by broad features due to the hydrogen-bonded amide groups. The FT infrared absorption and Raman spectra of nylon 6 are shown in Fig. 8.5. They serve to illustrate the complementary nature of the two techniques. Thus, wherever possible, analysis should be performed involving both spectroscopies. FT Raman results were also obtained from glass-reinforced nylon samples, but composite materials are discussed in detail later in this chapter.

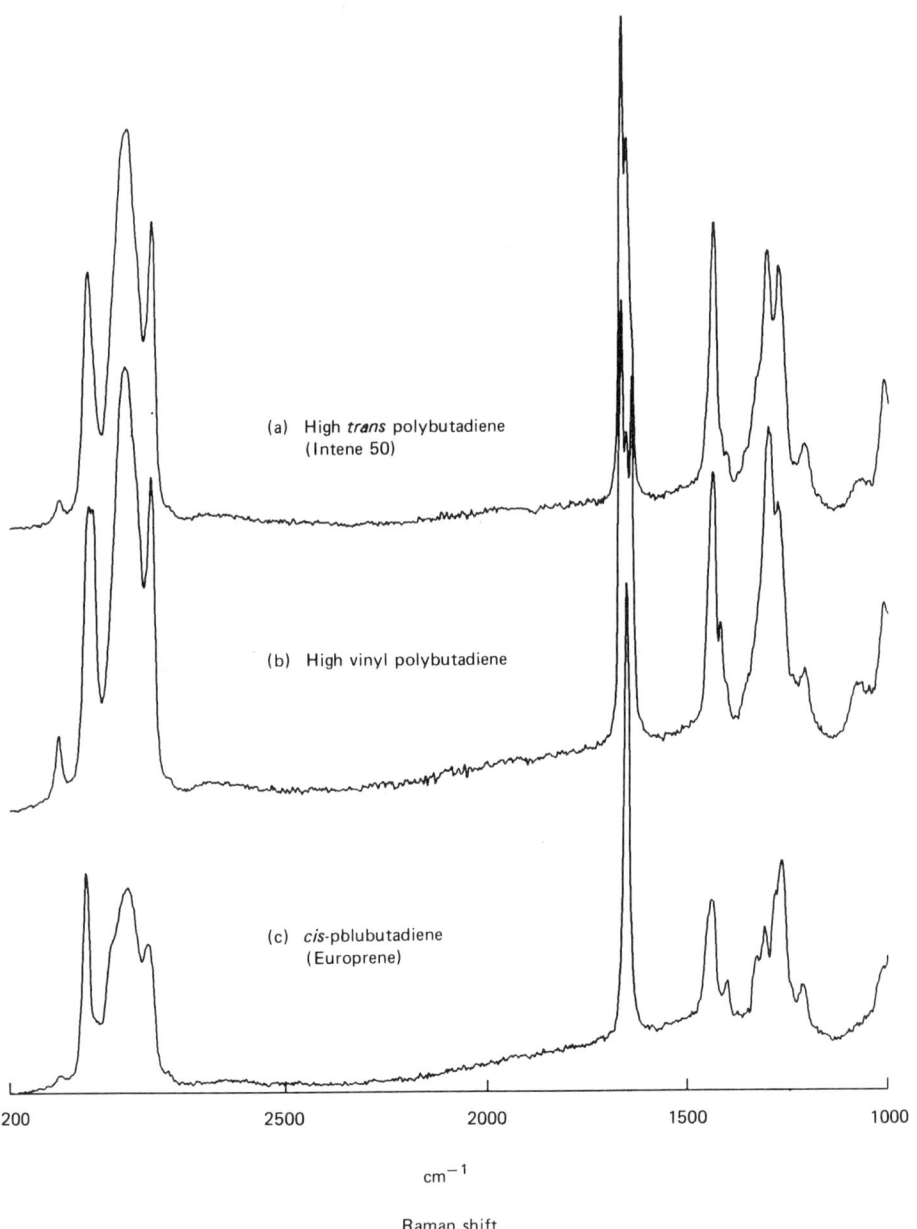

Fig. 8.4 — The FT Raman spectra of several different polybutadiene rubbers.

Another area in which FT Raman can play a useful role is in the identification of contaminants in a polymer matrix [11]. Figure 8.6 shows the FT Raman spectrum of an 80 μg inclusion taken from a polymer film. This demonstrates the tiny amount of

Sec. 8.3] **Crystallinity and tacticity studies** 181

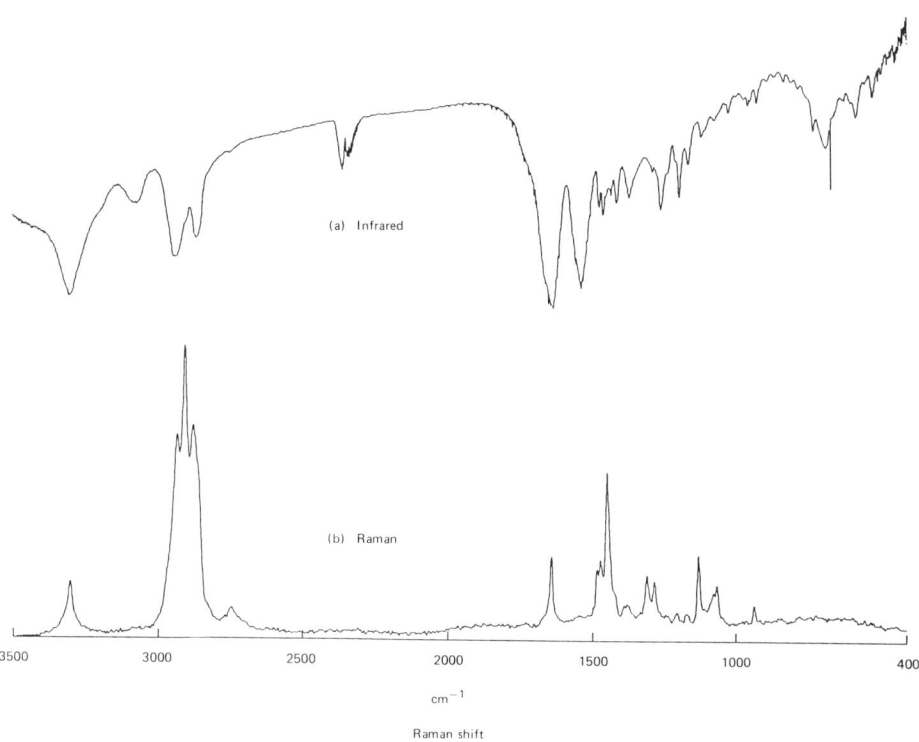

Fig. 8.5 — The FT Raman and FTIR spectra of Nylon 6.

sample required for such analyses. Assignment of the FT Raman spectrum with reference to library data identifies the inclusion as poly(ethylene vinylacetate) foam. This analysis would not have been possible using conventional Raman methods because of overwhelming sample fluorescence, and infrared microscopy would require extensive sample preparation.

8.3 CRYSTALLINITY AND TACTICITY STUDIES

Both Raman and its complement infrared absorption spectroscopy can give useful information on the tacticity, crystallinity and orientation of a polymer. Different tactic forms of a polymer frequently give rise to easily distinguishable Raman spectra. The spectroscopic differences between syndiotactic and isotactic polypropylene have been documented [15]. The standard method of determining tacticity is ^{13}C NMR; however, the advantage of the Raman method is that it can be applied to specimens without sampling. Broad Raman bands of low peak intensity are characteristic of materials which have a low stereochemical purity, such as atactic samples or those with mixed head-to-head and head-to-tail fragments. The wide range of

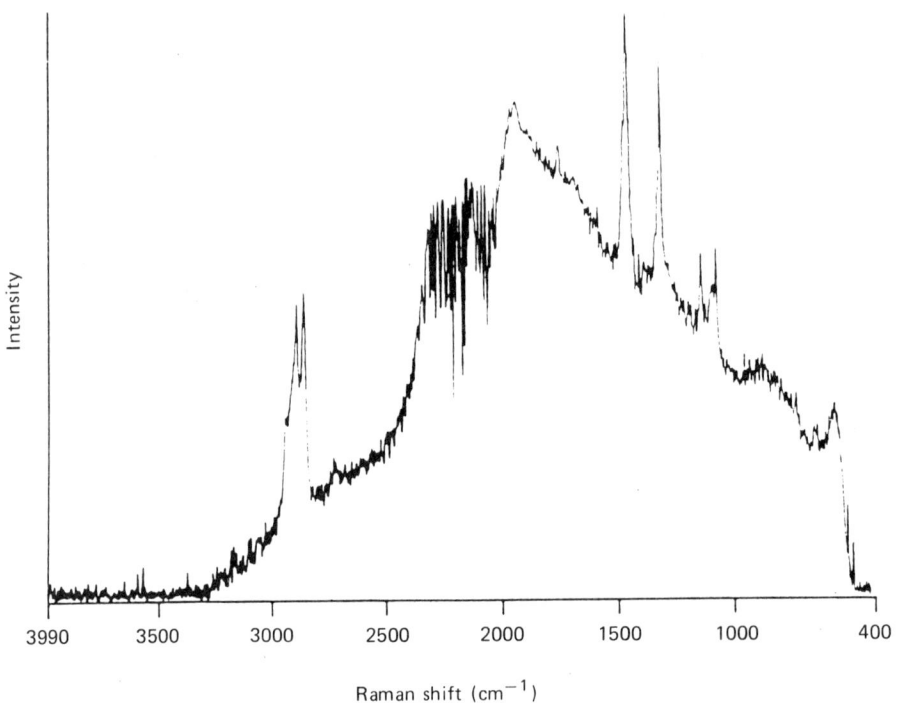

Fig. 8.6 — The FT Raman spectrum of an 80-μg poly(ethylene vinylacetate) impurity contained in a polymer film. Uncorrected data. Reproduced with permission from *Spectrochimica Acta*, **46A**, K. P. J. Williams and S. M. Mason, copyright 1990 Pergamon Press PLC.

significantly different chemical species present all have slightly different vibrational frequencies. The FT Raman spectrum of isotactic polystyrene is distinguishable from that of the atactic isomer (see Fig. 8.7) but the differences are mainly due to the crystallinity of the isotactic isomer [13]. Highly ordered species give rise to characteristically sharp and hence intense Raman spectral bands compared with those of disordered material. The tendency of amorphous materials to be poor Raman scatterers and hence to produce weak spectra further emphasizes the difference between the two spectra.

Hallmark *et al.* [16] studied the change in the FT Raman spectrum of poly (di-*n*-hexylgermane) as a function of temperature. The spectrum changes dramatically as the thermochromic transition temperature is traversed. Splittings in bands due to CH_2 bending motions were correlated with the packing of the hexyl side-chains. Above the thermochromic transition temperature, 12°C, the band profile is considered with an ordered *trans* planar zigzag conformation. Below 12°C, however, the packing of the side-chains becomes disordered, the spectral bands broaden and new bands due to gauche linkages are observed. Hallmark *et al.* concluded that the interaction between the side-groups and the backbone is then altered, introducing disorder into the germanium backbone structure. This disrupts the electronic overlap and causes a shift in the maximum of the UV/visible absorption.

Sec. 8.3] Crystallinity and tacticity studies

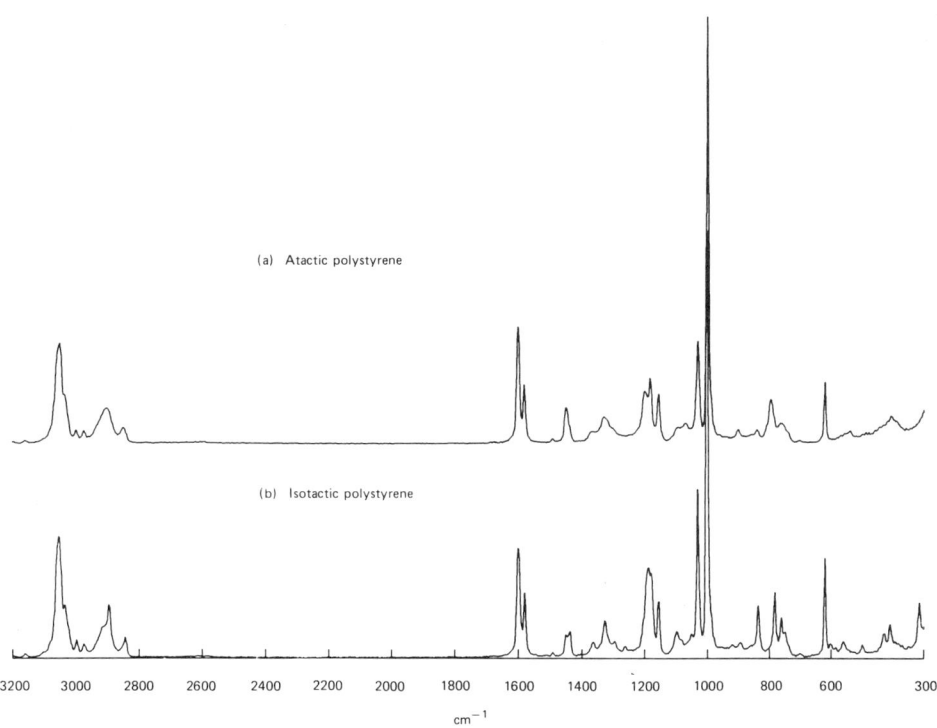

Fig. 8.7 — The FT Raman spectra of atactic and isotactic polystyrene.

Polymer molecules are also capable of internal rotations, and the consequent gyrations they perform produce very complex vibrational characteristics and hence broad Raman and infrared bands. On crystallization, the polymer chains become clamped into rigid and highly regular repeat units within a crystalline matrix. As a result sharp infrared and Raman bands are produced. These crystalline bands can be used to investigate the crystal habit, the number of chains and their symmetry within the unit cell.

Methods of obtaining the degree of crystallinity in a sample are well established [3]. A detailed description of a Raman method has been made by Strobl and Hagedorn [17]. The principle is the same as the methods used in Wide Angle X-ray Diffration (WAXR), NMR and IR. A spectrum/picture is recorded and then resolved into the contributions from the ordered and disordered phases. The mass fractions in the different orders can than be determined from the weights of the respective components either directly or after calibration. Strobl and Hagedorn use a three-phase model for partially crystalline polyethylene consisting of an orthorhombic crystalline phase, a meltlike amorphous phase and a disordered, anisotropic phase (the interfacial material). The method may, therefore, be used to estimate the proportion of interfacial material in a sample, as well as its crystallinity. The band intensity of pure crystalline material is estimated from chain-extended material, and that of the pure amorphous material from the melt. The integrated intensity of band at 1416 cm^{-1} (the A$_g$ component of the CH$_2$ bending vibration) serves as a measure of the amount of orthorhombic crystalline material present in the sample. Bands at 1303 and 1080 cm^{-1} are used to estimate the amorphous content. This method has been successfully applied to some other polymers [3]. FT Raman spectroscopy has recently been applied to determining the crystalline content of poly(aryl ether ether ketone) [18], following initial work by Louden using conventional Raman methods [19]. The FT Raman spectrum of nylon 6 has also been shown to change with crystalline content [14] and this may form the basis of a method of measuring its crystallinity.

Conventional Raman spectroscopy can be used to determine the lamellar thickness of polymer crystals, whether they be independent as in solution-crystallized materials or stacked as in the melt-crystallized case. This method relies on recording the frequencies of the longitudinal acoustic modes (LAMs). The frequency of these modes is inversely proportional to lamellar thickness and has been extensively reviewed [3,20]. However, the most intense mode, LA Mode 1, lies in the frequency range $\Delta v = 5–30$ cm^{-1}, making it impossible to record using current FT Raman instruments and filter technology.

Changes in crystalline habit as a function of temperature or pressure can be followed using Raman spectroscopy. A simple example is polyethylene, which contains two chains per unit cell. Each mode of each chain can occur in-phase or out-of-phase with its partner, giving rise to two bands in the Raman spectrum. This is known as correlation splitting [21], and can be monitored as a function of temperature and pressure. Jin Wang *et al.* [22] used conventional methods to follow changes in the Raman spectrum and hence crystallinity of a synthetic polyisoprene as a function of pressure. FT Raman techniques have also been used to study the development of cold crystallization in natural rubber [23]. Figure 8.8 shows the FT Raman spectra of natural rubber as a function of cold soaking time. A similar study of the crystallization process in chloroprene rubber has also been made [24].

Polytetrafluoroethylene undergoes a phase change at 19°C due to a change in its helical pitch from 13/6 to 15/7 [25]. This phase change has been studied using FT Raman spectroscopy [13]. In principle, different liquid crystal polymer phases may also be studied, but in practice this has to date been limited because of persistent sample fluorescence. Polyphosphazines have been studied successfully [26]. These materials exhibit thermotropic behaviour which is highly dependent on their thermal history and the nature and packing of their side-chains around the flexible P–N backbone. Figure 8.9 shows the structure of poly(bis-4-methoxy-phenoxy phosphazine) (PBMOPP) which has been studied in the crystalline and mesophase. A difference spectrum between these two states was used to identify band shifts in the aromatic ring breathing mode and the aromatic C–H stretch on crystallization. These shifts reflect the reduction in the mobility of the aromatic ring side-chains in the crystalline material. In addition, differences in the relative band intensities of several modes were observed. The C=C stretching modes increase in intensity on crystallization, which is consistent with the view that the arrangement of the substituent groups determines the nature of the material.

Ellis *et al.* [27] have also identified changes in the FT Raman spectrum of the thermotropic, main-chain liquid crystalline polymer, poly(heptamethylene

Sec. 8.3] Crystallinity and tacticity studies

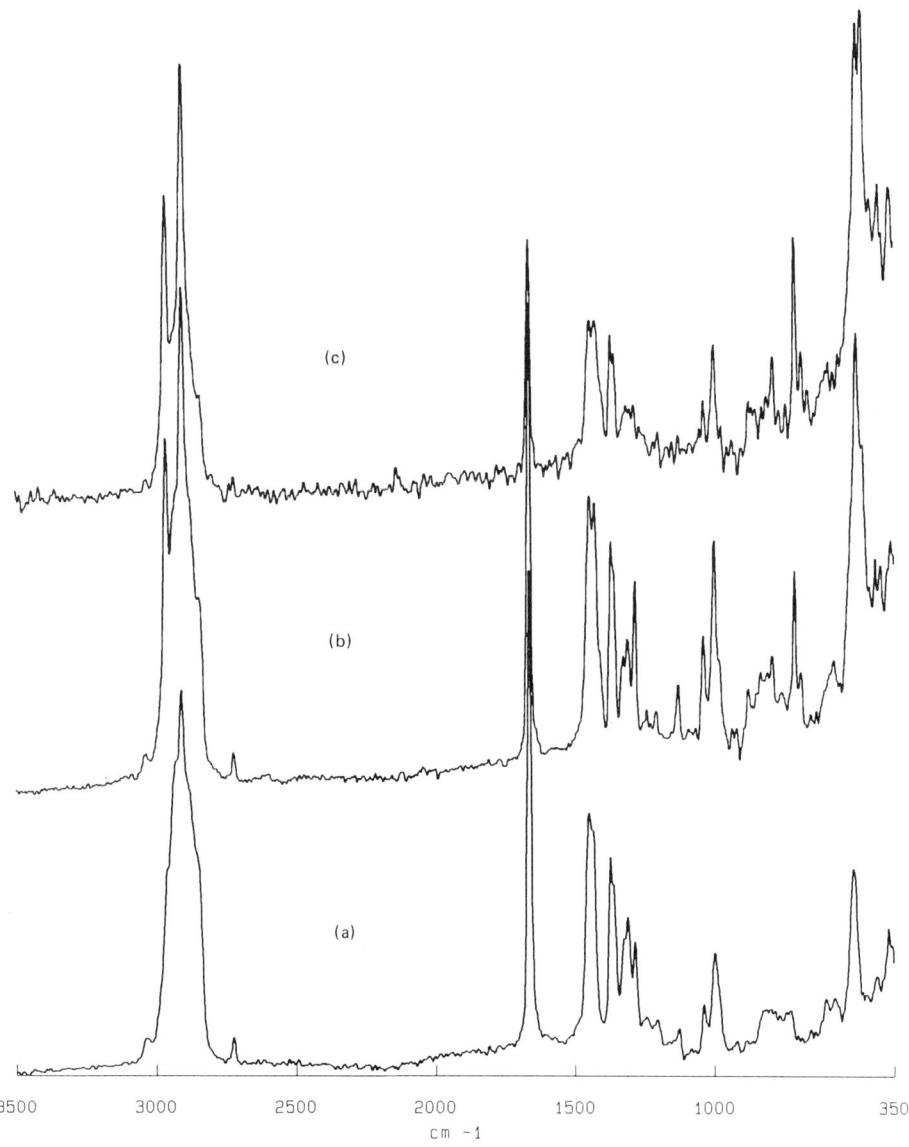

Fig. 8.8 — The FT Raman spectrum of natural rubber as a function of cold soaking time: (a) control, (b) 3 weeks' and (c) 3 months' cold soaking.

terephthaloyl-bis-4-oxybenzoate) (PHMTOB) as a function of temperature and thermal history. Main chain liquid crystalline polymers consist of rigid mesogen units linked by flexible spacers. The structure of PHMTOB is shown in Fig. 8.10(a). When heated above its crystalline melting point, long-range order persists and within distinct temperature regions 'liquid crystalline' phases are said to exist. PHMTOB exhibits a layered 'smectic' type phase which is shown in Fig. 8.10(b). Spectral

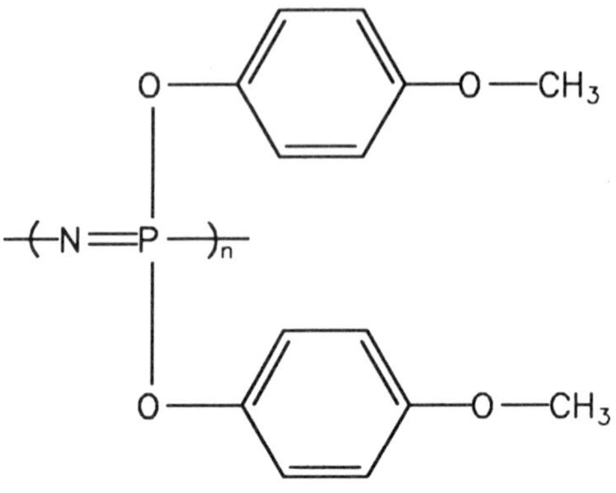

Fig. 8.9 — The structure of poly(bis 4-methoxyphenoxyphosphazine).

changes have been attributed to the differences in the intermolecular interactions or conformation of the polymer chains between the crystalline, liquid crystalline and amorphous phases.

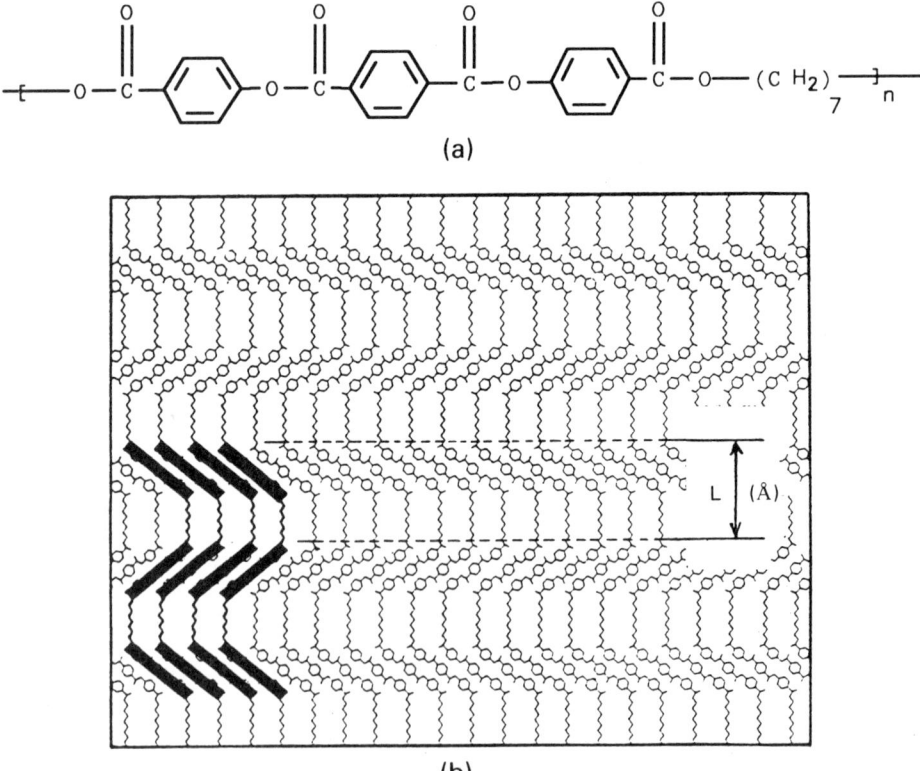

Fig. 8.10 — (a) The structure of poly(heptamethylene terephthaloyl-bis-4-oxybenzoate) and (b) the liquid crystalline, smectic phase (courtesy of G. Ellis).

FT Raman spectra can be recorded *in situ* at temperatures up to 200°C using the hot cell in Fig. 5.22. However, above 150°C a heating background is observed in the spectrum which increases with increasing temperature until it obliterates the Raman signal. Thus, only phase changes which occur below about 180°C are easily accessible.

Preliminary FT Raman results on poly(vinyl chloride) gels have also been reported [13]. Investigations to date into the structure of these gels and the nature of the crosslink or junction points have concentrated on the solids formed by evaporating the solvent from the gel phase. This is, of course, somewhat illogical. Hence Jawhari [28] used FT Raman to investigate 'wet' PVC gels at a series of temperatures both above and below the gel–solution transition temperature looking for changes in the spectra as the gel melts.

8.4 ELASTOMERS

The analysis of elastomers, and natural rubber in particular, using FT Raman spectroscopy serves to illustrate many of the advantages of the technique [29]. The potential of Raman for the analysis of 'real' rubber systems has long been appreciated [30]. However, conventional studies have been severely hampered by sample fluorescence. The study of polyisoprene and many synthetic rubbers had been strictly limited to highly purified samples [31,32]. The literature spectra are of relatively poor quality and have only been obtained after solvent extractions and techniques such as 'burning out' the fluorescence by exposing the sample to the laser excitation source for periods of up to 12 hours.

Near infrared FT Raman methods have allowed the study of unmodified samples often taken directly from the bale. Figure 8.11 shows the FT Raman spectra of natural rubber and several commercial synthetic elastomers. Natural rubber contains many contaminants (such as proteins, beetles and bark) and is usually compounded with fillers, anti-degradents and oils in its various applications. Figure 8.12 shows the FT Raman spectra of 'best' and 'worst' quality natural rubber. In Fig. 8.13 the effect of compounding the rubber with a synthetic oil is demonstrated. The only additive to date to prevent the acquisition of useful FT Raman spectra is carbon black. The laser radiation is absorbed by the sample and massive sample heating then results. This problem can be reduced by spinning the sample, but to date no useful results have been obtained, presumably because the Raman scatter is massively attenuated by the carbon.

The reduction in sample fluorescence achieved by exciting in the near infrared has also allowed the examination of crosslinked materials, both sulphur-vulcanized and free-radical-cured [12]. Vulcanization reactions are poorly understood and FT Raman spectroscopy should provide a useful insight into the mechanism of crosslinking. Figure 8.14 shows the FT Raman spectra of (a) raw, (b) compounded and (c) conventionally vulcanized natural rubber. The strong bands due to elemental sulphur are obvious in spectrum (B). No strong S–S band is obvious in the vulcanized material, which is highly anomalous as this group should give rise to an intense Raman band. New bands do, however, appear in the 1550–1700 cm^{-1} region of the

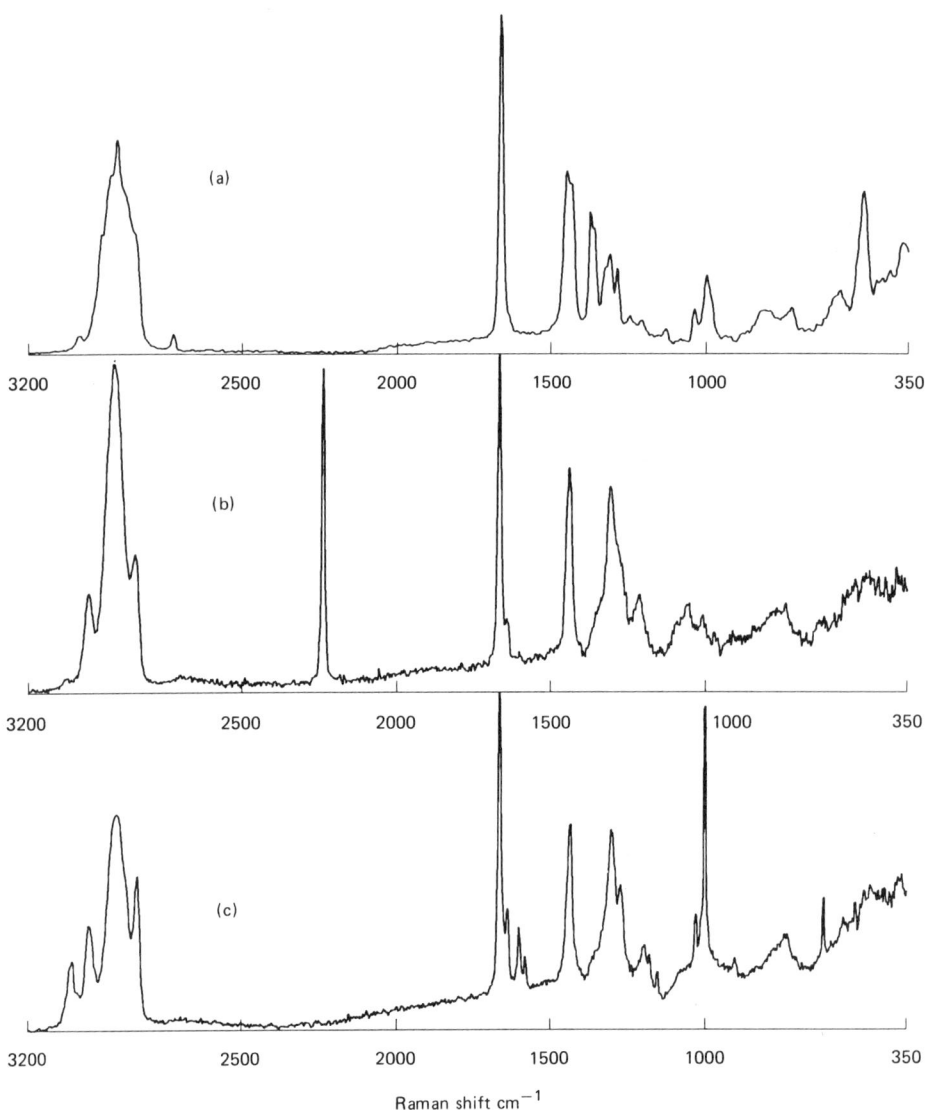

Fig. 8.11 — The FT Raman spectra of (A) natural rubber, (B) nitrile rubber and (C) styrene/butadiene rubber.

vulcanizate. It has been suggested that these are due to diene and triene species formed as a result of main chain modifications [31]. It should be noted that an unusually high percentage of sulphur (10%) was used in this mixture to emphasize the differences between the spectra.

Miller et al. [33] have used FT Raman spectroscopy to study the physical properties of reaction-injection-moulded polyurethanes such as the ones commonly

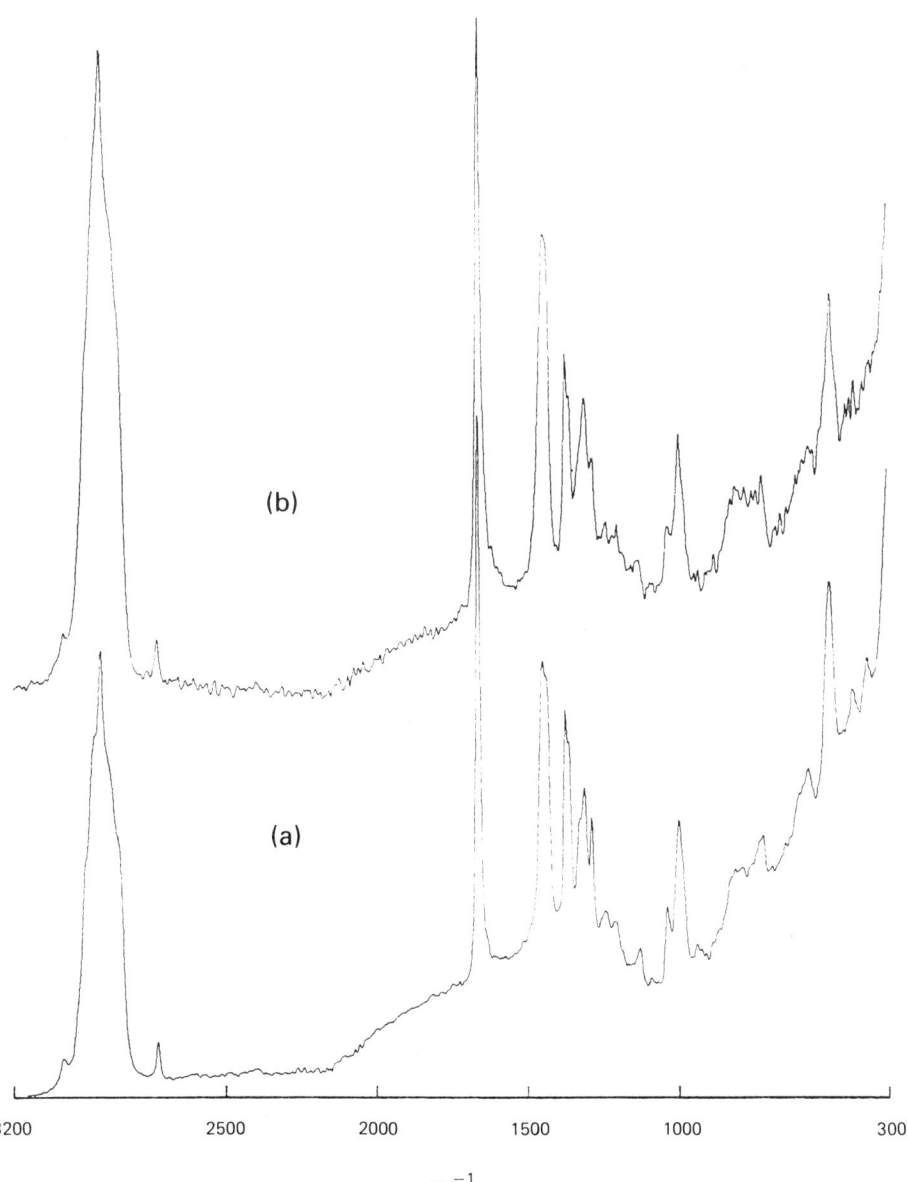

Fig. 8.12 — The FT Raman spectra of (a) 'best' quality natural rubber, grade SMRL, and (b) 'worst' quality natural rubber, grade SKIM, both courtsey of the Malaysian Rubber Producers' Research Association.

used in the automotive industry. A series of samples was acquired at various distances into the moulding with reference to the injection point. The flexive modulus of these samples was then monitored and found to vary, presumably

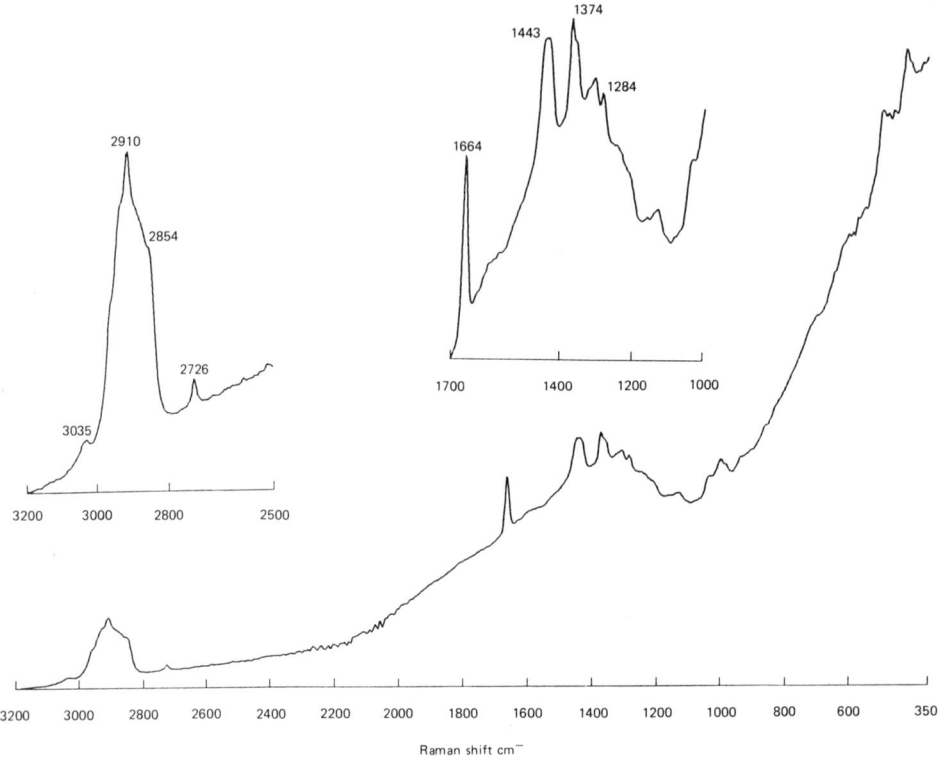

Fig. 8.13 — The FT Raman spectrum of natural rubber mixed with a synthetic oil.

because the extent of the polymerization process and/or the stoichiometry are injection-time-dependent. The samples were then studied by near infrared diffuse reflection and FT Raman methods. The information derived enabled the authors to comment on the chemistry and also the possibility of phase separation in the polymer. The latter is characterized by hydrogen bonding between segments of the polymer chain. Bands shifts will, therefore, occur in the FT Raman spectrum in those bands associated with the amide and carbonyl groups. Miller *et al.* concluded by suggesting that FT Raman might well be used routinely for rapid, non-destructive quality evaluation of polyurethane mouldings.

8.5 PAINTS

FTIR is well established as a useful technique for the study of paints but there are obvious areas where the results obtained are not satisfactory. Infrared studies of paint films tend to be restricted to reflection techniques because of strong absorption and scattering by pigments. The results of many reflection experiments have in turn proved difficult to interpret whilst quantitative measurements have been imprecise

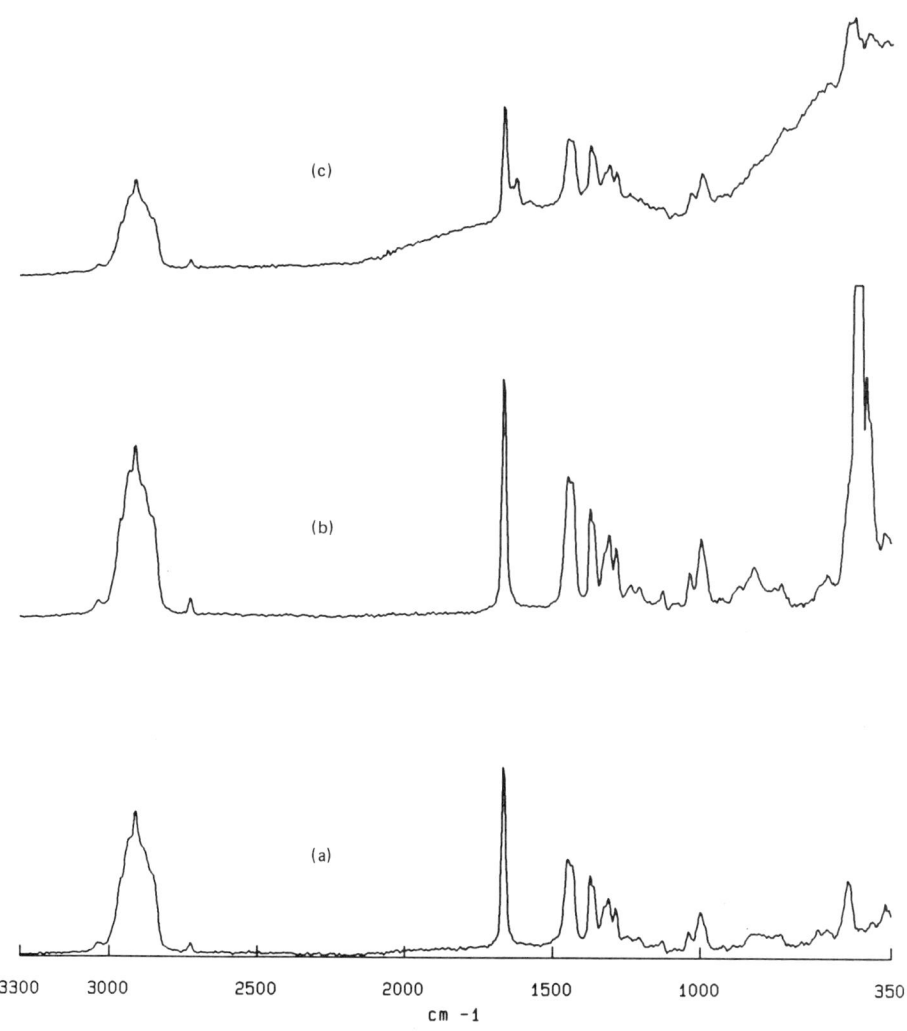

Fig. 8.14 — The FT Raman spectrum of (a) raw, (b) compounded and (c) vulcanized natural rubber.

without the use of complex modelling routines. In addition, many coating materials are water-based and their spectra therefore suffer from strong absorptions in the infrared, which prevent useful vibrational data from being acquired. NMR is traditionally used to examine the structure and reactions of coating materials; however, difficulties are encountered because the sample form becomes problematic.

FT Raman spectroscopy has been used to study several paint systems including the polymerization of water-borne emulsions which form a polymer latex and the curing processes and weathering reactions that occur in alkyd resins [34]. Because of chemical and environmental concerns, the polymer latex is becoming an increasingly important paint system. They can be formed by emulsion polymerization which allows the particle size and morphology to be controlled. The mechanism of these reactions is not well understood and studying them using infrared spectroscopy has always been hampered by the presence of water. Attempts at ATR IR studies were also misleading because the latex tended to deposit on the ATR crystal. FT Raman is a non-contact technique and results suggest that very dilute suspensions, down to well below 1% w/w solid can be studied in favourable cases. Small-path-length cells are used to study these emulsions in order to limit water self-absorption.

The emulsion polymerization of a butyl acrylate/methyl methacrylate/allyl methacrylate copolymer latex has recently been studied using FT Raman methods. The structure of the monomer units involved in the formation of this latex is shown in Fig. 8.15. The highlighted acrylic and methacrylic C=C bonds react to form the polymer chains, whilst the allylic double bonds give rise to crosslinking. Raman spectroscopy is particularly attractive because the bands due to the C=C stretching vibrations are strong and by monitoring their disappearance the extent of polymerization can be determined.

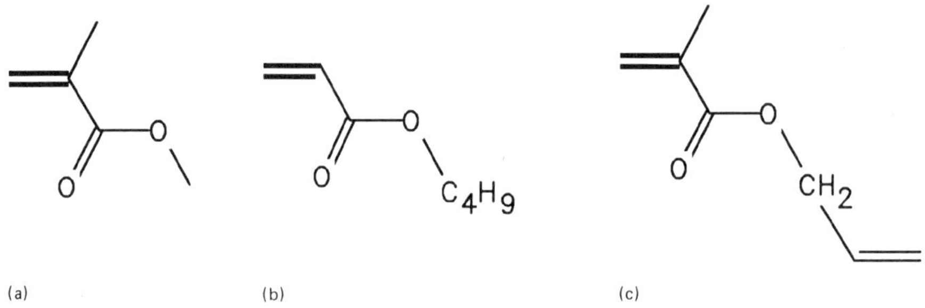

Fig. 8.15 — The chemical structures of (a) butyl acrylate, (b) methyl methacrylate and (c) allyl methacrylate copolymer.

The majority of oil-based paints and coatings are based on alkyd resins. The general structure of an alkyd resin is shown in Fig. 8.16. FT Raman has been used to

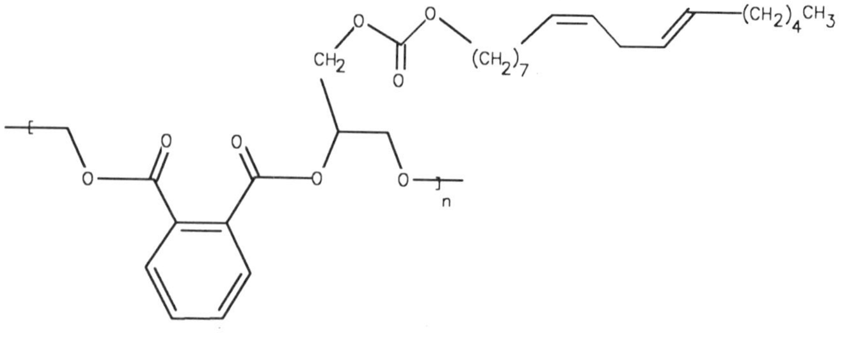

Fig. 8.16 — The general structure of an alkyd resin.

study '*o*-phthalate' alkyd resin, a soya-oil-based resin which contains about 70% fatty acid. The curing reaction was monitored over several weeks. Figure 8.17 shows the variation in the FT Raman spectrum with time. It can be seen that the level of unsaturation increases for a short period and then decreases as the cure proceeds. The increase in the band head at 1655 cm^{-1} is possibly due to isomerization. The addition of titania to produce white paint gives rise to new bands at 442 and 608 cm^{-1} and also improves the signal-to-noise ratio in the Raman spectrum, owing presumably to increased scattering efficiency. The cure appears to proceed as in the raw resin but at an increased rate and this has been confirmed from experience in the paint industry. Evidence for the isomerization of the unsaturation bands which occurs early in the cure is clearer in this system, because of improved signal-to-noise ratio.

The degradation reactions which occur whilst paints and coatings are in service are commercially important. The photodegradation of a number of systems has been studied using FTIR, although the results obtained are generally of poor quality [35]. In addition, TiO_2, which of course limits FTIR studies, is also thought to play an important part in the degradation reactions. It is postulated that even though it offers some protection from photodegradation by absorbing and scattering radiation, it also acts as an oxidation catalyst in photolysis reactions.

ATR can provide some useful results from weathered systems, but penetration of the infrared radiation is limited to a depth of about 2 μm, and therefore it tends only to give information on those reactions which occur at or very close to the surface early in the process. In contrast, FT Raman spectroscopy probes the bulk sample. Recently, FT Raman spectra have been compared with those obtained from ATR from unpigmented, red and white acrylic polyurethane coatings used on phosphated steel panels of motor vehicles [34]. The differences in the FT Raman spectra caused by weathering are small and unfortunately close to the noise limit. However, as instrumentation is constantly improving more information will undoubtably become available on these systems.

Many abrasion- and chemical-resistant paints and coatings are based on epoxy resins, which are also used in adhesives and sealing agents. FT Raman studies have been carried out on the curing reactions which occur in some of these materials [13]. Preliminary kinetic data has also been obtained on a curing commercial epoxy adhesive.

8.6 THERMOPLASTICS AND COMPOSITES

Poly(aryl ether sulphone) (PES) and poly(aryl ether ether sulphone) (PEES) and their copolymers are high quality, 'engineering' thermoplastic resins which are commonly used as the matrix element in composites. The structures of these materials is shown in Fig. 8.18. They are used extensively in the electronic, automotive and aerospace industries. ES and EES form random copolymers and both FTIR and FT Raman spectroscopies have been used to monitor the ratio ES:EES in samples of unknown composition, by constructing a 'standard curve' of band area versus concentration for a series of samples of known composition [36].

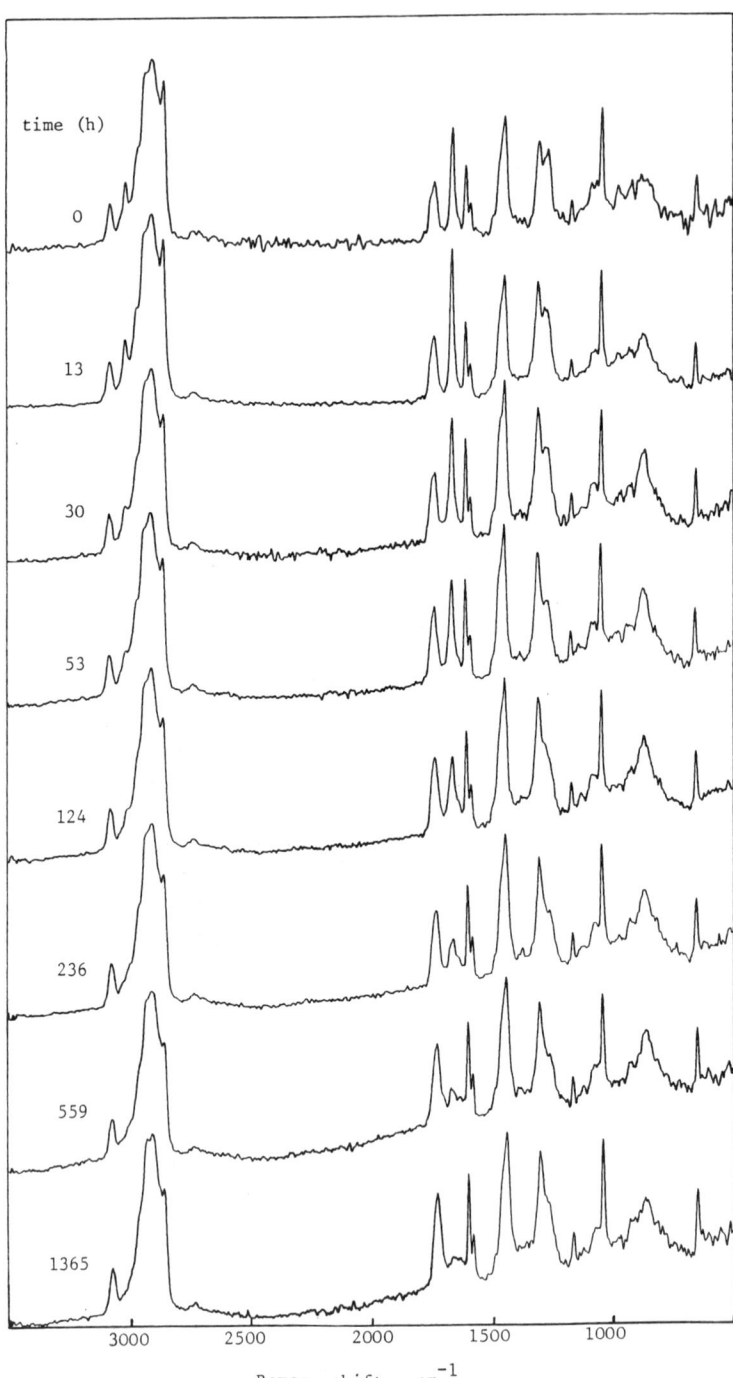

Fig. 8.17 — The FT Raman spectrum of '*o*-phthalate' alkyl resin as a function of time. Courtesy of J. K. Agbenyega.

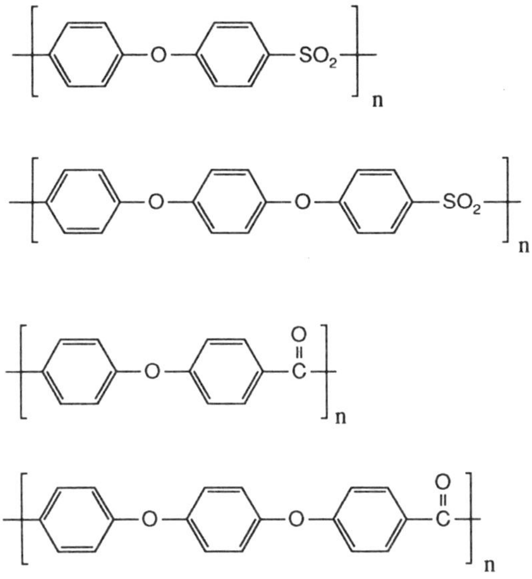

Fig. 8.18 — The structure of (a) poly(arly ether sulphone), (b) poly(aryl ether ether sulphone), (c) poly(arly ether ketone) and (d) poly (aryl ether ether ketone).

This procedure has been described in Chapter 6. In principle, this polymerization reaction could be monitored continuously. *In situ* measurements would also be possible using fibre-optic coupling to allow remote sampling (see Section 5.4.9).

Glass-reinforced PES has been studied using FT Raman and the results compared with those obtained using specular reflectance FTIR techniques. It was reported that the advantage of the FTIR method was that it could be done on a microscopic level, allowing the interface between resin and reinforcing agent to be monitored. Microscope attachments are, however, now available with some commercially available FT Raman instruments (see Section 5.7). Pioneering experiments on microscope attachments optimized for use in the near infrared were carried out by Messerschmidt and Chase [37]. They reported the FT Raman spectrum of a single Kevlar fibre. Further work by Bergin [38] demonstrated that spectra could also be obtained from very weak Raman scatterers such as a highly coloured polyurethane elastomer.

Poly(aryl ether ketone) (PEK) and poly(aryl ether ether ketone) (PEEK) are directly analogous to PES and PEES (see Fig. 8.18). They too are used in high performance composites, particularly in elevated temperature applications, because they are able to maintain their performance.† The FT Raman spectra of these materials and their composites have been reported [18].

Johnson and Wunder have published FT Raman results on polyimide composites [39], materials used as high temperature coatings in the electronic industry. FTIR has

---

† PEEK does, however, discolour under heat treatment and Raman studies show an increased level of fluorescence.

been used extensively to study these systems, but there have been problems investigating the bulk polymer and in producing high quality spectra from reinforced composites.† FT Raman spectroscopy has proven to be a useful complement since it has allowed composite materials to be studied *in situ*. The materials which are typically used as reinforcing agents, glass and quartz, are weak Raman scatterers and allow the matrix of the resin to be studied unhindered. Polyimides are prepared from solution. These polymerization reactions have been followed using FT Raman spectroscopy. Solvent loss, the disappearance of the prepolymer and the appearance of polyimide can all be monitored simultaneously.

Variations in the curing of polyimides can result in ranges of thermal and mechanical properties which may well affect in-service performance. Those groups, such as C=C, which are involved in the curing process give rise to strong Raman bands and this has allowed these reactions to be studied even in composite materials. Understanding the mechanism of curing reactions should enable more reproducible and reliable polyimide composites to be produced in the future, and FT Raman could be used to monitor the manufacturing process.

To date the examples given here of the types of FT Raman study possible on composite materials have been limited to the investigation of the matrix element. Aromatic polyamide fibres, such as Kevlar, have been shown to have high moduli and to be suitable for reinforcing certain resins. The conventional Raman spectra of Kevlar has been reported [40]. The study of certain grades of Kevlar and similar materials has, however, been severely limited by sample fluorescence. FT Raman studies are not limited by fluorescence and with the use of microscopes have been extended to single fibres [37,38]. In a study by Young [41], changes in the conventional Raman spectrum of fibres of poly(*p*-phenylene benzobisthiazole) (PBT) were monitored as a function of applied stress. The structures of this material is shown in Fig. 8.19. The heat-treated fibres deform elastically up to

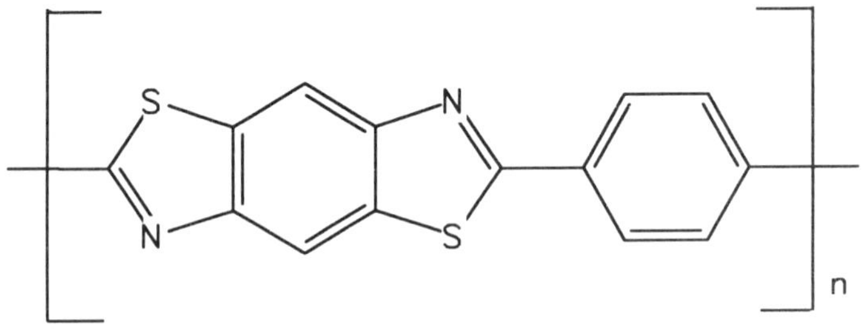

Fig. 8.19 — The structure of poly(*p*-phenylene benzobisthiazole) (PBT).

---

† The use of ATR on these materials is problematic as it is very hard to prepare a suitably flat surface to mate with the reflecting beam. The use of transmission is limited because the reinforcing materials frequently strongly absorb and can cause massive scatter to the detriment of the spectral quality.

fracture. A linear shift in the Raman band at 1480 cm^{-1} is observed as the strain is increased to fracture. Shifts of up to 12 cm^{-1} are common in these studies. The Raman band shifts reflect the strain experienced by the molecules in the fibre which can be related to the overall applied stress. These and other high modulus fibres, such as Kevlar, have also been studied when embedded in epoxy resin matrices in tension and compression. Similar measurement could now be possible on an increased range of samples using FT Raman techniques.

8.7 CONJUGATED AND ELECTRICALLY CONDUCTING POLYMERS

In recent years, in response to a growing demand from the optics and electronic industries, many novel polymeric materials have been developed. A variety of applications of FT Raman in the study of these materials are beginning to appear in the literature. FT Raman is being recognized as an ideal tool, primarily because of the reduced sample fluorescence and absorption, but also because it is non-destructive and allows samples to be studied *in situ*.

Many electrically conducting polymers such as polypyrrole and polyaniline have been synthesized. The modification of polydiacetylenes has also been studied in an attempt to produce crystalline conductive polymers. Hankin and Sandman have published conventional and FT Raman results from a series of chemically modified arylamine-substituted polydiacetylenes [42]. These materials react with halogens to produce very different structures. Figure 8.20 shows the structure of the two arylamine-substituted polydiacetylene starting materials, poly(1,6(di-*N*-carbazolyl)-2,4-hexadiyne), (poly(DCH)) and poly(1,1,6,6-tetraphenyl-1,2-hexadiynediamine), (poly(THD)). When poly(DCH) is reacted with bromine the FT Raman

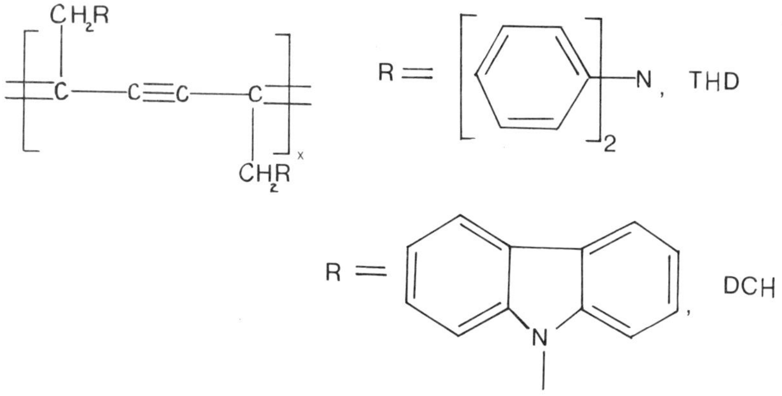

Fig. 8.20 — The structure of poly(1,6(di-*N*-carbazolyl)-2,4-hexadiyne) (poly(DCH)) and poly(1,1,6,6-tetraphenyl-1,2-hexadiynediamine) (poly(THD)).

spectrum of the product is consistent with that of a crystalline brominated material with a mixed polyacetylene structure. In contrast, when poly(THD) is reacted with either bromine or iodine, bands appear in the Raman spectrum which are characteristic of graphitic or glassy carbon and there is no evidence for the preservation of the conjugated polymer backbone.

The FT Raman spectra have already proved particularly useful in the elucidation of the polymer structure. Spectra excited from visible wavelengths are frequently dominated by large backgrounds. These backgrounds are caused by fluorescence and/or absorption of the laser frequency, leading to sample heating and/or sample degradation. Absorption of the visible radiation occurs because of electronic excitation of the conjugated polymer backbone. In the extreme case where the Raman exciting radiation wavelength approaches that of an absorption maximum in the electronic spectrum, massive enhancement of certain modes in the Raman spectrum can occur, i.e. the spectrum becomes resonance-enhanced. In these cases only those modes which may couple with the electronic transition are enhanced, and these are usually only the totally symmetric modes [43]. The advantages of this selective enhancement are that dilute samples, or those which are very weak Raman scatterers, can be studied. However, because the resonance-enhanced spectrum does not look like that of the unenhanced one the analytical value of the technique can be suspect. In other cases, the fact that only the symmetric modes of an otherwise highly complex spectrum are enhanced can be extremely useful.

Resonance enhancement of Raman active vibrational modes depends upon the frequency of the exciting radiation. Electrically conducting polymers such as conjugated polypyrrole, polyaniline and doped polyacetylene have been studied using resonance Raman excited in the visible [44], because they have absorption maxima in that region of the electromagnetic spectrum. Near infrared excitation is expected to afford new pieces of information on those conducting polymers which have strong absorptions in the near infrared region. Furukawa *et al.* [45,46] have recently published initial resonance enhanced FT Raman spectra of doped polyacetylene conducting polymers. When polyacetylene is doped with electron donors (e.g. sodium) or electron acceptors (e.g. iodine) it becomes highly electrically conducting. The electronic absorption spectrum changes and absorption maxima appear in the near infrared. This allows the doped regions of the polymer to be probed using resonance-enhanced FT Raman spectroscopy excited at 1064 nm. The position of the Raman band due to the *trans* C=C vibration (between 1450 and 1550 cm^{-1}) is sensitive to the number of conjugated *trans* C=C segments producing it. Furukawa *et al.* have shown that doped and non-doped regions of the polymer contain different amounts of conjugation and hypothesize on the significance of this observation [45]. They also report two distinct changes in the FT Raman spectrum of polyacetylene with increasing dopant concentration (sodium) [46]. The first of these changes, which occurs at low dopant levels, has been associated with an abrupt increase in the electrical conductivity, and the second, which occurs at much higher concentrations, with the appearance of a Pauli susceptibility in the material.

Williams and Gerrard have used FT Raman to study the degradation of polyvinylchloride (PVC) systems [47]. As PVC degrades, hydrogen chloride is lost and runs of conjugated carbon double bonds are formed. It has been shown that the extent of polymer degradation can be monitored through the formation of these polyenes. Further, these polyenes, if of long enough sequence length, have strong absorptions in the visible. Thus, when conventional Raman studies are carried out with excitation in the visible, the resonance condition can be satisfied and resonance-enhanced spectra can be recorded [48]. Spectral enhancement of the modes associated with the polyene backbone have been estimated in these cases to be of the order of 10^6. Thus, the very early stages of degradation can be monitored in this way when the concentration of polyenes is still incredibly small.

FT Raman spectra of fully processed PVC samples can also be recorded. The materials investigated by Williams and Gerrard were a PVC-clad steel coupon used in the roofing industry and PVC-backed polyurethane foam, a floor covering. Degradation of these samples was induced by heating, and was characterized by the formation of new bands. These bands were attributed to the C=C and C–C stretching vibrations of a polyene chain.

Spectra recorded from a yellow PVC coupon were compared with those from an unpigmented, white coupon. The yellow dye appears to inhibit the degradation of the PVC and spectral changes are only observed after several hours at elevated temperatures, after which plasticizer loss will have occurred. This observation is probably significant since previous studies have convincingly shown that plasticizer loss markedly accelerates the degradation process [49]. The FT Raman spectra from the inner and outer surfaces of the PVC-backed polyurethane foam indicated greater polyene formation at the surface in contact with the foam. This is thought to be due to amine derivative additives in the foam, which in turn accelerate the dehydrochlorination process. Thus the degradation process has been shown to be a function not only of the polymer, but also of the additives mixed with the polymer.

As discussed above in relation to polyacetylenes, the $\nu_{(C=C)}$ band position is sensitive to the length of the conjugated chain producing it. The $\nu_{(C=C)}$ band position observed in these PVC samples indicates a polyene chain length of approximately twenty carbon–carbon double bonds. Such a polyene has an intense visible absorption ($\lambda_{max}=480$ nm) which the authors claim tails into the near infrared. Williams and Gerrard contend that the C=C and C–C stretching vibration observed in these PVC samples are therefore resonance-enhanced.

Poly(di-*n*-alkylgermanes) have strong absorptions in the visible and ultraviolet regions of the electromagnetic spectrum. Thus, Raman excitation using visible wavelengths causes resonance enhancements of those bands associated with the backbone. The spectra are dominated by a strong band at about 1174 cm^{-1} due to the Ge–C vibration. Interpretation of the CH_2 stretching vibrations is hindered because of poor signal-to-noise ratios in this region. Hallmark *et al.* [16] avoided problems of resonance enhancement by using near infrared Raman excitation. They were able to study the CH_2 vibrations in detail and ascertain the packing of the alkyl side-groups as a function of temperature. This has been discussed in Section 8.2

8.8 MISCELLANEOUS STUDIES

Enhancement of weak Raman scatterers can also be achieved via an interaction with a metal surface. Surface-enhanced Raman spectroscopy (SERS) and its applications are described in detail in Chapter 10. Visible Raman studies have been carried out on the interactions between adhesives and metal and polymeric surfaces [50]. These types of study may also carried out in the near infrared. It is expected that the lack of fluorescence and sample heating will afford superior spectra and allow a greater range of samples to be investigated. Similarly, the *in situ* study of electropolymerization reactions should be possible.

Submicrometre-thick films, including organic films and molecular composites as well as polymer laminates, can be studied using FT Raman in conjunction with waveguide technology [51]. Figure 8.21 shows the experimental arrangement used to obtain spectra from thin polymer films. When the conditions for total internal reflectance are satisfied, an evanescent wave traverses the quartz substrate and energy moves into the thin film [52]. This excites the Raman scatter, which in this case is collected by a fibre-optic probe. Of particular interest because of their non-linear optical properties are 'guest/host' films. These consist of highly oriented conjugated chromophores in a thin film polymer matrix, e.g. naphthalene or 2-nitro-5-(*N*-methyl-*N*-octadecylamino)benzoic acid in a cellulose acetate matrix. Visible studies of these types of materials are limited because of fluorescence and very extensive sample heating, leading to degradation.

FT Raman spectroscopy has also been shown to be a powerful tool in the direct observation of very low levels of dyestuffs in polymeric materials. This was first demonstrated by Rabolt *et al.* [53]. The Raman scattering cross-section of a typical dyestuff is much greater than that of polymer fibre. However, most common dyes are fluorescent when exposed to visible radiation and therefore conventional Raman studies have been severely limited, although several resonance spectra have been reported. FT Raman spectra, free from fluorescence and resonance effects, have allowed the investigation of the complete structure of dyes, pure and when bound in polymer matrices [54]. Bourgeois and Church have published FT Raman and infrared spectra of red, blue and undyed polyacrylic fibres. The FT Raman spectra have been discussed in Section 7.2.3 and are shown in Fig. 7.11. The diffuse reflectance infrared spectra are essentially those of the acrylic fibre, yielding little information about the dye. In contrast, the Raman spectra show strong bands due to the dye. Future work may well elucidate the nature of the polymer–dye interaction. Care must be taken when carrying out subtractions to ensure that the spectra are exactly superimposed, otherwise spurious bands are observed. This has been discussed by Chase [55].

Polymers are commonly used in drug delivery systems. In many cases the drug release occurs via a leaching or diffusion process through the polymer matrix. Properties, such as molecular weight and degree of crosslinking, can be modified to control the rate of drug release. A number of drug–polymer systems have been studied using FT Raman spectroscopy [56]. Qualitative and quantitative characterization of these systems has been possible and information on the nature of the interaction between the components is again available using computer subtraction routines. The diclofenac–sodium alginate system is shown in Fig. 7.14. The spectrum of the polymer support, sodium alginate, is weak compared with that of the diclofenac and the drug can be studied at concentrations down to below 5% w/w.

Another method of drug release is the erosion of the polymer by slow solution or by hydrolysis. A number of synthetic polymers have now been developed which

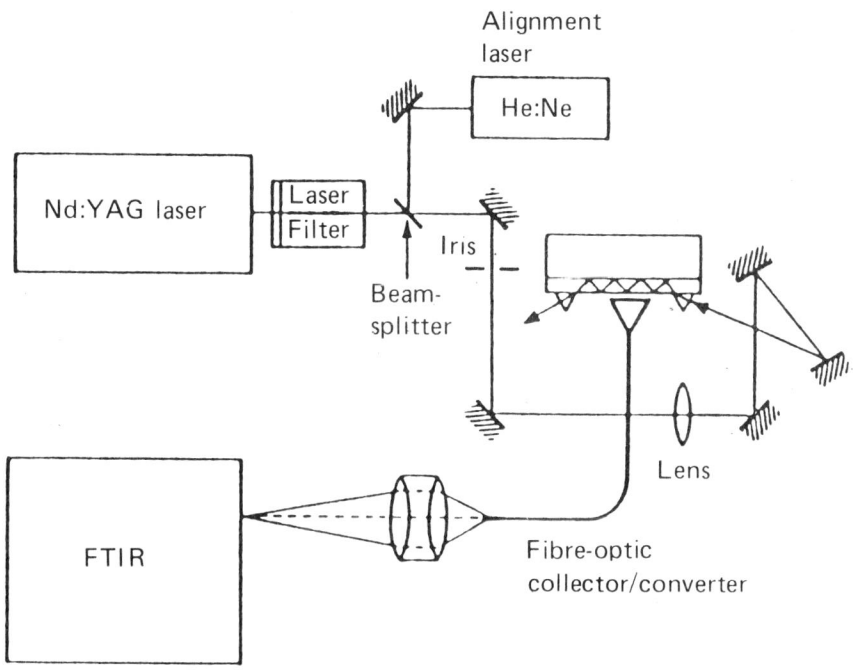

Fig. 8.21 — The experimental arrangement used in experiments carried out by Rabolt et al. [51]. Reproduced with permission from C. G. Zimba et al. J. Phys. Chem., **94**, 939. Copyright 1990 American Chemical Society.

possess a so-called biodegradable bond which is hydrolytically cleaved in biological fluids. Since release in these cases invariably involves the immersion of the drug/ support system in simulated body fluids the Raman method is appropriate because there is little interference from water. FT Raman spectroscopy has been used to follow the hydrolysis reaction of poly(sebacic anhydride) [56]. Other biopolymer systems are described in Chapter 9.

REFERENCES

[1] R. Signer and J. Weiler, *Helv. Chim. Acta*, 1932, **15**, 649.
[2] J. L. Koenig, *Appl. Spectrosc. Rev.*, 1971, **4**, 233.
[3] D. L. Gerrard and W. F. Maddams, *Appl. Spectrosc. Rev.*, 1986, **22**, 251.
[4] B. Schrader, *Angew. Chem. Internat. Edit.* 1973, **12**, 884.
[5] H. Kogelnik and S. P. S. Porto, *J. Opt. Soc. Am.*, 1963, **53**, 1446.
[6] T. R. Gilson and P. J. Hendra, *Laser Raman Spectroscopy*, Wiley, London (1970).
[7] H. Baranska, A. Labudzinska and J. Terpinski, *Laser Raman Spectroscopy — Analytical Applications*, Ellis Horwood, Chichester, 1987.
[8] T. Hirschfeld and D. B. Chase, *Appl. Spectrosc.*, 1986, **40**, 133.
[9] D. B. Chase, *Analyt. Chem.*, 1987, **59**, 881A.
[10] P. J. Hendra and H. M. Mould, *Int. Laboratory*, 1988, **18**, 34.
[11] K. P. J. Williams and S. M. Mason, *Spectrochim. Acta*, 1990, **46A**, 187.
[12] K. D. O. Jackson, M. J. R. Loadman, C. H. Jones and G. Ellis, *Spectrochim. Acta*, 1990, **46A**, 217.
[13] J. K. Agbenyega, G. Ellis, P. J. Hendra, W. F. Maddams, C. Passingham, H. A. Willis, and J. Chalmers, *Spectrochim. Acta*, **46A**, 197.
[14] W. F. Maddams and I. A. M. Royaud, *Spectrochim. Acta*, 1990, **46A**, 747.
[15] G. V. Fraser, P. J. Hendra, D. S. Watson, M. J. Gall, H. A. Willis and M. E. A. Cudby, *Spectrochim. Acta*, 1973, **29A**, 1525.
[16] V. M. Hallmark, C. G. Zimba, R. Sooriyakumara, R. D. Miller and J. F. Rabolt, *Macromol.*, 1990, **23**, 2346.
[17] G. R. Strobl and W. Hagedorn, *J. Polym. Sci. (Polym. Phys. Ed.)*, 1978, **16**, 1181.
[18] G. Ellis, A. Sanchez, P. J. Hendra, K-N. Kruger, J. M. Chalmers and J. G. Eaves, in press.
[19] J. D. Louden, *Polym. Comm.*, 1986, **27**, 82.
[20] D. J. Cutler, P. J. Hendra and G. Fraser, in *Developments in Polymer Characterization Volume 2*, Ed. J. V. Dawkins, Applied Science Publishers, 1980.
[21] D. I. Bower and W. F. Maddams, *The Vibrational Spectroscopy of Polymers*, Cambridge Solid State Science Series, Cambridge University Press, Cambridge, 1989.
[22] J. Wang, M. Yang, D. Lin, J. Sang, Y. Zhoa, Z. Liu and Q. Cui, *Polymer*, 1989, **30**, 524.
[23] C. H. Jones and P. J. Hendra, *Proc. European Symp. Polym. Spectrosc.*, Cologne, 1990, in press.
[24] P. J. Wallen, *Spectrochim. Acta*, 1991, **47A**, in press.
[25] J. L. Koenig and F. J. Boerio, *J. Chem Phys.*, 1969, **50**, 2829.
[26] G. Ellis, C. Marco, M. A. Gomez, J. G. Fatou and R. C. Haddon., *Polym. Bull.*, 1991, **25**, 351.
[27] G. Ellis, J. Lorente, C. Marco, M. A. Gomez, J. G. Fatou and P. J. Hendra, *Spectrochim. Acta*, 1991, **47A**, in press.
[28] T. Jawhari, *PhD Thesis*, Southampton University, UK, 1989.
[29] G. Ellis, C. H. Jones, P. J. Hendra, K. D. O. Jackson and M. J. R. Loadman, *Kautschuk & Gummi Kunststoffe*, 1990, **43**, 118.
[30] K. Kurosaki, *Int. Poly. Sci. Technol.*, 1988, **15**, 601.
[31] S. W. Cornell, J. L. Koenig, *Macromolecules*, 1969, **2**, 546.
[32] S. W. Cornell, J. L. Koenig, *Macromolecules*, 1968, **2**, 540.
[33] C. E. Miller, D. D. Archibald, M. L. Myrick and S. M. Angel, *Appl. Spectrosc.*, 1990, **44**, 1297.
[34] G. Ellis, M. Claybourn and S. E. Richards, *Spectrochim. Acta*, 1990, **46A**, 227.
[35] L. M. Briggs, D. R. Banes and R. O. Cater III, *Int. Engng. Chem. Res.*, 1987, **26**, 667.
[36] G. Ellis, A. Sanchez, P. J. Hendra, H. A. Willis, J. M. Chalmers, J. G. Eaves, W. F. Gaskin, K-N. Kruger, in press.
[37] R. G. Messerschmidt, D. B. Chase, *Appl. Spectrosc.*, 1989, **43**, 11.
[38] F. J. Bergin, *Spectrochim. Acta*, 1990, **46A**, 153.
[39] C. Johnson and S. L. Wunder, *SAMPE JOURNAL*, 1990, **26**, 19.
[40] L. Penn and F. Milanovich, *Polymer*, 1979, **20**, 31.
[41] R. J. Young, *Polymer Int.*, 1991, in press.
[42] R. J. Day, I. M. Robinson, M. Zakikhani, R. J. Young, *Polymer*, 1987, **28**, 1833.
[43] K. Nakamoto, *Infrared and Raman Spectra of Inorganic and Coordination Compounds*, 4th Edition, Wiley, New York, 1986.
[44] I. Harada, *Proc. Twelfth Int. Conf. Raman Spectrosc. (ICORS), Columbia, South Carolina*, Ed. J. R. Durig and J. F. Sulivan, Wiley, p. 736, 1990.
[45] Y. Furukawa, H. Ohta, M. Tasumi, *Proc. Twelfth Int. Conf. Raman Spectrosc. (ICORS), Columbia, S. Carolina*, Ed. J. R. Durig and J. F. Sulivan, Wiley, p. 736, 1990.
[46] Y. Furakawa, H. Ohta, A. Sakamoto and M. Tasumi, *Spectrochem. Acta*, 1991, in press.
[47] K. P. J. Williams and D. L. Gerrard, *Euro. Polym. J.*, 1990, **26**, 1355.
[48] A. Baruya, D. L. Gerrard and W. F. Maddams, *Macromol.*, 1983, **16**, 578.

[49] H. J. Bowley, D. L. Gerrard, K. P. J. Williams and I. S. Biggin, *J. Vinyl Tech.*, 1986, **8**, 176.
[50] F. J. Boerio, P. P. Hong, P. J. Clark and Y. Okamoto, *Langmuir*, 1990, **6**, 721.
[51] C. G. Zimba, S. Turrell, J. P. Swalen, V. M. Hallmark and J. F. Rabolt, *J. Phys. Chem.*, 1990, **94**, 939.
[52] R. Iwamoto, M. Miya, K. Ohta and S. Mima, *J. Chem. Phys.*, 1981, **74**, 4780.
[53] V. M. Hallmark, C. G. Zimba, J. D. Swalen and J. F. Rabolt, *Spectroscopy*, 1987, **2**, 40.
[54] D. Bourgeois and S. P. Church, *Spectrochim. Acta*, 1990, **46A**, 295.
[55] D. B. Chase, *SPIE (Fourier Transform Spectroscopy)*, 1989, p. 173.
[56] M. C. Davies, J. S. Binns, C. D. Melia and D. Borgeois, *Spectrochim. Acta*, 1990, **46A**, 277.

9
Biological applications

9.1 INTRODUCTION

Many reviews have been published outlining some of the biological applications of conventional Raman spectroscopy using visible lasers [1,2]. Vibrational spectra contain useful information about chemical composition as well as the secondary and tertiary structures of large biological molecules. For example, some evidence for the basic amino acid sequence and its three dimensional structure (α-helix or β-pleated sheet, etc.) is available from the infrared and Raman spectra of proteins. It has been shown that many of the building blocks of life, proteins, polypeptides, nucleic acids and saccharides, can be isolated and studied using conventional and resonance Raman techniques. However, the impurities and chromophores present in biological samples often fluoresce and some can degrade or denature when irradiated at visible wavelengths, preventing useful data from being acquired. FT Raman spectroscopy using near infrared excitation should in principle prove particularly useful for studying systems where laser-induced decomposition and fluorescence have hampered previous studies. Goral and Zichy obtained excellent quality FT Raman spectra from nucleic acids and their components, polypeptides, and a variety of saccharides [3]. These spectra are generally of superior quality data because many earlier conventional studies have been limited to using low laser powers to minimize sample damage.

Studies of samples of tissue, organs and vegetable matter have been particularly problematic. Fluorescence spectroscopy has been used extensively in pathological applications because the technique is non-destructive and can be employed clinically using fibre-optic probes, allowing disease progression to be monitored *in vivo*. However, FTIR and FT Raman techniques have the potential to provide much more detailed molecular level information on the changes which occur in tissue as a result of disease. An additional advantage of FT Raman is that the sample penetration of near infrared radiation is such that it can be used to probe biological substituents several hundred micrometers below the tissue surface. Problems encountered in previous visibly excited Raman studies on body tissue and vegetable matter resulting from sample heating can also be reduced by defocusing the laser beam as the viewed area is much larger. Like fluorescence spectroscopy, FT Raman is non-destructive and ideal for remote sampling (see Section 5.4.9). This could in principle allow the development of an FT Raman probe for *in vivo* diagnosis of, for instance, dental disease or alternatively the incorporation of the probe into an endoscope for the investigation of internal organs.

9.2 MEMBRANES

Levin *et al.* have published several papers which outline the advantages and disadvantages of FT Raman spectroscopy as a technique for studying biological assemblies in general, and membranes in particular [4–7]. Many biological aggregates are easily degraded by laser excitation, are very weak Raman scatterers and suffer from massive sample fluorescence. The problems of degradation and fluorescence can be eased by using near infrared excitation, but are certainly not eliminated. Many techniques have been explored by Levin *et al.* in order to boost the Raman signal, and spectral quality is improving all the time as the sensitivity of the instruments improves. One problem, however, is proving difficult to overcome. The overtone region of water, the usual medium for biological cellular assemblies, unfortunately falls in the Stokes Raman scattering range of a spectrum excited using an Nd^{3+}:YAG laser operating at 1064 nm. Thus, samples absorb scattered radiation, and problems of temperature control have been encountered; these are particularly important in the biological field. In these cases, internal temperature calibrations are made and, in an attempt to minimize the effects, the laser spot can be defocused and fibre optics employed to deliver the incident laser excitation and to collect the Raman scatter. The use of deuterium oxide as an alternative solvent to water also has a beneficial effect since the v_{OD} absorptions are shifted away from the Stokes Raman spectral region. Problems encountered because of water self-absorption have also been covered in Chapter 6. Possible future developments to overcome this problem are discussed in Chapter 12.

Membranes are an important constituent of all mammalian cells and are composed of lipid molecules arranged in lyotropic liquid crystalline phases. A lipid molecule contains two portions: a hydrophilic head and a hydrophobic tail, usually a long alkyl chain. A typical ordered bilayer packing geometry for these molecules is shown in Fig. 9.1(b). Membranes mediate the passage of materials in and out of cells. Of interest to researchers is the packing and dynamic properties of the lipid molecules which aggregate to form membranes and their interactions with materials such as antibiotics and other drugs. In particular, information is required on the process of interdigitation. Interdigitation causes a decrease in the bilayer thickness of model membrane systems and an ordering of the lipid chain. The resulting packing geometry is shown in Fig. 9.1(c). Model lipid systems have been studied in both the gel and liquid crystalline phases using FT Raman methods [5–7]. Real membranes, such as bovine synaptic plasma membrane, have also been studied, although the procedure could not be described as routine [4]. The sample was cooled to prevent heating and it was necessary to accumulate spectra for two and half hours to achieve a spectrum with an adequate signal-to-noise ratio.

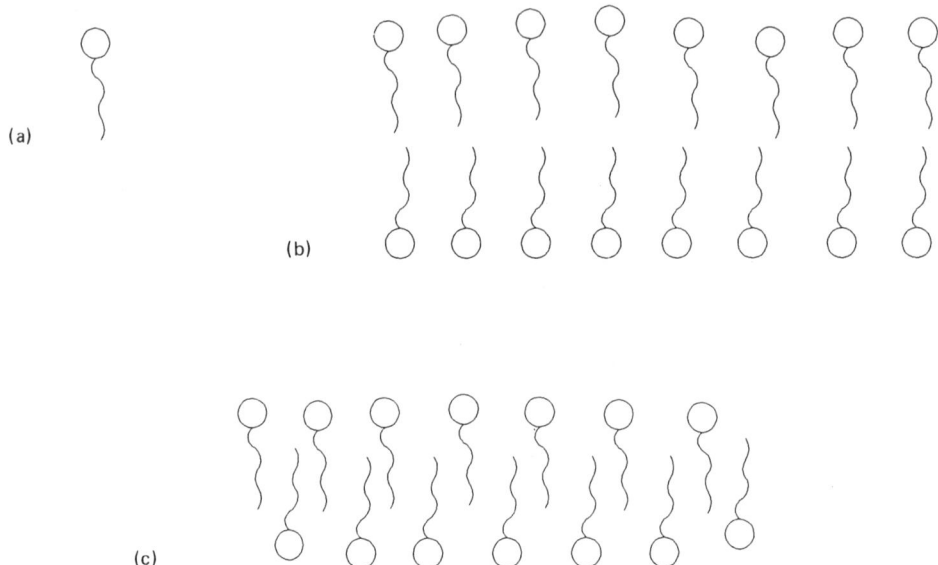

Fig. 9.1 — A stylized diagram of (a) a lipid molecule, (b) a bilayer assembly and (c) an interdigitated model membrane.

FT Raman spectroscopy has also been applied to the study of a dipalmitoylphosphatidylcholine model membrane system and its interaction with the polyene antibiotics, amphotericin A and B, and nystatin [5]. The structures of these molecules are shown in Fig. 9.2. These drugs are known to be active against fungal infections; however, the mechanism of the interaction at a molecular level is not understood. The effect of cholesterol on the interaction between the antibiotic and both the bilayer and multilamellar structures was also investigated. Studies on these systems using conventional Raman with excitation in the visible at 514.5 nm were limited because of massive sample fluorescence. In contrast, FT Raman spectra of good signal-to-noise ratio were obtained. By analysing the C–H stretching region, the ratio of the intensities of the symmetric in-plane methyl stretch at 2850 cm^{-1} and the asymmetric out-of-plane stretch at 2880 cm^{-1} was used to monitor the packing properties of the membrane system as the antibiotics were added. The ratio 2850/2880 cm^{-1} decreased, indicating a significant ordering of the chains, as the drugs were added. Cholesterol did not appear to influence the packing and/or ordering of the membrane. These observations were interpreted by Levin *et al.* in terms of the mechanism of formation and the structure of the lipid membrane, which is thought to have undergone an interdigitation process.

In another study, Levin *et al.* investigated the effect of unsaturation on the packing and dynamic properties of phosphatidylcholine membrane assemblies [6].

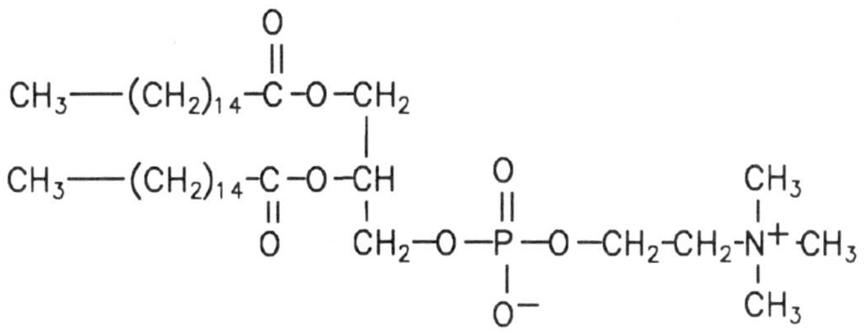

(a)

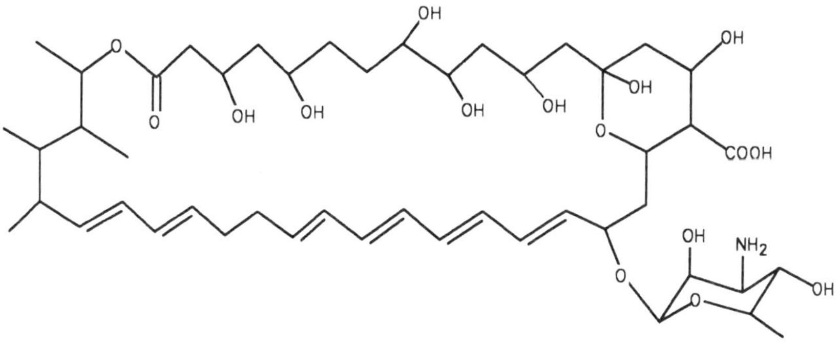

(b)

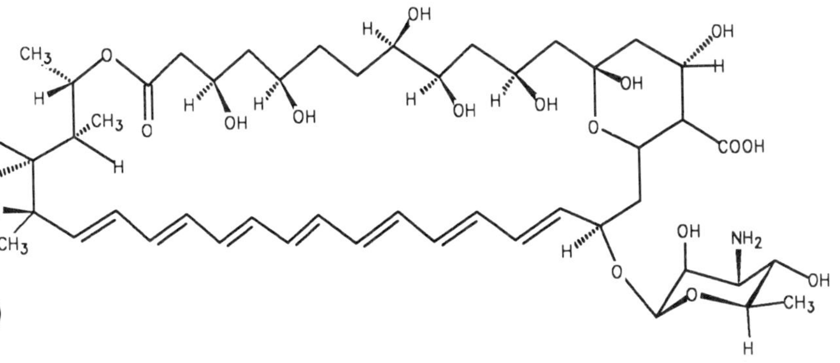

(c)

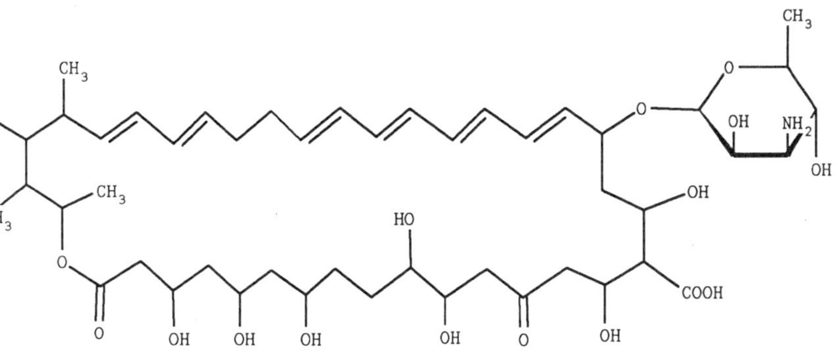

(d)

Fig. 9.2 — The molecular structure of (a) dipalmitoylphosphatidylcholine and the antibiotics (b) amphotericin A, (c) amphotericin B and (d) nystatin.

Raman studies of model membranes containing saturated lipids have been carried out. However, studies on models comprised of unsaturated alkyl chains are limited because of the tendency of these materials to undergo peroxidation, producing highly fluorescent degradation products. It has been shown recently that the functional activity of the integral membrane protein, rhodopsin, is modulated by conformational changes which occur in the membrane. Understanding the packing of membranes will enable the mechanism by which they modify protein behaviour to be elucidated. The packing of unsaturated phosphatidycholine molecules containing one, four and six double bonds was monitored using the v_{CH} I_{2850}/I_{2880} ratio mentioned earlier. Levin found that the ratio decreased with increasing unsaturation, indicating increased bilayer order. The C=C symmetric stretching mode also showed an increase in band half-width with increasing unsaturation. This was interpreted as being due to a higher degree of rotational freedom.

9.3 ANIMAL TISSUE

FT Raman spectroscopy has already found a number of other specific applications in the medical, biochemical and food/nutrition areas of research. The acquisition of relatively fluorescence-free spectra using near infrared radiation has opened up many new areas of research, and the rapid accumulation of FT Raman data of adequate quality will allow extensive studies to be carried out from which reliable conclusions may be drawn. This point is well illustrated by the research carried out by Nie et al. on human eyes [8]. Conventional Raman spectra have been obtained from young, healthy human eye lenses, but as the eye ages, spectra become increasingly difficult to record because of the appearance of various fluorophors in the lens. These fluorophors may be of a photochemical, metabolic or pathological origin. Techniques to limit the effects of sample fluorescence, such as time-resolved studies, SERS, resonance and non-linear Raman and photobleaching have been attempted with varying success. Nie et al. have studied a large variety of eyes and have catalogued the results such that they have been able to distinguish those processes which occur naturally in the lens because of ageing from those associated with disease, in this case the development of lenticular opacity (cataracts). Differences between the FT Raman spectra relating to the formation of cataracts have been interpreted in terms of changes in certain protein conformations. The ease of use of the FT Raman technique should enable libraries of eye spectra to be accumulated and other abnormalities which occur in the lens, such as lipid peroxidation and protein glycation, to be studied. FT Raman spectroscopy also allows the study of brunescent human eyes, which absorb visible radiation, as well as pigmented animal lenses [9]. Certain diurnal (daytime) animals, such as squirrels and chipmunks, have highly pigmented lenses in order to protect the retina from near UV-induced damage and to enhance the sharpness of their vision by reducing chromatic aberrations in the lens, particularly from shorter wavelength light. These pigments have prevented previous Raman studies. In addition to the study of lenses, pathological developments in other parts of the eye may also be investigated using FT Raman methods. Preliminary spectra have been observed from human and chicken eye sclera [9]. The sclerotic membrane is the hard firm opaque coating outside the iris which forms the white of the eye. A comparison of the two spectra indicates that chicken sclera contain a band at 963 cm^{-1} absent in that of humans. This band has been attributed to the symmetric P–O stretch in the inorganic salt apatite, which is commonly found in bones and teeth.

Rava, Baraga and Feld have carried out an FT Raman study of healthy and diseased human arteries [10]. Many people die each year because of heart attacks, nearly all attributable to coronary atherosclerosis. Several treatments for atherosclerosis are available, but no reliable method exists to determine how each patient will react to the particular therapy chosen. The *in situ* biopsy of blood vessels is difficult and thus detailed information on the state of the diseased vessel is limited. It has been demonstrated that FT Raman spectroscopy can provide information on the molecular level changes which are occurring as a result of the atherosclerosis occurring in the human aorta. Changes in the spectrum associated with an initial build-up of cholesterol in the inner layer of the aorta (the intima) can be identified. Other differences between the spectra of healthy and atherosclerotic tissue are attributed to accumulation of calcium and the formation of calcium hydroxyapatite crystals in the tissue several hundred micrometres below the surface. This is thought to occur in the advanced stages of the disease. Information on the increased levels of cholesterol and calcium should eventually enable physicians to prescribe the correct treatment for each patient.

Visible Raman studies of bone tissue have been severely hampered by sample fluorescence. Nie *et al.* [9] have obtained FT Raman spectra from chicken bone of excellent quality. They illustrated the problem of fluorescence by comparing the FT Raman spectrum with one recorded using krypton 647.1 nm excitation. This diagram has been reproduced as Fig. 9.3. The basic structural unit of bone is an osteon composite consisting of inorganic salts embedded in organized lamellae of collagen. Over 70% of bone is comprised of the calcium phosphate salt, hydroxyapatite. FT Raman could be used as a probe to monitor the changes associated with normal bone mineralization and identify those which occur as the result of disease.

Similar studies could also be carried out on teeth, which have traditionally given very fluorescent spectra. FT Raman examination of a human tooth yields different spectra from the various regions within the tooth [9]. Figure 9.4 shows spectra obtained from (a) the tooth enamel (the hard outer layer of the tooth), (b) the dentine inside the tooth and (c) the root. The enamel spectrum is dominated by bands which can be attributed to the mineral apatite (591, 961 and 1071 cm^{-1}). The dentine and root spectra indicate the presence of a much larger proportion of organic material. The C–H stretching bands at 2932 and 2880 cm^{-1} are more intense, and amide I and III bands at 1670 and 1243 cm^{-1} respectively have been identified. These bands correlate with those found in the FT Raman spectra of bone, indicating a similar protein conformation in the two materials.

9.4 POLYPEPTIDES

Polypeptides have been studied by a number of different groups [3,11–13] but the general conclusion is clear. Isolated amino acids and crystalline homopolypeptides

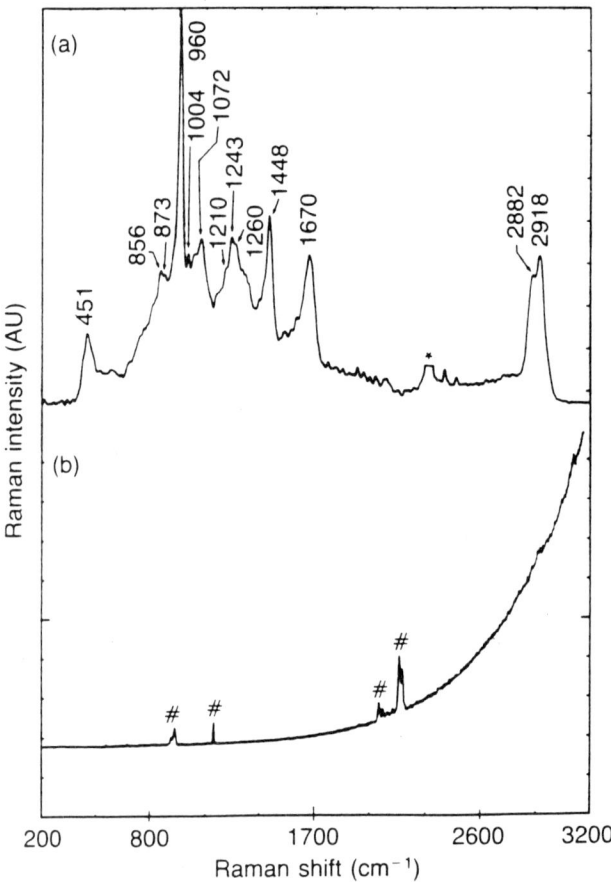

Fig. 9.3 — Comparison of the near infrared and visible excited Raman spectra of a chicken leg bone. (a) FT Raman spectrum and (b) conventional Raman spectrum, # represents a plasma line and * indicates an artifact. Uncorrected data. Reproduced with permission from *Spectroscopy*, **5**, S. Nie *et al.*, copyright 1990 Pergamon Press PLC.

give FT Raman spectra of excellent quality. Figure 9.5 shows the FT Raman spectrum of L-tryptophan and as a comparison that of polytryptophan. These spectra are easily identified and can be recorded routinely in only a few minutes' acquisition time. In contrast to the infrared spectra of polypeptides, it is the side-group modes which tend to dominate the Raman spectra rather than those of the common polyamide backbone. Consequently, the Raman spectra of different polypeptides are far more characteristic than their infrared counterparts. Conventional Raman studies have provided similar information from purified homopolymeric polypeptides, but it is clear that the use of near infrared excitation will remove the need for extensive purification routines. Many studies have appeared over the years on polypeptides, particularly in the solid state, since the vibrational spectra are rich in

Sec. 9.4] Polypeptides 211

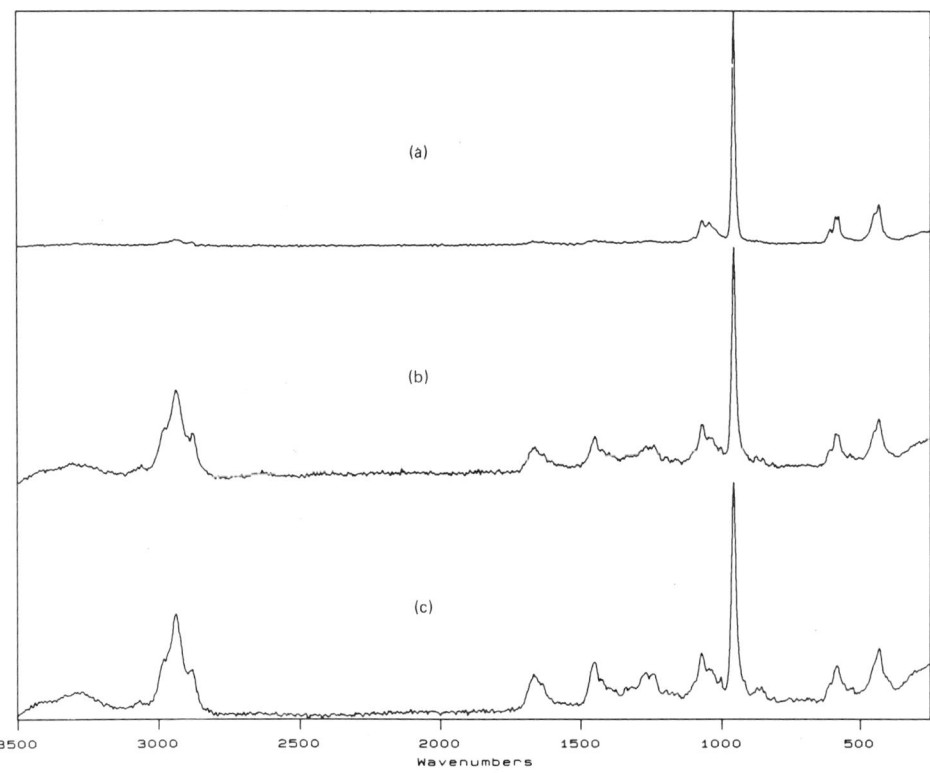

Fig. 9.4 — The FT Raman spectra of human teeth: (a) the enamel, (b) the dentine and (c) the root. Uncorrected data. Teeth courtesy of Douglas and Timothy Allen.

information on the conformation of the molecules and their crystallinity. FT methods are bound to be of value in this field. The quality of spectra obtainable from dilute aqueous solutions has always been modest because polypeptides are poor scatterers. In the near infrared the problem is further exacerbated by solvent absorption.

FT Raman has also been used to study heteropolypeptides, i.e. proteins. The spectra obtained tend to contain many overlapping bands, which makes assignments difficult and limits their analytical usefulness. These materials are inevitably of low crystallinity and are often highly hydrogen-bonded. Consequently, spectral bands tend to be relatively broad. However, some information on the composition and structure of these materials is available. Naturally occurring proteins, such as the globular protein bovine serum albumin, have been investigated [12]. Although the amide I and III modes can be identified, the spectrum is very complex and band assignments are difficult. However, deconvolution routines and chemometric methods may assist in this process in the near future. Figure 9.6 shows the FT Raman spectrum of human nail which contains the protein keratin [3]. Keratin is composed of α-helical polypeptide chains which are linked by the disulphide bridges of cystine

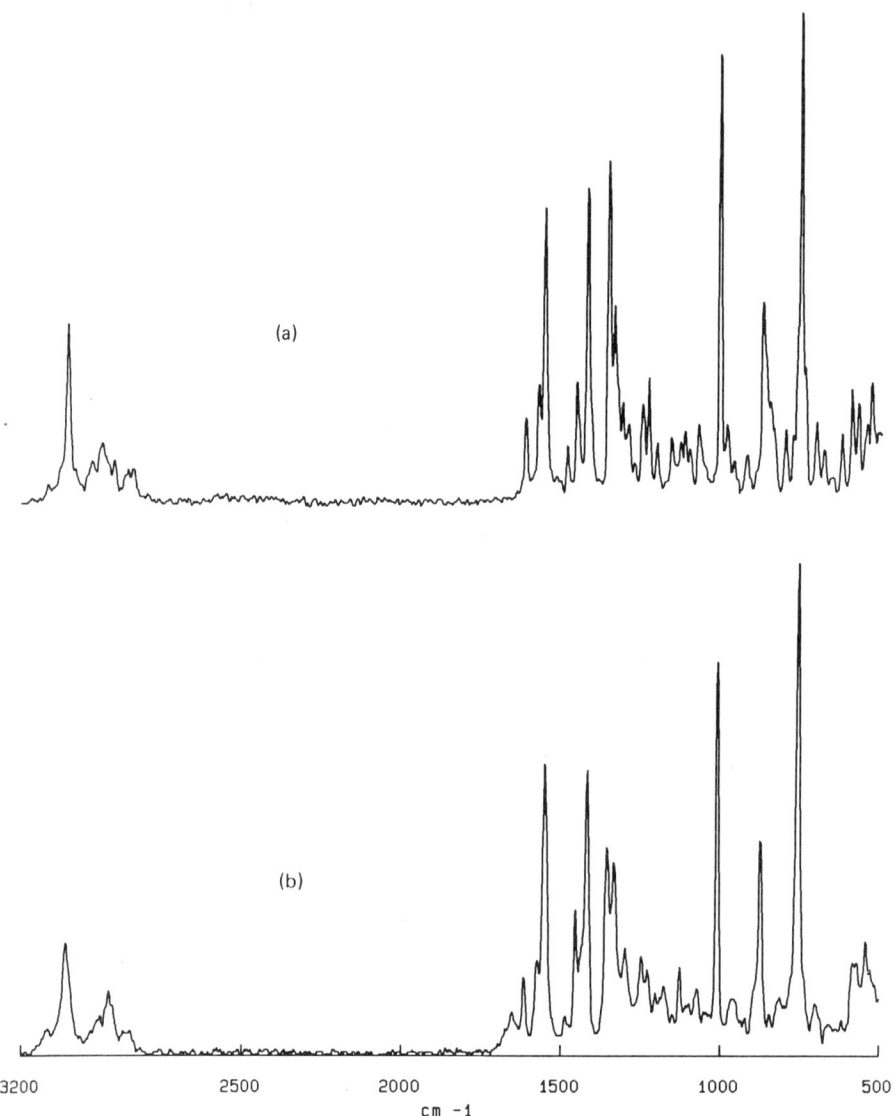

Fig. 9.5 — The FT Raman spectrum of (a) L-tryptophan and (b) polytryptophan.

molecules. This compound is also found in hair, birds' feathers and horses' hooves. When hair is 'permed' (i.e. a permanent wave or kink is introduced) the disulphide bridges are first reduced, the hair is formed into the desired shape and then they are reformed in an oxidizing reaction. A novel application of FT Raman spectroscopy could be to monitor this procedure using the disappearance and return of the band near 490 cm^{-1} associated with the S–S stretch.

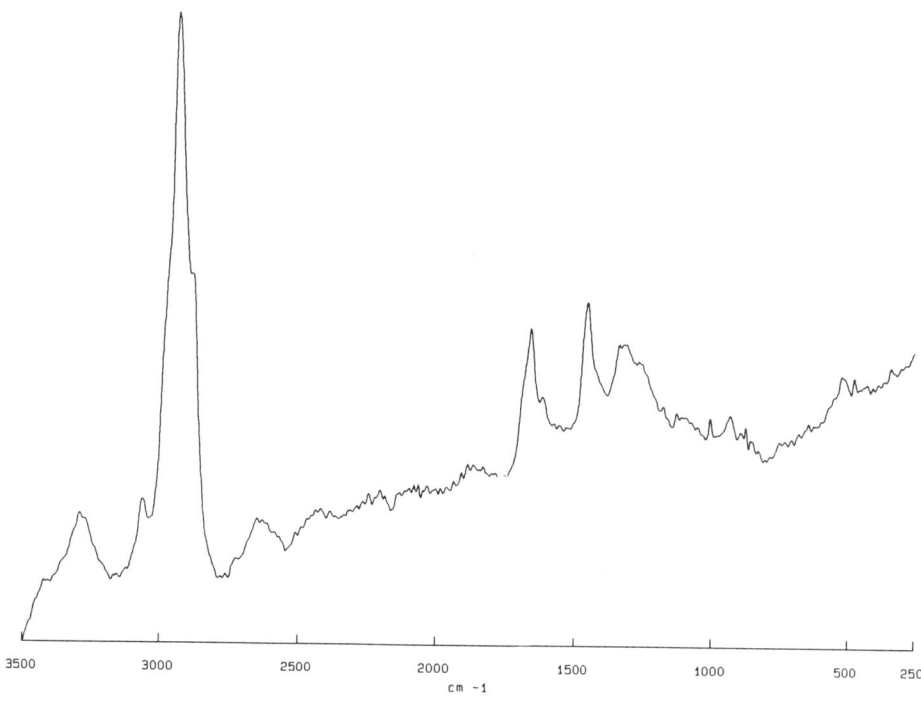

Fig. 9.6 — The FT Raman spectrum of human nail. Uncorrected data.

9.5 LIGHT-SENSITIVE SYSTEMS

One area of research where near infrared excited FT Raman spectroscopy has already caused much interest is in the study of reactions which are light-sensitive or where investigation using a ultraviolet or visible source precipitates photochemistry. Cobalamins are one class of compounds whose investigation has been hampered by this problem. Cobalamins, such as simple bioactive methyl cobalamin and the more complex co-enzyme B_{12}, contain a thermostable Co–C bond. It is the cleavage of this Co–C bond which is thought to be the key step in the biologically important catalysis reactions which are dependent upon the presence of co-enzyme B_{12}. The Co–C band is, however, highly photolabile and it has not been identified in previous Raman studies [14]. Nie *et al.* have recorded the FT Raman spectra from a series of photolabile organocobalt B_{12} model compounds [15] and several bioactive cobalamins [16]. These spectra are thought to indicate the presence of the Co–C bond. Using near infrared excitation, bands at 493 cm^{-1} and 463 cm^{-1} can be observed in the FT Raman spectrum of CH_3-cobalamin and CD_3- cobalamin respectively. The sharp 493-cm^{-1} band in methylcobalamin has also been monitored as a function of recrystallization conditions and it appears to be insensitive to crystalline form. Nie *et al.* have also studied the factors, such as environment and steric and electronic properties, which may influence the Co–C bond stretching frequency. To do so they recorded spectra from a large number of cobaloximes since these model compounds respond in an analogous manner to cobalamins. The results have been interpreted in terms of their relevance to the mechanism of homolytic cleavage of the Co–C bond [17]. The study of the Co–C bond in holoenzyme-B_{12} co-enzyme complexes is now anticipated. Little information is currently available on these complexes since no X-ray or NMR characterization has been carried out. Structural determinations are of limited precision because of the large size of the system.

Rhodopsins are retinal chromophores which are highly photosensitive, and absorption of visible radiation tends to cause extensive photodegradation. Resonance Raman spectroscopy is limited because of this problem. FT Raman spectroscopy with near infrared excitation almost invariably eliminates photoinduced decomposition, because of the lack of electronic absorptions of these retinal chromophores in the near infrared. Johnson and Rubinovitz [18] have described the spectrum of bacteriorhodopsin prepared from cultures of *Halobacterium halobium*. The spectrum is dominated by vibrations of the retinylidene chromophore, the protein-bound pigment. Nie *et al.* have proposed [14] that spectral subtraction routines can be employed to reveal the protein vibrational lines and thus study interactions between the protein and the chromophore. This is of interest because photoisomerization processes involve significant changes in protein conformation. Johnson and Rubinovitz noted that the intensity ratios of the retinylidene modes were almost identical in the 1064-nm excited spectrum, as they are in the resonance Raman spectrum excited at 514.5 nm and the pre-resonance spectrum excited at 752 nm. They concluded, therefore, that the near infrared excited spectrum was also pre-resonance enhanced [18].

Chlorophylls, the pigments found in photosynthetic organisms, are also light-sensitive and suffer from problems associated with sensitized photoxidation and fluorescence. Johnson and Rubinovitz have presented a preliminary spectrum from the photosynthetic reaction centre found in *Rhodobacter sphaeroides* [18]. The reaction centre of this photosynthetic bacteria contains four bacteriochlorophyll and two bacteriopheophytin chromophores as well as a carotene pigment. The FT Raman spectrum is highly complex, containing many unresolved overlapping bands from the three components. However, the lower frequency region of the spectrum is potentially rich in structural information about protein conformation. In another article Mattioli *et al.* [19], report the FT Raman spectrum from bacteriochlorophyll-a in the reaction centre of the photosensitive bacterium, *Rhodospirillum rubrum*. Bacteriochlorophyll can undergo several different electronic transitions including Q_x, Q_y and Soret. Traditionally, resonance Raman studies have exploited the Q_x and Soret absorptions, which lie in the visible region. The Q_y absorption, the lowest allowed singlet transition, occurs in the near infrared. The FT Raman spectrum of the bacteriochlorophyll-a is thought to be in pre-resonance with this state. The FT Raman spectrum is dominated by modes due to the chlorophyllic pigments. The pattern of band intensities is very similar to that of the non-fluorescent resonance Raman spectrum of the copper analogue of bacteriochlorophyll-a, excited in the Q_y absorption band. However, it is distinctly different from resonance studies excited in the Q_x and Soret transitions. Information on the interactions between carbonyl groups and the central magnesium atom is available by monitoring the keto and acetyl carbonyl modes which appear at 1687 and 1658 cm^{-1} respectively. Other bands at 1595, 1442 and 1358 cm^{-1} are also known to be magnesium-co-ordination-sensitive. The band at 1595 cm^{-1} in the FT Raman spectrum is interpreted in terms of two bound tetrahydrofuran molecules. The mode would be expected to shift to 1610 cm^{-1} if only one ligand were coordinated.

Mattioli *et al.* have also reported that spectra can also be obtained from the reaction centre of the bacterium *Rhodobacter sphaeroides* [19]. However, the most comprehensive study of this system to date is due to Noguchi *et al.* [20]. They studied bacteriochlorophyll-a from *Chromatium vinosum* in various solid films and in chromophores from a blue–green mutant of *Rhodobacter sphaeroides* demonstrating the use of FT Raman spectroscopy for the study of the *in vivo* state. They noted the sensitivity of the Q_y absorption to molecular interactions of the bacteriochlorophyll-a. This absorption can shift over 100 nm (between 770 and 870 nm) depending on its environment. Therefore, the Raman spectra resonant with the Q_y band are expected to provide information on molecular interactions. Hydrated bacteriochlorophyll exists in three forms, which have different absorption maxima. The blue–green mutant of *Rhodobacter sphaeroides* does not contain carotenoids and has only one type of light-harvesting pigment–protein complex, bacteriochlorophyll-a, which has an absorption maximum at 870 nm. The amount of bacteriochlorophyll-a present in the mutant is much larger than in the normal reaction centre, which also contains bacteriopheophytin-a. Thus, the FT Raman spectrum of the mutant, which is of excellent quality, is predominantly that of the bacteriochlorophyll which absorbs at 870 nm. ^{15}N substitution was also used to aid band assignments. It was concluded that the band positions in the C=C stretching region of the spectra could be interpreted in terms of the co-ordination number of the central magnesium atom, and the relative intensities of the bands between 1200 and 1000 cm^{-1} were found to be characteristic of the different bacteriochlorophyll states.

9.6 WOOD AND PLANT TISSUE

The FT Raman spectra of several woods have been reported [3,9,21,22]. Plant cells can be distinguished from those of animals by the presence of a thick cell wall. The most important components of the cell wall are cellulose and lignin. Cellulose is a polymer of glucose which contains β-1,4-glycosidic bonds, and lignin is a highly complex polymer of phenyl propane residues. Traditional methods of investigating the structure of woody tissue have involved extraction and isolation processes which almost inevitably alter the morphology of these molecules. Irradiation of woody tissue with visible laser light invariably causes massive sample fluorescence. The methods employed to reduce this problem have included immersing the sample in deuterated water and acquiring data over several hours (up to 8 hours for some samples). In many samples, however, the high levels of fluorescence encountered prevent useful data from being recorded. Nie *et al.* have reported the spectra of several native North American trees [9]. By comparison with work carried out on model cellulose and lignin compounds, they have assigned some of the bands in the spectra. With the advent of the near infrared Raman microscope, Nie *et al.* confidently predict that studies on the orientation of cellulose and lignin within the cell walls of different species should be possible. Atalla and Agarwal have already provided some information on the orientation of these materials using a conventional Raman microprobe [24].

Goral and Zichy extended their study of wood samples, comparing the spectra they obtained with those from certain wood products, such as Xerox paper and tissue paper [3]. The spectra of these woody materials were also compared with the spectrum of cotton fibres. Wood contains between 40 and 50% cellulose whilst cotton fibres are essentially pure cellulose. Kenton and Rubinovitz also describe the application of FT Raman spectroscopy to wood and other forest products [21]. The results of a feasibility study into recording spectra from coatings on paper were also presented. A number of different paper samples were studied. In one case, spectra were recorded from filter paper which had been partially covered with red ink. By subtracting the spectrum of clean white paper from that of paper impregnated with dye, a spectrum of the pure ink was obtained.† Traditional methods of analysing inks and dyes always required that they were first extracted from the paper. A spectrum was also obtained from a high quality paper sample. Bands were identified due to the fillers, titanium oxide and clay, and the adhesives, poly(vinyl acetate) and poly(styrene butadiene).

In addition, Kenton and Rubinovitz also compared two hard woods, oak and balsa, with two soft woods, pine and redwood [21]. The classifications hard and soft do not relate to the physical properties of the wood, but rather to the manner in which the trees produce their seeds and the type of lignin that they contain. Soft woods bear cones whereas hard woods do not. Soft wood lignin consists of guaiacyl units, whilst hard wood contains a mixture of guaiacyl and syringyl units. Guaiacyl lignin is a polymer based on *trans*-coniferyl alcohol, whereas syringyl lignin is based on the precursor *trans*-sinapyl alcohol. These precursor alcohols are shown in Fig. 9.7. Several differences between the FT Raman spectra of the hard wood samples and those of the soft woods can be identified. Figure 9.8 shows the FT Raman spectra of (a) apple, a hard wood, and (b) Douglas fir, a soft wood. A weak carbonyl band at 1736 cm^{-1} is observed in the spectra of hard woods. This band has been attributed to the quantity of holocellulose present in the sample. Holocellulose is the term used to describe the combination of cellulose and hemicellulose in a material. Kenton and Rubinovitz concluded that the ratio of holocellulose to lignin was higher in hard woods than in soft woods. Bands at 1120 cm^{-1}/1093 cm^{-1} and 895 cm^{-1}/593 cm^{-1} were also ratioed to confirm this analysis. However, the main difference between the two sets of spectra is a band at 1269 cm^{-1}, which was strong in the spectra of the soft woods. This band was attributed to the presence of rosin. However, Kenton and Rubinovitz also proposed that the signal seen at this frequency may be due to the ring breathing mode associated with guaiacyl lignin. Perhaps the most comprehensive study on wood to date was carried out by Evans following a survey of twelve different woods, six hard and six soft [23]. Evans' work generally confirms many of the conclusions of previous workers, that soft woods contain a higher proportion of lignin, whereas hard woods are richer in holocellulose, although he asserts that the band at 1273 cm^{-1} is due solely to the guaiacyl content of wood.

† This procedure has recently been applied to the *in situ* FT Raman study of TLC plates [22].

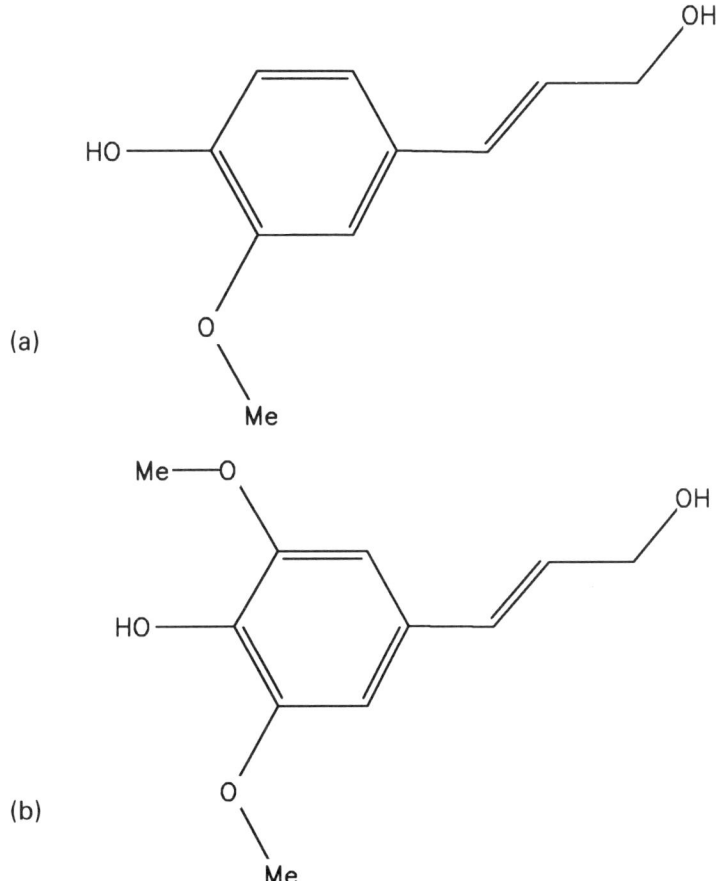

Fig. 9.7 — The molecular structure of (a) *trans*-coniferyl alcohol and (b) *trans*-sinapyl alcohol.

FT Raman spectroscopy has also been applied to the study of other plant extracts. Goral and Zichy published the FT Raman spectrum of sesquiterpene lactones [3]. These are extracts from the family of plants known as Compositae. They have the general structure shown in Fig. 9.9. The FT Raman spectra of helenalin and linifolin-A, where R=H and C_2H_5 respectively are shown in Fig. 9.10. As expected from their similar structure the spectra are almost identical, but subtle changes in the C–H stretching region due to the presence of the acetyl group can be observed. This is not, however, always the case, as mentioned in Chapter 7, where the example of the alkaloids was cited (see Fig. 7.7). Very subtle changes in the substituents in the series result in very large spectral variations.

9.7 PHARMACEUTICALS

A paper has appeared in the literature by Espinosa *et al.* on the low frequency FT Raman spectra of elliptocines [24]. Elliptocines are natural plant alkaloids which

218   **Biological applications**   [Ch. 9

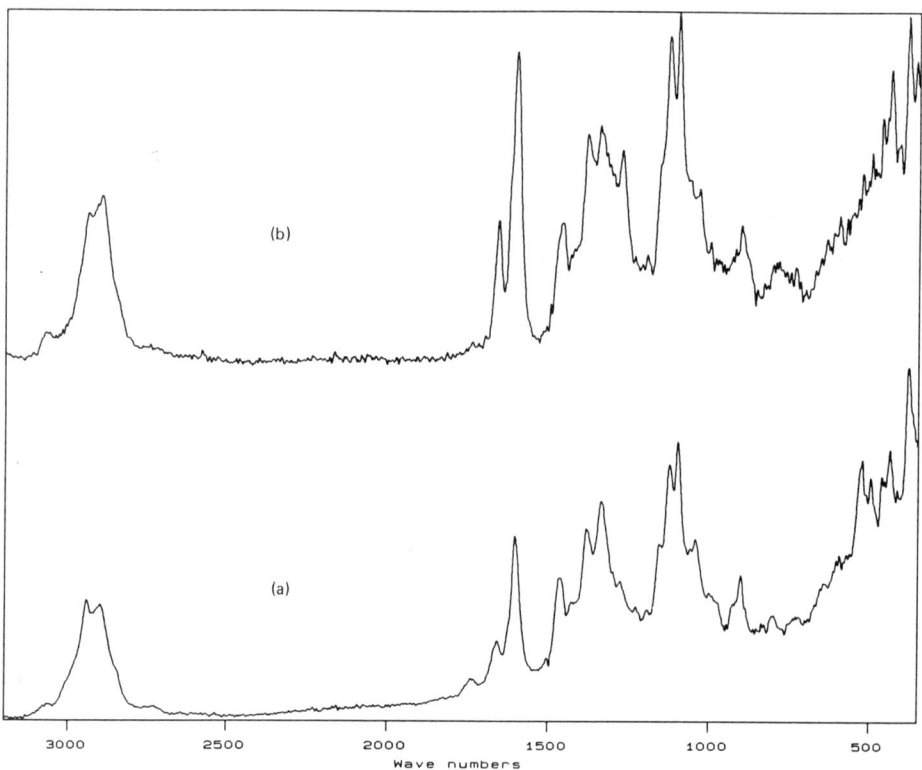

Fig. 9.8 — The FT Raman spectrum of (a) Apple wood and (b) Douglas fir. Diagram Courtesy of P. A. Evans.

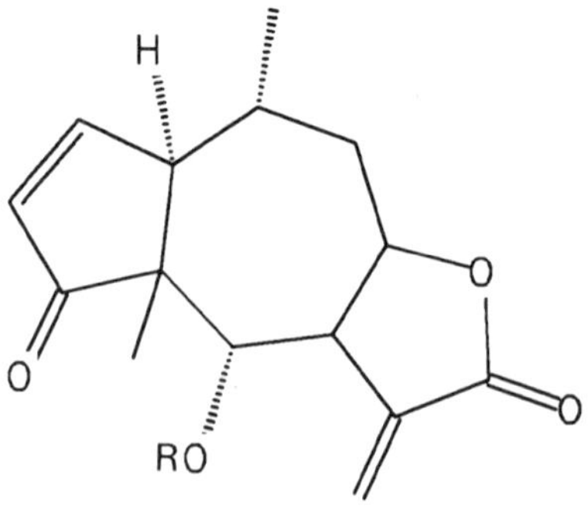

Fig. 9.9 — The general molecular structure of a sesquiterpene lactone.

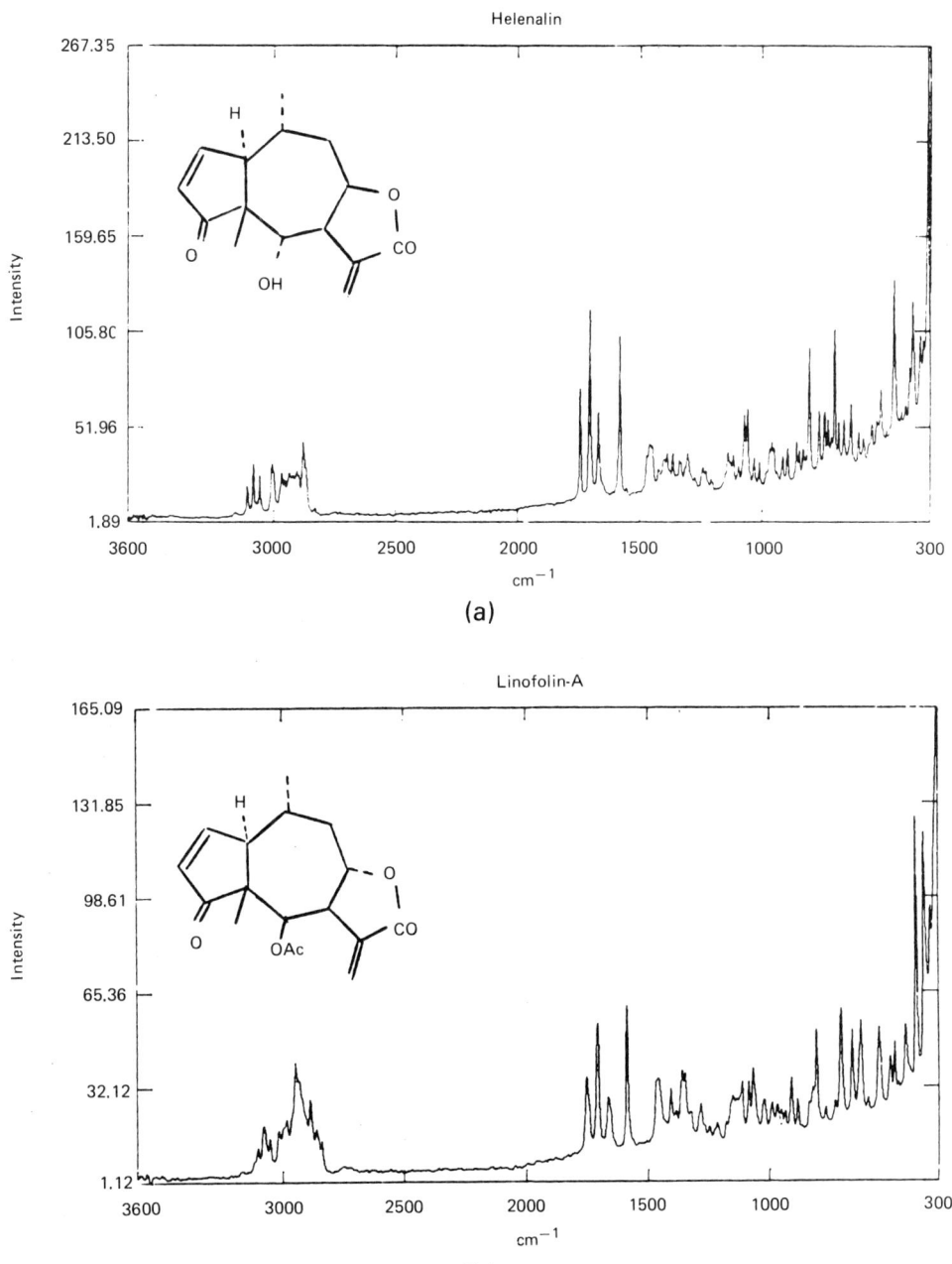

Fig. 9.10 — The FT Raman spectrum of (a) helenalin and (b) linifolin-A. Reproduced with permission from *Spectrochimica Acta*, **46A**, J. Góral and V. Zichy, copyright 1990 Pergamon Press PLC.

have the general structure shown in Fig. 9.11. Near infrared excited FT Raman spectra can be obtained from elliptocines with the substituents indicated in Fig. 9.11. Visible studies of these highly coloured materials are limited because of sample heating and fluorescence, a problem that persists with 1064-nm excitation in the case where X=OH. Elliptocines are of pharmaceutical importance because of their antitumour properties. It is believed that they intercalate between DNA base pairs, but little is known about the mechanism of this binding process. The low frequency region of the Raman spectra (400–10 cm^{-1}), which contains information about intermolecular hydrogen bonding, should allow the interactions between DNA and elliptocines to be monitored.

This preliminary paper assesses the potential of FT Raman to yield information in this low frequency region. The Raman spectra of the nucleoside thymidine and the corresponding nucleotide salt thymidine-5-monophosphate dipotassium obtained using Nd^{3+}:YAG (1064-nm) and Ar^+ (514.5-nm) laser excitation sources were compared. The particular set of filters used in this work allowed shifts down to 85 cm^{-1} to be recorded. This preliminary work on nucleosides and nucleotides identified several features in the near infrared excited spectrum which were not real and which the authors attribute to Nd^{3+}:YAG emission lines. These lines were quite intense and appeared at 110, 85, 60 and 50 cm^{-1} shift. However, real emission lines at these frequencies would be blocked by the Nd^{3+}:YAG rejection filters, and these lines are far more likely to be computational artifacts. The FT Raman spectra from the elliptocine samples were then recorded and some assignments made, with the confidence that the bands assigned were real. The authors then went on to record FT Raman spectra from two forms of DNA. These samples originated from calf thymus and sperm herring. Espinosa *et al.* concluded that they expected a band in the spectrum of elliptocine band near 115 cm^{-1} due to intermolecular hydrogen bonding to yield information about the intercalating reaction which occurs between DNA and elliptocines. Goral and Zichy have also made a study of nucleic acids and their components [3].

A number of other papers have also appeared in the literature describing the application of FT Raman spectroscopy to the study of pharmaceuticals. These applications have been described in detail in Chapter 7. Hodges and Akhavan have published the FT Raman spectra of the alkaloids codeine, heroin and morphine [26,27] and other illicit drugs such as amphetamine sulphate. The spectra are shown in Figs 7.7 and 7.8 respectively. The results they obtained indicate that FT Raman spectroscopy has the potential to be used in a screening environment. Davies *et al.* studied a number of polymeric biomaterials and drug delivery systems [28]. Diclofenac, indomethacin and promethacin give excellent quality spectra (see Figs 7.14, 7.12 and 7.13 respectively). These drugs can be monitored down to concentrations below 5% w/w in inert polymer matrices such as sodium alginate. Neville and Shurvell report the FT Raman spectra of diazepam and four closely related 1,4-benzodiazepines [29]. These drugs have tranquillizing properties and diazepam is better known by its trade name, Valium. Spectral studies can give vital information on the relative molecular interactions of a series of these compounds. This is important because it has been shown that the unacceptable side-effects of this family

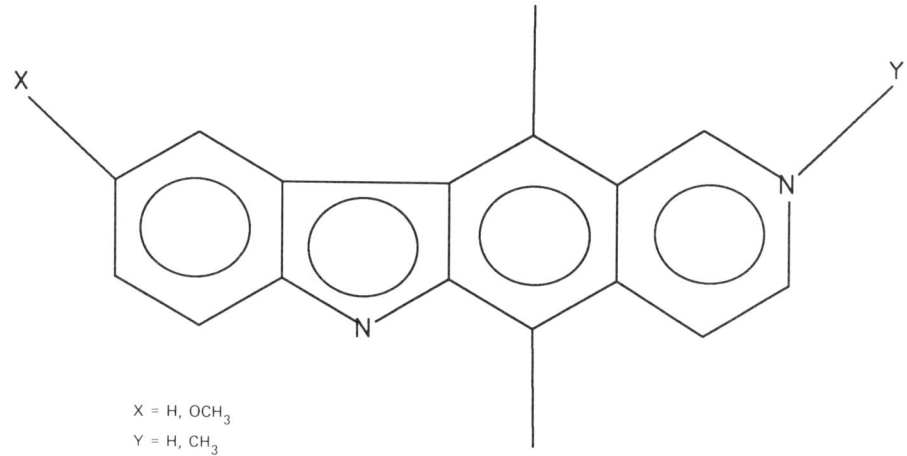

X = H, OCH$_3$
Y = H, CH$_3$

Fig. 9.11 — The general molecular structure of an elliptocine.

of drugs are mediated by different substituents. These materials are strongly coloured and tend to heat catastrophically in visible Raman studies.

9.8 FOODSTUFFS

The examination of food products is challenging to the analyst. Many 'wet chemical' tests remain the only way to characterize many products. For example, the unsaturation level in a fat-containing food is measured by titration with iodine to give an 'iodine number'. Sadeghi-Jorabchi *et al.* [30,31] have demonstrated that FT Raman spectroscopy can be used to give a quantitative measure of the level of unsaturation and also the ratio of *cis* to *trans* isomers. Conventional Raman spectra of oils and fats are highly fluorescent (see Fig. 1.4). FTIR spectra of a number of oils and fats have been recorded and the weak 1660 cm^{-1} band due to the $\nu_{C=C}$ used to determine the degree of unsaturation in the sample. This method is unreliable in the case of margarines which contain water and therefore suffer from a strong absorption at 1660 cm^{-1} due to the O–H bending vibration. FTIR spectra can yield an estimation of the concentration of the *trans* isomer, but no independent measure of the *cis* isomer content. Figure 9.12 shows the FT Raman spectra of olive oil, butter and 'Blue Band' margarine. The ratio of the band areas of the 1661-cm^{-1} mode (which is due to $\nu_{C=C}$ vibration) and the 1444 cm^{-1} mode (which is due to the CH$_2$ scissoring δ(CH$_2$) motion) correlates with the determined iodine numbers, such that a standard curve can be constructed. This procedure has been described in the Chapter 6 and can be applied to many food products in a non-destructive manner.

Figure 9.13 shows the FT Raman spectra of several different nuts which have been described by Goral and Zichy [3]. From the ratio of the 1661/1444 cm^{-1} bands the relatively healthy option is definitely the walnut!

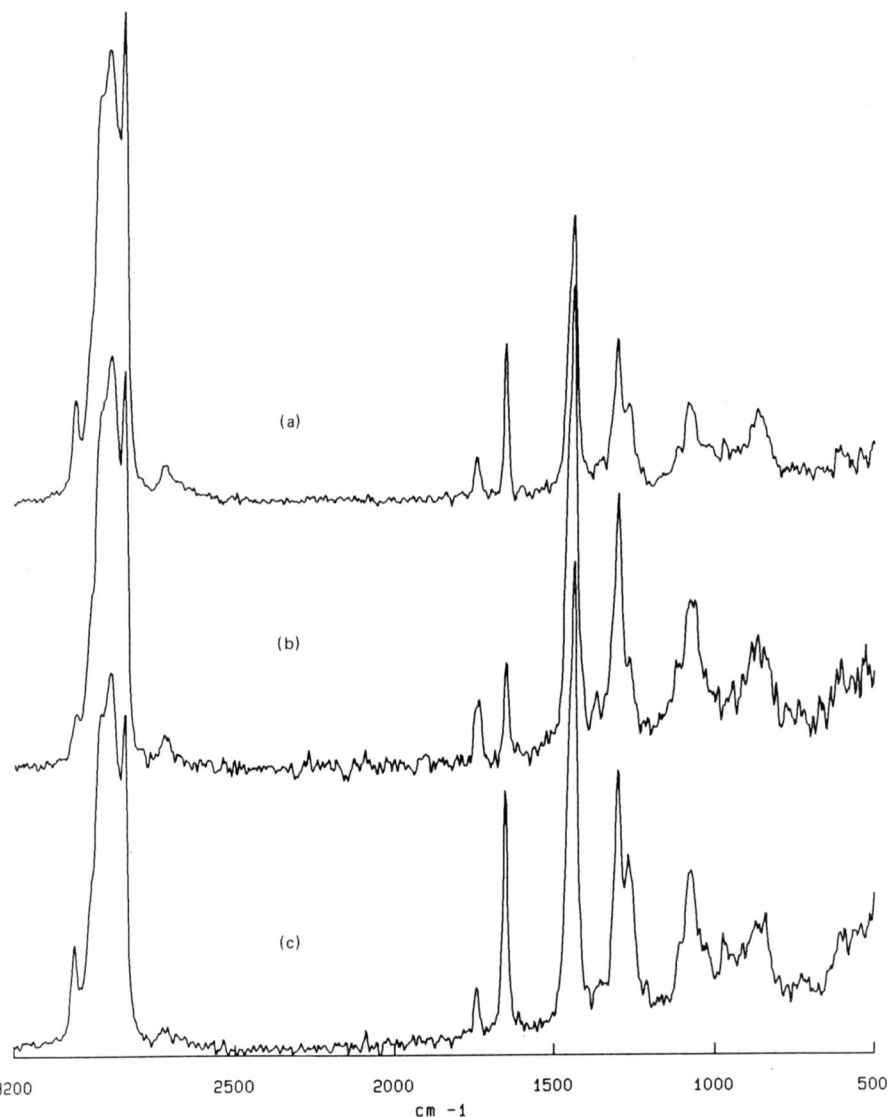

Fig. 9.12 — The FT Raman spectra of (a) olive oil, (b) butter and (c) 'Blue Band' margarine.

Derivatization of oils and fats to methyl esters followed by analysis using gas chromatography (GC) is the standard method of estimating the fatty acid composition of lipids and also determining the degree of unsaturation as well as the ratio of *cis* to *trans* isomers. Results available from this technique have been compared with those derived from FT Raman studies [31]. It has been noted from GC studies that hydrogenation, a process involved in converting an oil into a margarine, causes not

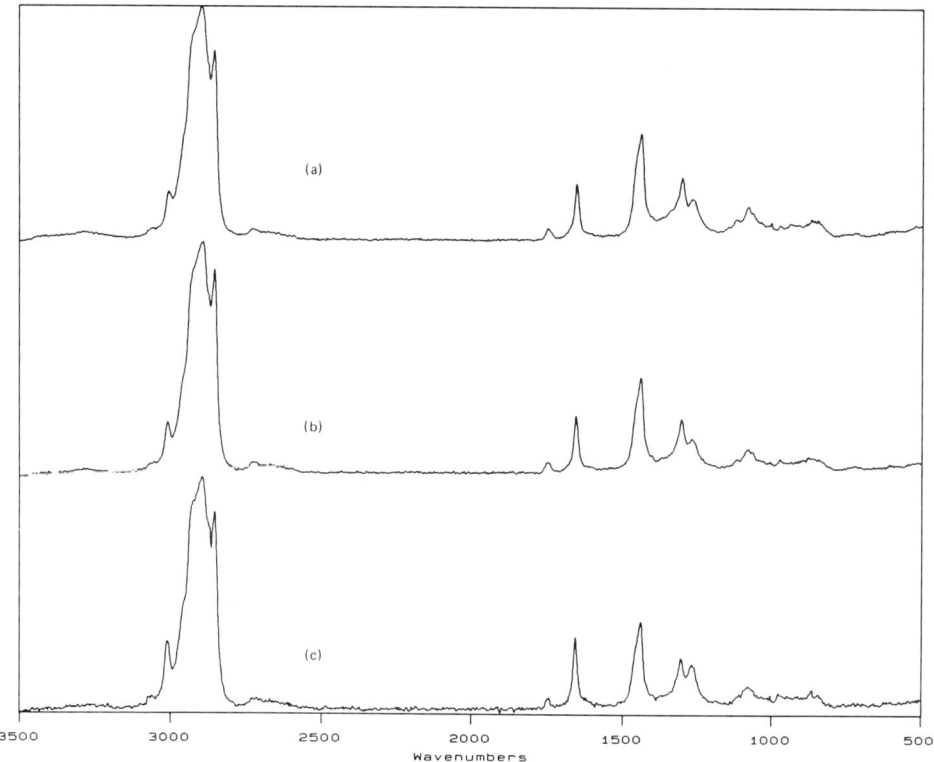

Fig. 9.13—The FT Raman spectrum (a) cashew nut, (b) brazil nut and (c) walnut. Uncorrected data.

only a decrease in the total degree of unsaturation, but also a decrease in the ratio of the *cis* polyunsaturate to *cis* monounsaturate content, with a concurrent increase in the concentration of *trans* monoenes. The results are confirmed by FT Raman studies. The band ratio $\nu_{C=C}$ (band area recorded between 1700 and 1601 cm^{-1}) to $\nu_{C=O}$ (band area between 1790 and 1713 cm^{-1}) decreases as the total level of unsaturation in the sample. The band recorded at 1265 cm^{-1} correlates with the concentration of the *cis* isomer in the sample. The decrease in the ratio of the *cis* polyunsaturate to *cis* monounsaturate is reflected in the shift in the *cis* $\nu_{C=C}$ band from 1657 to 1655 cm^{-1}. The appearance of a shoulder at 1666 cm^{-1} indicates the increase in the *trans* isomer content on hydrogenation. Problems are encountered, however, in reliably measuring the *trans* content at low concentrations, i.e. below 20%. The correlation between the GC results and those obtained using the FT Raman method was excellent for the standard lipid samples and mixtures; however, it is suggested that real systems will benefit from the application of the sophisticated multicomponent analysis routines, such as CIRCOM, which are now becoming available to assist particularly with the estimation of the isomer content. However, the advantages of the technique in an analytical environment are enormous. The speed and ease of the acquisition of data coupled with the lack of sample preparation make the technique very attractive.

Goral and Zichy have studied a number of carbohydrates [3]. Carbohydrates are an important group of compounds which include mono-, di- and polysaccharides, starch, glycogen and cellulose. Raman analysis of these compounds, and in particular naturally occurring saccharides, has traditionally been limited because of the problems associated with a weak signal and sample fluorescence. Conventional Raman spectra of crystalline monosaccharides, such as D-glucose, have been obtained using 514.5-nm excitation [32,33]. There are discrepancies between the intensity ratios of two D-glucose spectra shown in the literature. These may arise from differences in sample purity and instrument sensitivity, and FT Raman studies of this area are continuing. Figure 9.14 shows the FT Raman spectra of α- and β-D-

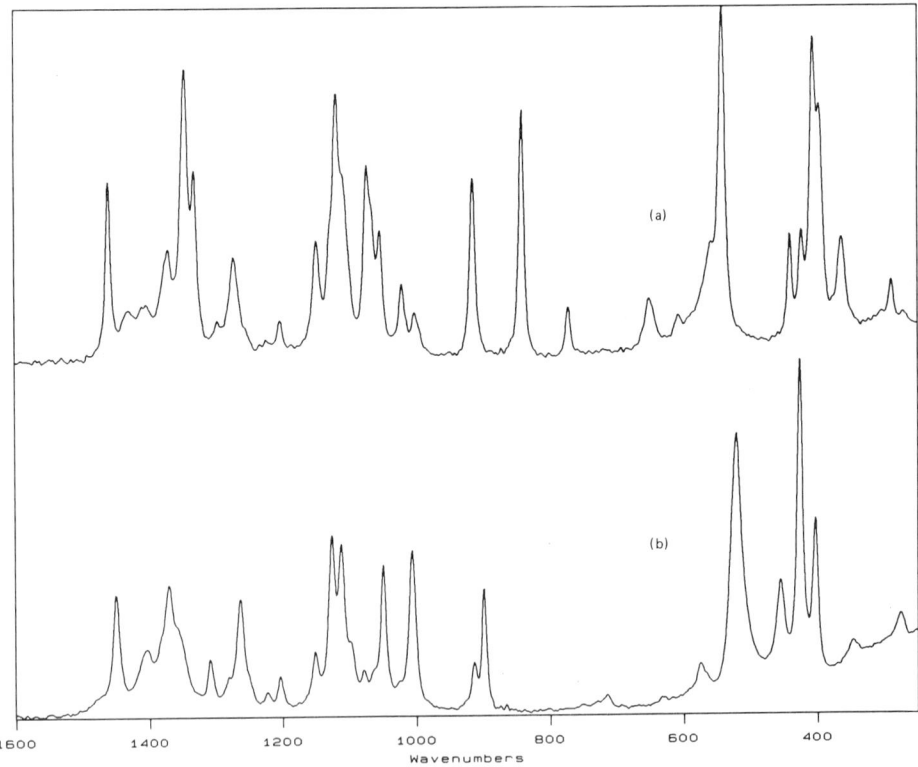

Fig. 9.14 — The FT Raman of (a) α- and (b) β-D-glucose. Uncorrected data.

glucose. Monosaccharides can exist in these two possible crystalline forms, which are shown in Fig. 9.15.

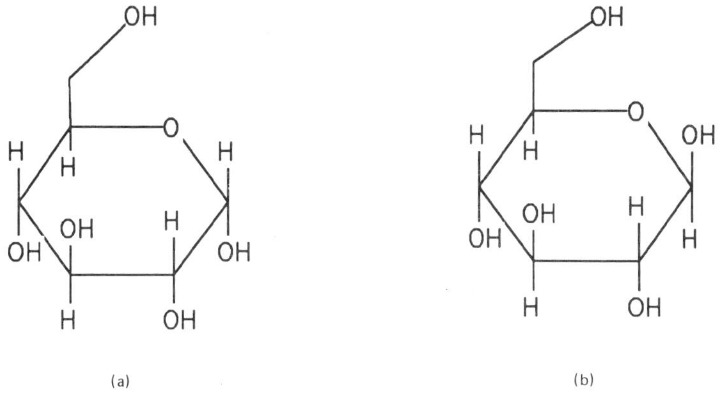

Fig. 9.15 — The α and β forms of crystalline D-glucose.

The FT Raman spectra of several disaccharides have also been recorded, including those of sucrose, maltose and lactose. The FT Raman spectrum of sucrose is shown in Fig. 9.16(a). Sucrose is an α-D-glucopyranosyl-β-D-fructofuranose and its spectrum contains bands characteristic of the anomers of both monosaccharide components — α-glucose and β-fructose at 847 and 868 cm^{-1} respectively. The polysaccharides cellulose, starch, agarose, dextrin, dextran, Sephadex and Ficoll have been studied using FT Raman spectroscopy [3]. Cellulose has been studied as a dried pulp and also in woods, papers and cottons as described above. Ficoll is a poly(sucrose) and many common features can be identified in the spectra of Ficoll and sucrose (see Fig. 9.16).

Comparing the spectrum of α-D-glucose with that of β-D-glucose in Fig. 9.14, it is most surprising that the spectra contain so many differences. Intuitively one would expect the spectra to be very similar because the two compounds have almost identical structures. However, large spectral changes as a result of minor structural changes are a general observation for saccharides. Thus, FT Raman spectroscopy appears to have great potential for use as a routine analytical technique for studying these compounds. However, on the whole the samples studied to date have been highly crystalline pure materials. Analysis of saccharides in real food products is the subject of further work. Some carbohydrate-containing foods have, however, already been studied, such as rice, semolina and tapioca. These materials give essentially the same spectrum, that of starch.

9.9 CONCLUSION

The application of FT Raman spectroscopy to the investigation of biological systems will enable many previously inaccessible specimens to be studied. The relative freedom from fluorescence and the ability to study samples without inducing unwanted photochemistry or laser-induced degradation opens up many new areas of

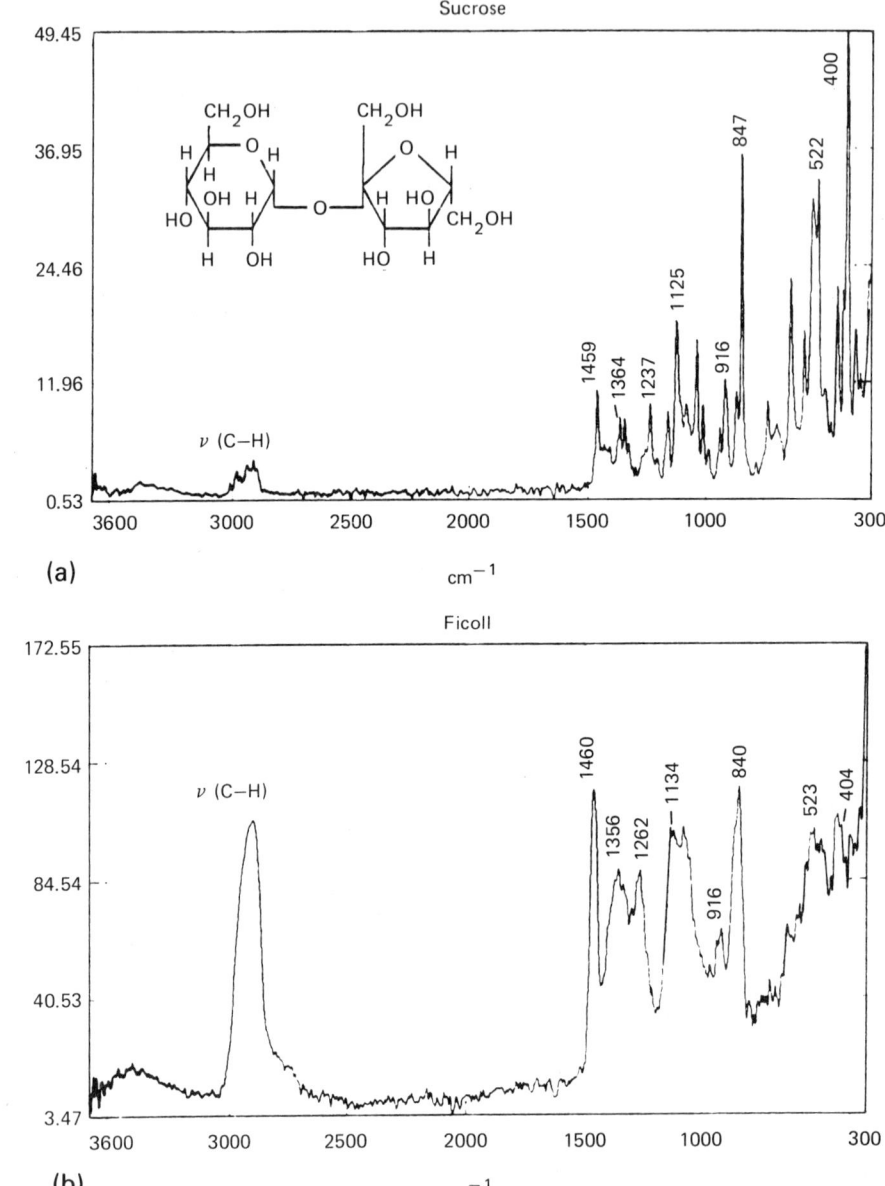

Fig. 9.16 — The FT Raman spectrum of (a) sucrose and (b) Ficoll, a polysucrose. Reproduced with permission from *Spectrochimica Acta*, **46A**, J. Góral and V. Zichy, copyright 1990 Pergamon Press PLC.

research. Improvements in sensitivity, which will inevitably occur, will further enhance the usefulness of the technique, since many of the biological species are very poor Raman scatterers. However, it is clear that the 1064-nm source is far from ideal for investigating dilute aqueous solutions, which are commonly of interest to the biologist and biochemist. These systems await the development of alternative excitation sources.

REFERENCES

[1] J. L. Koenig, Chapter 3, Raman Spectroscopy, in *Introduction to the Spectroscopy of Biological Polymers*, Ed. D. W. Jones, Academic Press, London, 1976.
[2] S. K. Freeman, *Applications of Laser Raman Spectroscopy*, Wiley-Interscience, New York, USA, 1974.
[3] J. Goral and V. Zichy, *Spectrochim. Acta*, 1990, **46A**, 253.
[4] I. W. Levin and E. N. Lewis, *Anal. Chem.*, 1990, **62**, 1101.
[5] I. W. Levin and E. N. Lewis, *SPIE (Fourier Transform Spectroscopy)*, 1989, **1145**, 99.
[6] E. N. Lewis, B. J. Litman and I. W. Levin, *SPIE (Fourier Transform Spectroscopy)*, 1989, **1145**, 602.
[7] I. W. Levin, M. T. Devlin and E. N. Lewis, *Proceedings of the Twelfth International Conference on Raman Spectroscopy, Columbia, South Carolina*, Ed. J. R. Durig and J. F. Sullivan, Wiley, New York, 1990, p872.
[8] S. Nie, K. L. Bergbauer, J. F. R. Kuck Jr. and N-T. Yu, *Exp. Eye Research*, 1990, **51**, 619.
[9] S. Nie, K. L. Bergbauer, J. F. R. Kuck Jr. and N-T. Yu, *Spectroscopy*, 1990, **5**, 24.
[10] R. P. Rava, J. J. Baraga and M. S. Feld, *Spectrochim. Acta*, 1991, **47A**, 509.
[11] D. Gani, P. J. Hendra, W. F. Maddams, C. Passingham, I. A. M. Royaud, H. A. Willis and V. Zichy, *Analyst*, 1990, **115**, 1313.
[12] P. J. Hendra, C. Passingham and I. A. M. Royaud, *Proceedings of the Twelfth International Conference on Raman Spectroscopy, Columbia, South Carolina*, Ed. J. R. Durig and J. F. Sullivan, Wiley, New York, 1990, p678.
[13] V. Hallmark and J. F. Rabolt, *Macromol.*, 1989, **22**, 500.
[14] S. Nie, L. G. Marzilli and N-T. Yu, *Proceedings of the Twelfth International Conference on Raman Spectroscopy, Columbia, South Carolina*, Ed. J. R. Durig and J. F. Sullivan, Wiley, New York, 1990, p878.
[15] S. Nie, L. G. Marzilli and N-T. Yu, *J. Am. Chem. Soc.*, 1989, **111**, 9256.
[16] S. Nie, P. A. Marzilli, L. G. Marzilli and N-T. Yu, *J. Chem. Soc. Chem. Commun.*, 1990, p770.
[17] S. Nie, P. A. Marzilli, L. G. Marzilli and N-T. Yu, *J. Am. Chem. Soc.*, 1990, **112**, 6084.
[18] C. K. Johnson and R. Rubinovitz, *Applied Spectrosc.*, 1990, **44**, 1103.
[19] T. A. Mattioli, A. Hoffmann, M. Lutz and B. Schrader, *C. R. Acad. Sci. Paris*, t.310, Serie III, 1990, p441.
[20] T. Noguchi, Y. Furukawa and M. Tasumi, *Spectrochim. Acta*, (1991) in press.
[21] R. C. Kenton and R. L. Rubinovitz, *Applied Spectrosc.*, 1990, **44**, 1377.
[22] P. M. Fredericks and C. de Bakker, *Proceedings of the Eighth Int. Conf. Fourier Transform Spectroscopy*, Lubeck-Travemunde, SPIE, 1991, in preparation.
[23] P. A. Evans, *Spectrochim. Acta*, 1991, **47A**, in press.
[24] R. H. Atalla and U. P. Agarwal, *J. Raman Spectrosc.*, 1986, **17**, 229.
[25] J. M. Espinosa, D. H. Christensen, G. O. Sorensen, O. Faurskov Nielsen, J. Aubard, M. Schwaller and G. Dodin, *Spectrochim. Acta*, 1991, in press.
[26] C. M. Hodges and J. Akhavan, *Spectrochim. Acta*, 1990, **46A**, 303.
[27] C. M. Hodges, P. J. Hendra, H. A. Willis and T. Farley, *J. Raman Spectrosc.*, 1989, **20**, 745.
[28] M. C. Davies, J. S. Binns, C. D. Melia and D. Bourgeois, *Spectrochim. Acta*, 1990, **46A**, 277.
[29] G. A. Neville and H. F. Shurvell, *J. Raman Spectrosc.*, 1990, **21**, 9.
[30] H. Sadeghi-Jorabchi, P. J. Hendra, R. H. Wilson and P. S. Belton, *JAOCS*, 1990, **67**, 483.
[31] H. Sadeghi-Jorabchi, R. H. Wilson, P. S. Belton, J. D. Edwards-Webb and D. T. Coxon, *Spectrochim. Acta*, 1991, **47A**, in press.
[32] C. Y. She, N. D. Dinh and A. T. Tu, *Biochim. Biophys. Acta*, 1974, **372**, 345.
[33] P. D. Vasko, J. Blackwell and J. L. Koenig, *Carbohydr. Res.*, 1971, **19**, 297.

10

The application of Fourier transform Raman spectroscopy to the study of surfaces

10.1 RAMAN SPECTROSCOPY OF SURFACES

All heterogeneous chemical processes take place at an interface, by definition. As such reactions form the basis of our knowledge of electrochemistry, corrosion, heterogeneous catalysis, semiconductor manufacture, etc. the study of surfaces has become of considerable significance in chemistry. As a result, many techniques have been developed to study surfaces [1]. The principal goal is always to find a method which provides information primarily from the surface and not the bulk. Vibrational spectroscopy has featured strongly in this field, notably electron energy loss spectroscopy (HREELS), various forms of infrared spectroscopy (transmission, reflection, RAIRS) and Raman spectroscopy. The dominating drawback with the majority of surface techniques is that they can only function under ultra-high-vacuum conditions (e.g. HREELS, Auger spectroscopy, low energy electron diffraction). As such, they are powerful tools for fundamental investigation of single crystal metal or oxide surfaces, but cannot study electrodes immersed in electrolyte or catalytic reactors. The only techniques which are capable of working at realistic temperatures and pressures are infrared and Raman spectroscopy.

Raman spectroscopy has principally been applied to two main groups of surfaces–electrodes and heterogeneous catalysts. It can provide unique information from electrochemical interfaces which present considerable experimental difficulties in infrared spectroscopy due to water absorption [2,3]. Electrochemical Raman experiments utilize an effect known as surface-enhanced Raman scattering (SERS), which will be discussed at length. Raman experiments on heterogeneous catalysts have the advantage that they are not complicated by bulk catalyst absorptions. Both of these fields have started to be studied by FT Raman spectroscopy and the results so far and future prospects will be reviewed in this chapter.

10.2 SURFACE-ENHANCED RAMAN SPECTROSCOPY

The SERS effect was first observed by Fleischmann *et al.* in 1973 [4,5]. They studied the adsorption of calomel (Hg_2Cl_2) on a roughened mercury electrode. Considering that the spectra obtained were only due to species directly adsorbed to the electrode, they were extraordinarily intense. By changing the potential it was possible to reproducibly adsorb and desorb calomel on the surface. The work was extended to pyridine adsorbed on a roughened silver electrode and even higher quality spectra were obtained. The spectra varied as a function of applied potential, with various pyridine species being adsorbed and desorbed at different potentials. Later researchers, notably Jeanmaire and Van Duyne [6], and Albrecht and Creighton [7], proposed that the intensity of the Raman spectra obtained could not simply be due to the increased surface area of the electrode caused by roughening or from a relatively thick layer of complexes on the surface. They suggested that there were other physical effects coming into play which massively enhanced the Raman signal. Over time this phenomenon became known as surface-enhanced Raman spectroscopy. Enhancements of the order of 10^6 in Raman intensity have been observed, especially at copper, silver and gold electrodes.

Although the SERS effect was first observed on roughened electrode surfaces, similar enhancements have been observed from other substrates. These include metal colloids, filter paper impregnated with these colloids, metal island films and metal particles on oxide supports. All of these substrates have one vital factor in common — they all have atomic scale roughness. In recent years, SERS publications have started to shift in emphasis. Whereas in the past most of the literature has centred on trying to understand the mechanism of the enhancement, now analytical chemists have been exploiting SERS as a new analytical technique and have started to expand its applications. SERS can provide unique information from interfaces and is therefore very applicable to studying corrosion, electrode processes (e.g. batteries), catalysts, etc. It also provides a method of studying very dilute solutions if the solute can be adsorbed to a metal surface.

As covered in Chapter 2, Raman scattering relies on the fact that the electron clouds of molecules can be polarized slightly when placed in an electric field. A dipole is induced in the molecule according to the following equation

$$\mu = \alpha E \qquad (10.1)$$

The intensity of the Raman scatter is related to μ^2. Thus to obtain an increase in Raman scatter there must be an increase in the polarizability, α, or the electric field, E. It is now generally agreed that the majority of the enhancement comes from an increase in the electric field (about 10^4), and the rest from surface roughness (about 10^2). Various models have been proposed for the enhancement in E. Comprehensive reviews can be found in refs. [8–11]. Perhaps the most widely accepted is the so-called electromagnetic or em model. Here it is thought that the incoming photons used to produce Raman scattering also excite surface 'plasmons'. These are oscillations in the density of the 'electron gas' in small metal particles or in the irregularities at surfaces. The plasmon resonance frequency is related to the size and shape of the particles, and the electrical and optical properties of the materials from which they are made. Once the plasmon is excited, it cannot re-radiate easily and the net result is a greatly enhanced electric field surrounding the metal particles. Thus Raman scattering from molecules in close proximity to this particle is greatly enhanced. Not only is the electric field of the incident exciting radiation enhanced, but that of the Raman-scattered radiation as well. This theory has also been successful in modelling phenomena related to SERS such as second harmonic generation (SHG) at surfaces.

The spectrum of a material obtained by SERS is not the same as its normal unenhanced Raman spectrum (see Fig. 10.1). Bands which are observed in the unenhanced spectrum may not be observed in the SERS spectrum. Conversely, the SERS spectrum may show bands which are absent from the normal Raman spectrum. It is clear that not all the Raman bands are enhanced, and in this sense SERS is similar to resonance Raman scattering. Also, the SERS spectrum of a material may vary with the SERS substrate to which it is adsorbed. As can be seen from Fig. 10.1, the SERS spectra from pyridine over a silver film and a silver colloid are different, and both differ in turn from liquid pyridine itself. Such differences may convey considerable information, but understanding the exact nature of the spectra is far from simple. For instance, if a molecule is adsorbed to the surface in such a way as to have high polarizability normal to the surface, and thus in the direction of the magnified electric field, its vibrations may be preferentially enhanced. Thus SERS spectra contain orientational information, although this can be hard to interpret, particularly as roughened SERS active surfaces do not have a regular morphology. It is also possible to obtain very intense spectra which show the resonance Raman effect in addition to SERS.

The various models proposed to explain the enhancement make predictions about the magnitude of the enhancement as a function of the wavelength of the incident exciting radiation (see Fig. 10.2). Some theories predict that the SERS effect should not be observed in the near infrared. However, there is now a growing collection of papers clearly demonstrating strong SERS enhancement in the near infrared on silver, copper, gold and platinum substrates. So far, studies have centred on electrodes and metal colloids and the more novel SERS substrates have yet to be investigated. The remainder of this discussion will centre on these two systems where success has already been demonstrated.

10.3 NEAR INFRARED FT RAMAN SERS SPECTRA OF ELECTROCHEMICAL SYSTEMS

The original SERS experiments carried out in the 1970s used electrode surfaces as the SERS substrate. To observe the enhancement, the electrode has to be specially pretreated to produce an atomic-scale roughness. This is done by first polishing the electrode with successively finer grades of alumina, then placing the electrode in the electrolyte and sweeping the electrode potential through several oxidation–reduction cycles (ORCs). These first oxidize the metal atoms at the surface to produce ions in the electrode double layer, and then reduce them back to metal atoms, randomly depositing them on the electrode surface (see Fig. 10.3). This produces an electrode surface of increased surface area with the topography required for the SERS effect to

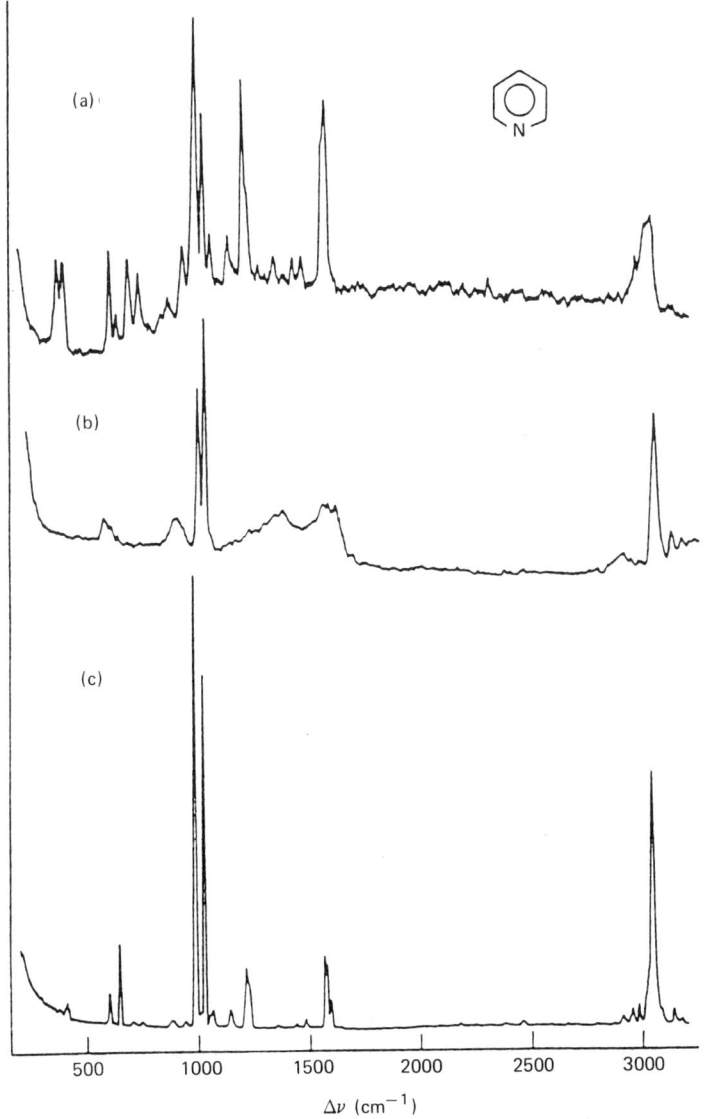

Fig. 10.1 — The SERS spectra of pyridine on (a) cold-deposited silver film and (b) silver colloid compared withh the spectrum of (c) liquid pyridine. Reproduced with thanks from ref. [10].

function. Great care must be taken not to contaminate the electrode or electrolyte as only a minute amount of contaminant will cause problems if it is absorbed on the electrode surface. Consequently, triply distilled water and Analar or spectroscopic-grade reagents must be used in scrupulously clean cells.

There are several good reasons for attempting to excite SERS spectra with a near infrared laser. Theories describing the mechanism of SERS enhancement are not yet

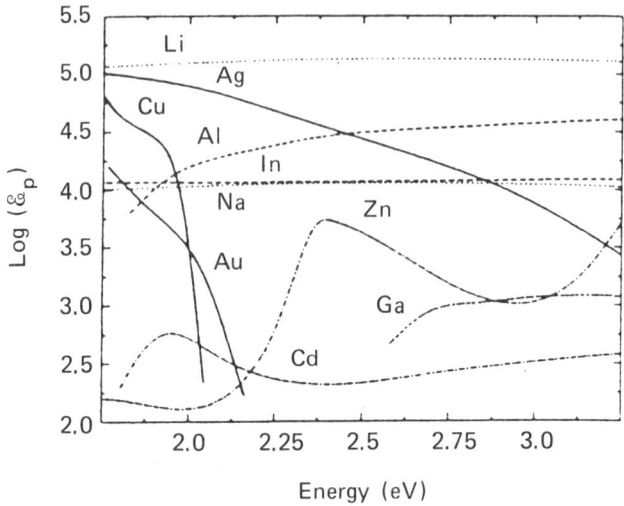

Fig. 10.2 — Predicted SERS enhancement for particles optimized for size and shape for different metals as a function of excitation frequency. Reproduced with permission from E. J. Zeman and G. C. Schatz, *J. Phys. Chem.*, **91**, 634. Copyright (1987) American Chemical Society.

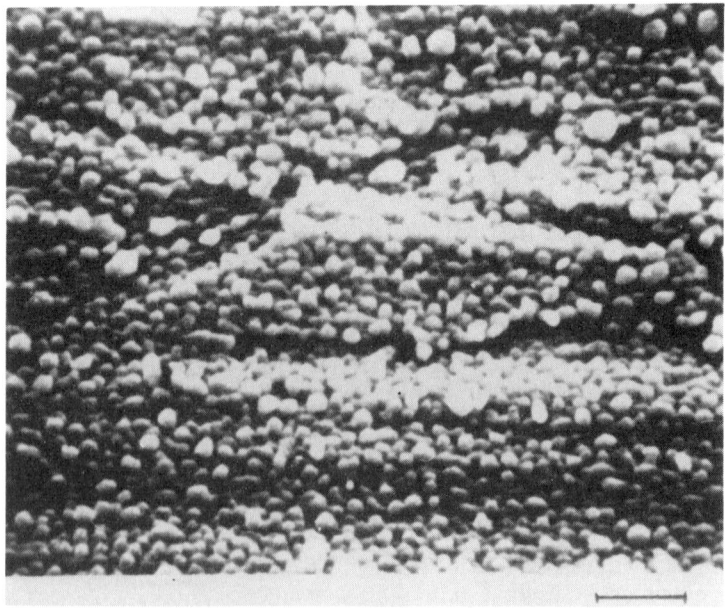

Fig. 10.3 — An electron micrograph showing the structure of an electrochemically roughened silver electrode, courtesy of Dr J. A. Creighton.

fully comprehensive. Hence, observation of SERS spectra and measurement of SERS enhancement factors in the near infrared provide useful experimental data with which to compare theoretical predictions. Also, in electrode studies using visible lasers, species adsorbed to the electrode have been shown to be both generated and removed by laser radiation [13]. As the photon energy in the near infrared is reduced, photochemical changes are less likely to be induced by the laser. SERS spectra frequently appear over an intense background which varies in intensity with the number of ORCs used in preparing the SERS surface. Again by moving to near infrared excitation these backgrounds might well be reduced, thus improving the signal-to-noise ratio in the spectra. A typical cell used for electrochemical SERS studies is shown in Fig. 10.4. A potential can be applied between the working

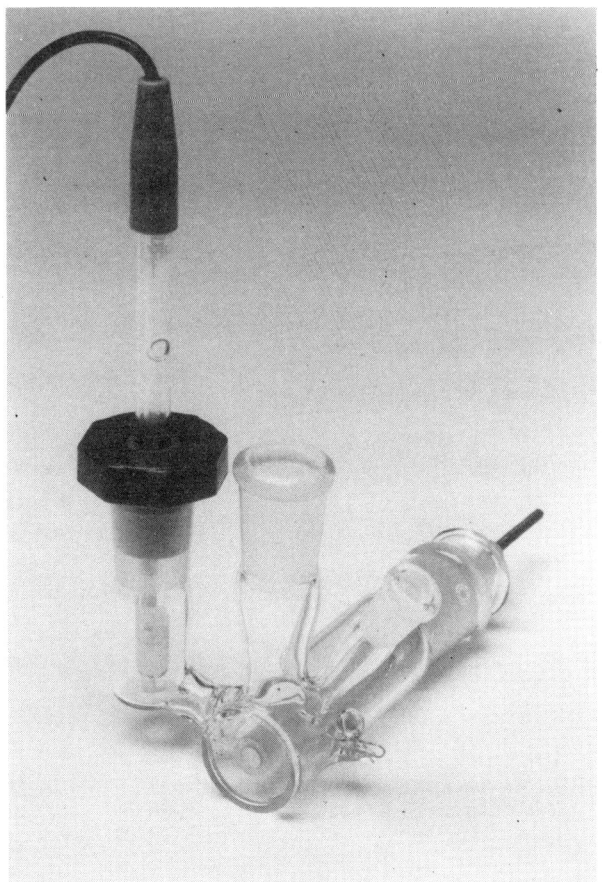

Fig. 10.4 — Typical electrochemical cell used for FT SERS experiments.

electrode and the secondary electrode, and this potential is measured with respect to a reference electrode (usually the saturated calomel electrode (SCE)). Spectra are recorded of adsorbates on the working electrode.

The first near infrared FT Raman spectra obtained showing the SERS effect were reported by three independent groups in the summer and autumn of 1988 and all used electrodes as SERS substrates. Chase and Parkinson [14] studied pyridine on silver and gold while Angel et al. [15] studied pyridine on copper and gold, and Fleischmann, Hendra et al. [16] studied pyridine on silver, copper and gold. All three groups used FT Raman systems with an Nd^{3+}:YAG laser, although papers on different systems have subsequently appeared which use a 785 nm diode laser and conventional monochromator and photomultiplier detection [17]. The SERS spectrum of pyridine adsorbed to electrode surfaces has been extensively investigated with visible excitation and was the obvious place to start a near infrared study.

Figure 10.5 shows a near infrared SERS spectrum of pyridine on silver. The first

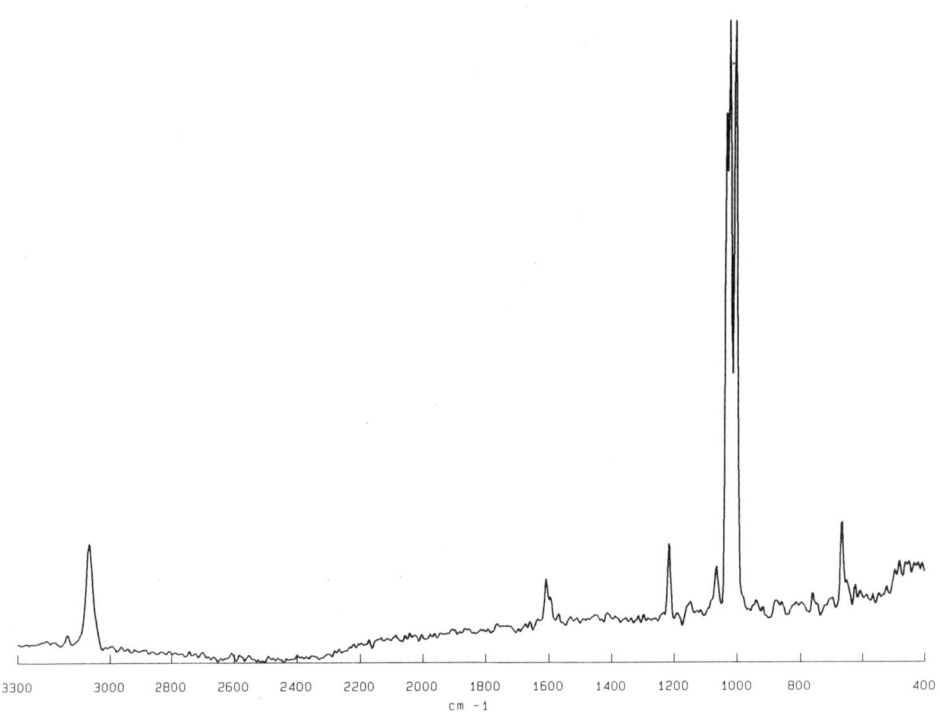

Fig. 10.5 — SERS spectrum of pyridine on silver electrode (pH 7.6, -0.2 V SCE) obtained using an Nd^{3+}:YAG laser and an FT Raman spectrometer. Even with near infrared excitation the spectrum still appears on a background. Courtesy of D. Sockalingum.

thing to note about the spectral bands is that they are very intense and excellent signal-to-noise ratios are attainable. It is clear that the SERS effect operates very effectively with 1064-nm excitation, with an estimated enhancement factor of 10^6 [15]. Pyridine has C_{2v} symmetry and thus all its vibrations are Raman active. The strongest bands in the spectrum are centred around 1000 cm^{-1} and are due to ring breathing motions of the pyridine molecule. These bands are so strong that in some cases they can exceed the dynamic range of some spectrometers. SERS spectra have been obtained from electrolyte pyridine concentrations of 0.05 M down to 8 μM. Also present in the spectra are deformation modes, and Fleischmann et al. have shown the CH stretching region.

The Raman bands appear over a broad diffuse background, and this has been subtracted from some of the published spectra. The exact nature of this background is as yet uncertain. Fleischmann et al. [16] have produced identical backgrounds from electrodes which have been polished and then roughened in KCl solution alone, thus showing that the background is not due to the pyridine adsorbate but is dependent on the nature of the metal. Angel et al. [15] report an intense background from a pyridine on copper electrode before ORC roughening, and observe that this background can be dramatically reduced after one or two ORCs. However, the reverse of this is observed for pyridine on gold where the intensity of both the SERS spectra and the background increase as more ORCs are performed. As well as producing Raman enhancements, rough metal surfaces have also been shown to alter the fluorescence levels recorded from different molecules. Similar effects to those described by Angel are described in the literature. For instance, it has been observed that fluorescence emission from very fluorescent molecules (i.e. ones which have a high fluorescence quantum efficiency) such as Rhodamine 6G can be reduced by a factor of ten on absorption to a roughened surface [11,18]. Counter to this, fluorescence from basic fuchsin has been seen to increase on adsorption to a rough silver surface, and the efficiency of this process depends on the distance of the molecules from the surface [19].

Many of the FT Raman spectra published to date have not been corrected for the instrumental response function. This is particularly evident in FT near infrared SERS spectra, whose backgrounds show oscillations which are principally due to the Rayleigh light rejection filter characteristics. The lack of corrected spectra means that care must be taken when comparing the spectra recorded by different laboratories. The use of different filters and detectors produce spurious changes in relative band intensity, and the spectral range over which the instrument is sensitive. For instance, Chase and Parkinson report that the 1215 cm^{-1} band of the pyridine/silver system is much weaker than in the visible, and the ratio of the 1007/1037 cm^{-1} bands is different from visible experiments. It is not clear whether these differences arise from fundamental changes in the enhancement or simply from instrumental response.

The spectra obtained from adsorbed pyridine vary with the electrode metal, applied potential and pH. Fleischmann, Sockalingum and Musiani have studied the effect of all three of these variables [20]. Figure 10.6 shows the spectra of pyridine adsorbed to different metal electrodes. These in turn differ from the pyridine/silver spectrum shown in Fig. 10.5. Lewis-coordinated pyridine is not found on copper and gold electrodes, as it is on silver, at any pH, as shown by the absence of a band at 1025 cm^{-1}. Also, Fig. 10.6 shows that it is possible to record weak but observable SERS spectra from platinum in the near infrared. Figures 10.7 and 10.8 show the potential dependence of the pyridine/silver system at constant pH, and the pH dependence at constant potential respectively. It is thought that there are three

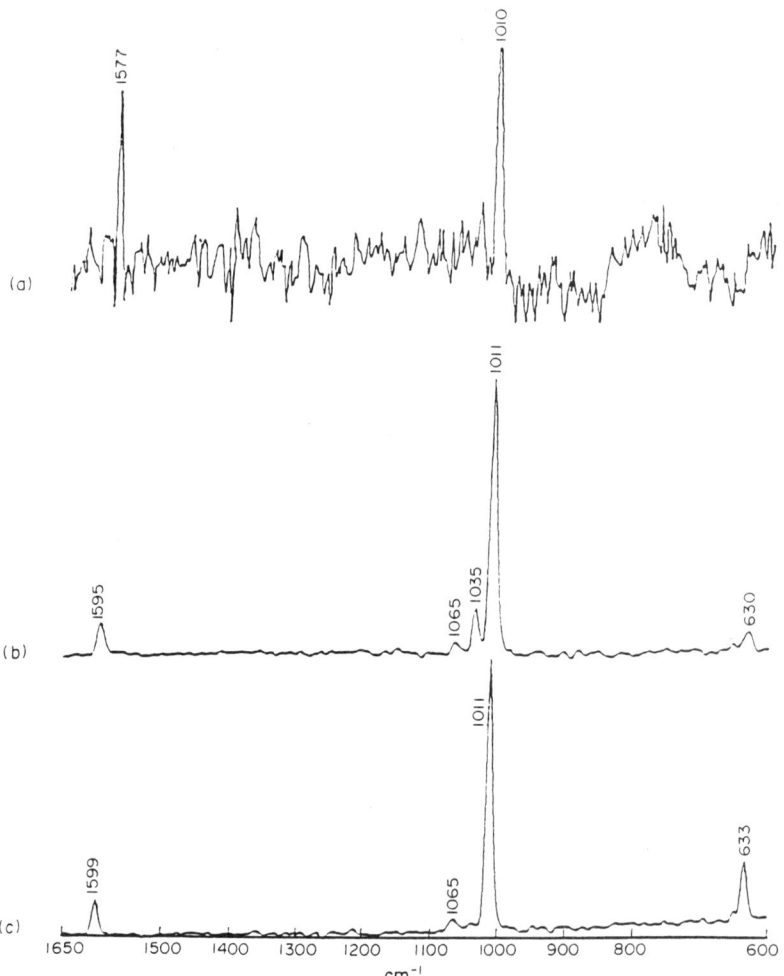

Fig. 10.6 — FT near infrared SERS spectra of 0.05 M pyridine in 0.1 M KCl solution adsorbed to (a) platinum, (b) gold and (c) copper electrodes. Reproduced with permission from *Spectrochimica Acta*, **46A**, M. Fleischmann *et al.*, copyright 1990, Pergamon Press PLC.

pyridine species (free pyridine, pyridinium cations and Lewis-coordinated pyridine) in two orientations which are being selectively adsorbed and desorbed on the electrode surface. In addition, there may be a range of different sites on the electrode surface. Thus, the spectra obtained are a complicated mixture of these species, and a detailed analysis is awaited.

One factor which is definitely different between visible SERS spectra and those in the near infrared is the effect of absorption of Raman light by water. Water is completely transparent at visible wavelengths, but at about 7400 cm^{-1} absolute in the near infrared the first overtone of the OH stretching vibration causes it to absorb

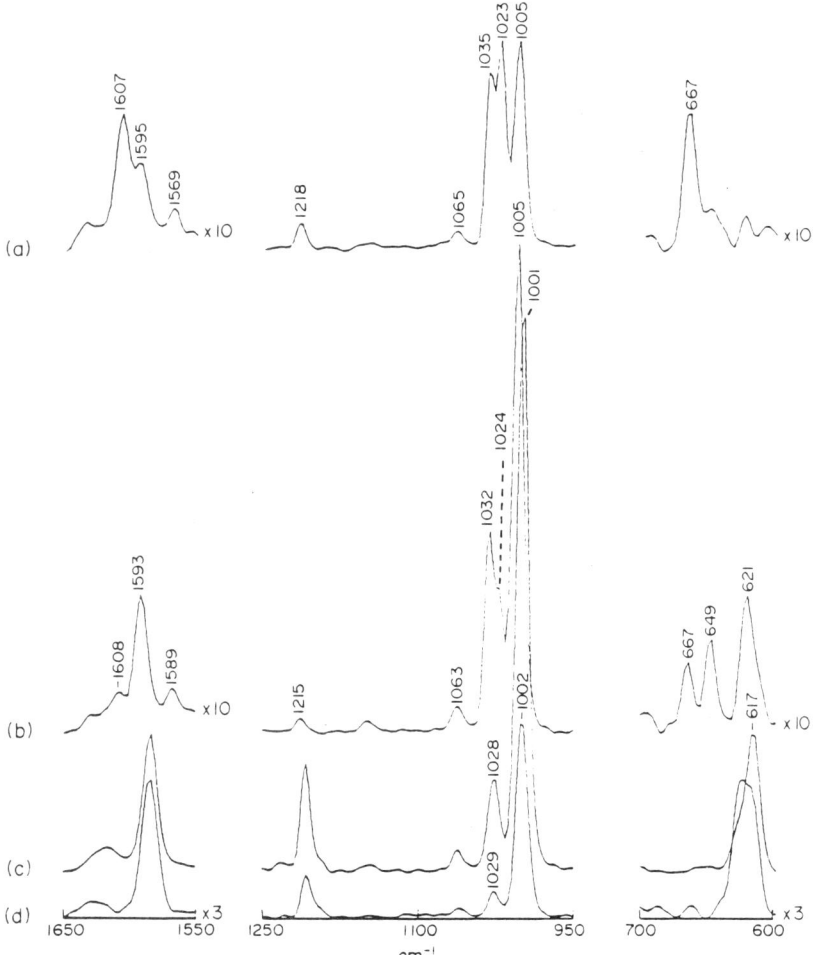

Fig. 10.7 — The effect of changing potential on the silver/pyridine system at a constant pH of 7.8. (a) −0.2 V/SCE, (b) −0.6 V/SCE, (c) −1.2 V/SCE, (d) −1.4 V/SCE. Reproduced with permission from *Spectrochimica Acta*, **46A**, M. Fleischmann *et al.*, copyright 1990, Pergamon Press PLC.

strongly. This corresponds to 2000 cm^{-1} Stokes shift away from the Nd^{3+}:YAG laser line. The near infrared absorption spectrum of water is shown in Fig. 6.13. Excess bulk aqueous electrolyte between the electrode and the spectrometer collection optics will cause the spectrum to be severely attenuated, particularly bands at greater than 2000 cm^{-1} shift. Thus, to observe the CH stretching vibration unattenuated it is necessary for the electrode to be hard up against the glass front of the cell. In this case the water absorption is minimal and in Fig. 10.5 it only manifests as a weak indentation on the background around 2500 cm^{-1} shift. Water absorption is not a drawback of the FT technique itself but is a result of using an Nd^{3+}:YAG laser. It could be alleviated by shifting to a laser line around 900 nm, and changing the filters and detector. Alternatively, D$_2$O can be used as a solvent.

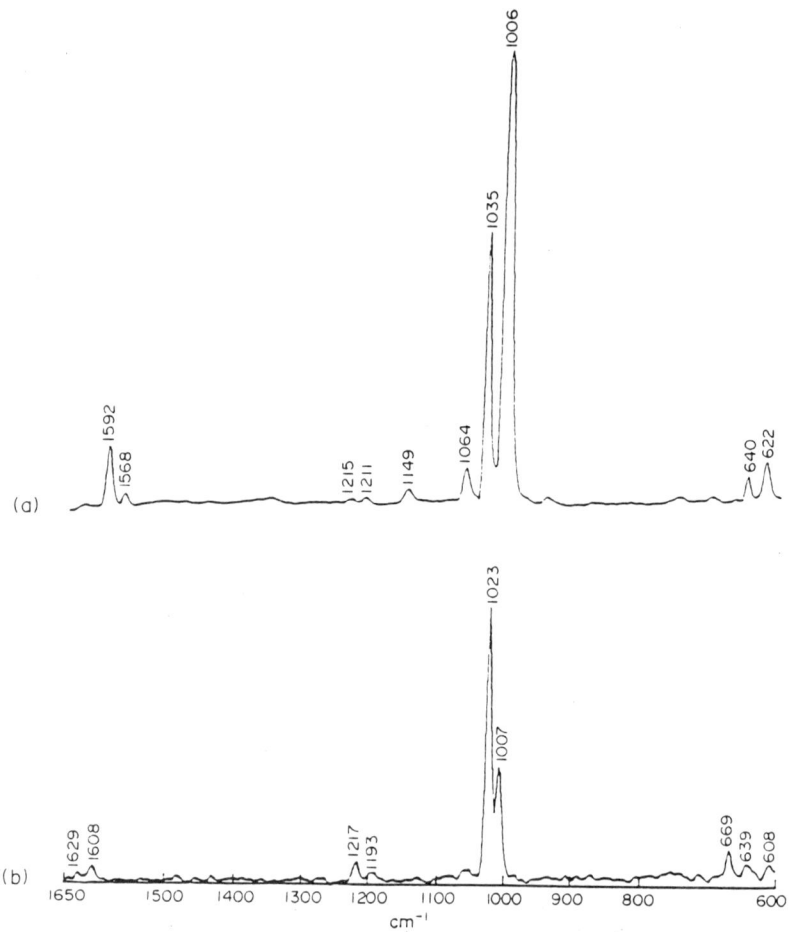

Fig. 10.8 — The effect of changing pH on the pyridine/silver system at a constant potential of −0.6 V/SCE. (a) pH 10.6, (b) pH 3.4. Reproduced with permission from *Spectrochimica Acta*, **46A**, M. Fleischmann *et al.*, copyright 1990, Pergamon Press PLC.

FT near infrared SERS has also been used to study the absorption of ferro- and ferricyanide ion to the gold electrode in lithium and caesium ion containing electrolytes [20]. By changing the electrode potential it is possible to change the oxidation state of the ion. The spectra of lithium and caesium electrolytes are significantly different, implying that the cations must be co-adsorbed with the ferro/ferricyanide anions, perhaps in pairs or clusters. This can be seen in Fig. 10.9. SERS has long been viewed as a useful tool for studying corrosion of metal surfaces, a system of great practical importance. The absorption of benzotriazole (BTA) and 2-aminopyrimidine (2-AMP) to the copper electrode has been studied with FT near infrared SERS in chloride containing electrolytes. Figure 10.10 shows SERS spectra of both pure inhibitor (A and B) and mixtures of the two (C and D). Despite the fact that the concentration of 2-AMP is much greater than that of BTA, the spectra are dominated by bands due to the latter showing that BTA is preferentially absorbed.

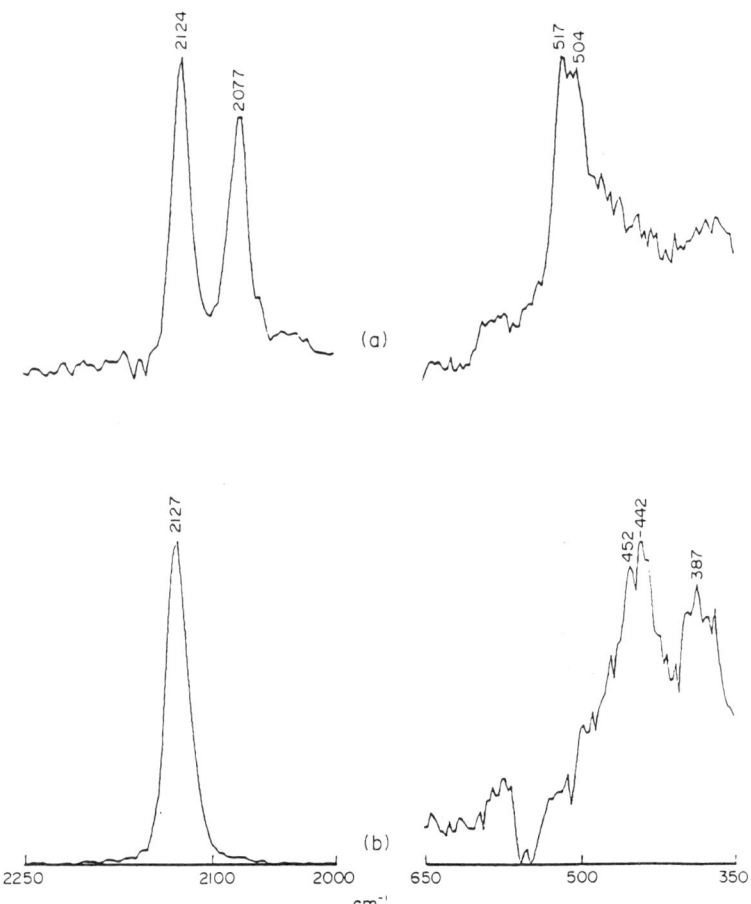

Fig. 10.9 — FT near infrared SERS spectra of ferro/ferricyanide ions (0.05 M) adsorbed to a gold electrode with differing supporting electrolytes. (a) 0.1 M CsCl, (b) 0.1 M LiCl. Reproduced with permission from *Spectrochimica Acta*, **46A**, M. Fleischmann *et al.*, copyright 1990, Pergamon Press PLC.

Also, in all the spectra there is a band at 290 cm^{-1} which it is thought arises from the Cu—Cl stretching frequency. The chloride ion is normally considered as a species responsible for corrosion, but the FT Raman spectra show that it is clearly co-adsorbed with the inhibitors. When the cell was left in open circuit for 70 minutes, SERS spectra were still obtained with only slightly reduced intensity showing that both the inhibitor and chloride ion remain adsorbed to the surface.

Near infrared SERS is being investigated by some authors to establish its utility as a method of monitoring environmental contaminants [15]. The hope is to perform remote sampling of the Raman light sample by using a fibre optic probe which can be lowered into deep water. Mixtures of 3-picoline (methylpyridine) and 3-chloropyridine adsorbed to a copper electrode have been studied with varying electrode

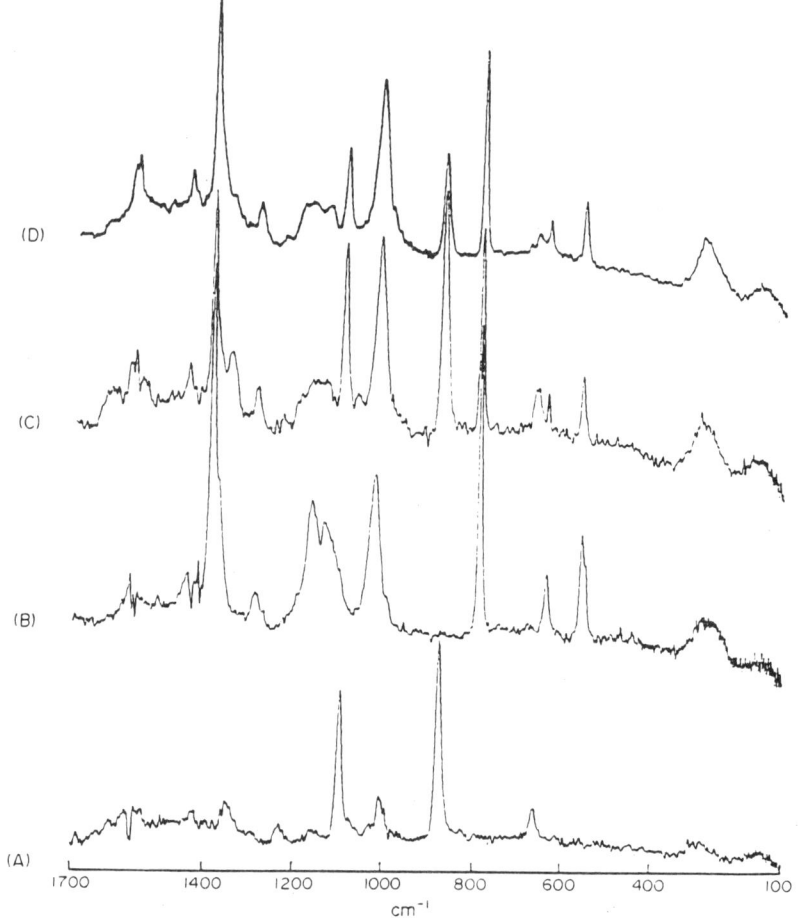

Fig. 10.10 — FT near infrared SERS spectra of corrosion inhibitors in 1 M KCl solutions adsorbed to a copper electrode at -0.7 V/SCE. (A) 0.1 M 2-AMP, (B) 0.1 M BTA, (C) 0.1 M 2-AMP and 0.001 M BTA, (D) 0.05 M 2-AMP + 0.005 M BTA. Reproduced with permission from *Spectrochimica Acta*, **46A**, M. Fleischmann *et al.*, copyright 1990, Pergamon Press PLC.

potential [21]. Spectra were recorded of 1-mM solutions of the analytes, and of mixtures containing 0.5 mM of each. The optimum potential for absorbing both species was 0.999 V relative to the SCE, though the spectrum obtained was not a simple sum of the two SERS spectra of the individual materials. Chloropyridine was found to be selectively adsorbed (or enhanced) at potentials more negative than picoline.

10.4 NEAR INFRARED FT RAMAN SERS SPECTRA FROM METAL COLLOIDS

As mentioned previously, since their first observation on metal electrodes, SERS spectra have also been obtained from molecules adsorbed to other substrates. In particular, aqueous metal colloids have proved to be very useful SERS substrates as they allow the convenient investigation of very dilute aqueous solutions by Raman spectroscopy. This is not only of interest to analytical chemists, but also widely used in the Raman spectroscopic study of *in situ* biochemical processes where it is not possible or meaningful to obtain more concentrated solutions of the analyte. In these studies resonance SERS spectra are also often obtained.

Metal colloids are obtained by reducing a dilute solution of a metal salt with a reducing agent such as alkaline sodium borohydride solution [22,23]. A distribution of particle size is produced by this process, and by subsequent particle aggregation. These different particle sizes can be separated by allowing the colloids to sediment in a glass column for several days. This process can be laborious but, recently, it has been demonstrated that it is possible to obtain SERS spectra from commercially available gold colloids [24]. An aliquot of the analyte solution is added to the colloid before submitting it to Raman spectroscopic investigation. As in all SERS techniques, the ability to observe spectra relies on the analyte being absorbed onto the SERS substrate. In an electrochemical experiment one can change the electrode potential to favour adsorption, but this of course is not the case with colloidal solutions. However, analyte concentration and solvent type are still controllable variables, but care must be taken with pH changes so as not to re-oxidize the colloidal particles.

At low concentrations the limiting factor to the observation of SERS spectra from colloids is the analyte adsorption equilibrium. At the other end of the scale, the colloidal particles have a finite total surface area, and once this is completely covered with analyte no increase in signal strength will be seen with increasing concentration. As a result the intensity of bands can be difficult to relate to the solution concentration. In visible Raman experiments, the detection limit for an analyte which adsorbs strongly onto silver colloid is of the order 10^{-9} M. Since the Raman spectrometer need only view a small sample (as little as 2 μl), spectra can be obtained from femtomole amounts. With the great success of FT near infrared SERS experiments with electrochemical systems, the next step was to establish whether similar enhancements could be obtained from colloids. Work published to date has been performed by Angel *et al.* [25].

FT near infrared SERS spectra have been obtained from dilute solutions of pyridine and chloropyridine in the presence of colloidal copper particles [25]. The copper colloids were prepared by reduction of copper sulphate in sodium citrate solution by the action of sodium borohydride in sodium hydroxide solution. The colloidal solution darkened on addition of the analyte and the SERS intensity was observed to increase over the period of an hour. This observation is not new and is generally attributed to the aggregation of the colloidal particles. A transmission electron microscope (TEM) image of a gold colloid aggregate is shown in Fig. 10.11 to demonstrate the point. Copper colloids present some experimental difficulties as they readily oxidize and frequently produce black precipitates rapidly on addition of the analyte.

The FT near infrared SERS spectra of a 0.2-mM solution of pyridine on copper colloid is shown in Fig. 10.12. As with electrochemically produced SERS spectra, the enhanced spectrum from the colloid is not the same as that of pure unenhanced

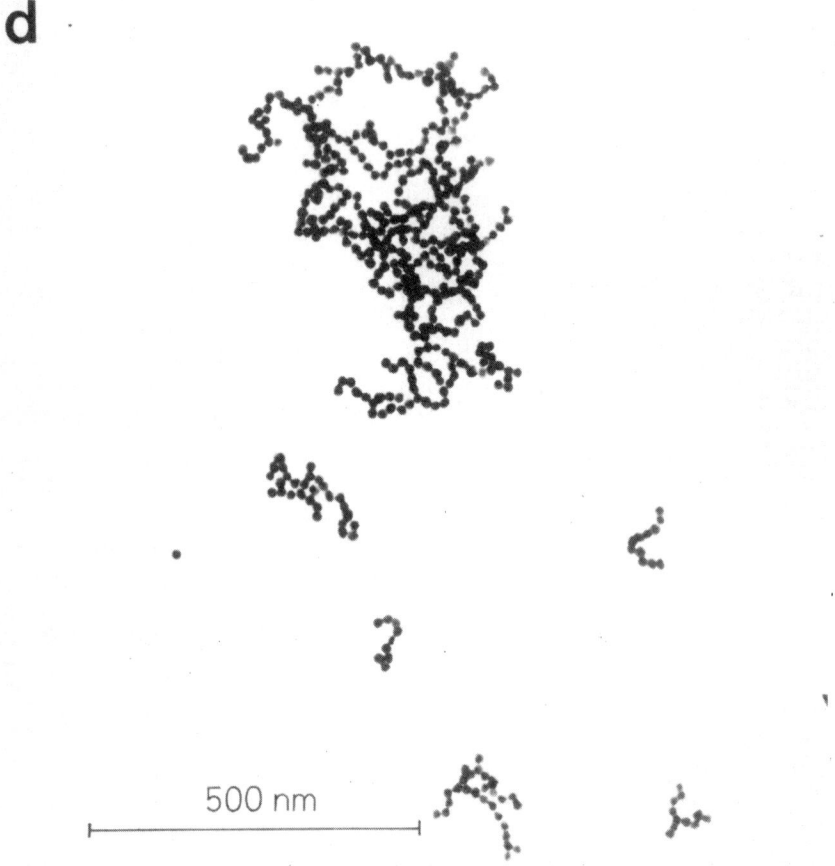

Fig. 10.11 — A transmission electron micrograph of an aggregated gold colloid, courtesy of Dr J. A. Creighton.

pyridine. The SERS enhancement is estimated at 2×10^5. It was also observed that the intensity of the SERS spectra increased on addition of aliquots of 0.1-mM potassium chloride solution. Very intense SERS spectra were also observed from 0.1-mM solutions of 3-chloropyridine (Fig. 10.13). These spectra were actually more intense than that of neat chloropyridine at the same laser power!

Rubidium(II)(phenanthroline)$_3$Cl$_2$ is a deeply coloured and highly fluorescent material. This complicates recording its Raman spectrum using visible laser excitation. A resonance Raman spectrum can be obtained with band intensities perturbed from their true values. Exciting Raman spectra with near infrared light at 1064 nm does not produce a resonance spectrum and the true Raman spectrum is obtained. FT near infrared SERS spectra have been obtained from this material adsorbed to both gold and copper colloid particles and are shown in Fig. 10.14. The spectra obtained from the gold and copper colloids are different, with changes in band intensity and frequency being apparent. All the spectra obtained from the

Sec. 10.4] **Near infrared FT Raman SERS spectra from metal colloids**   243

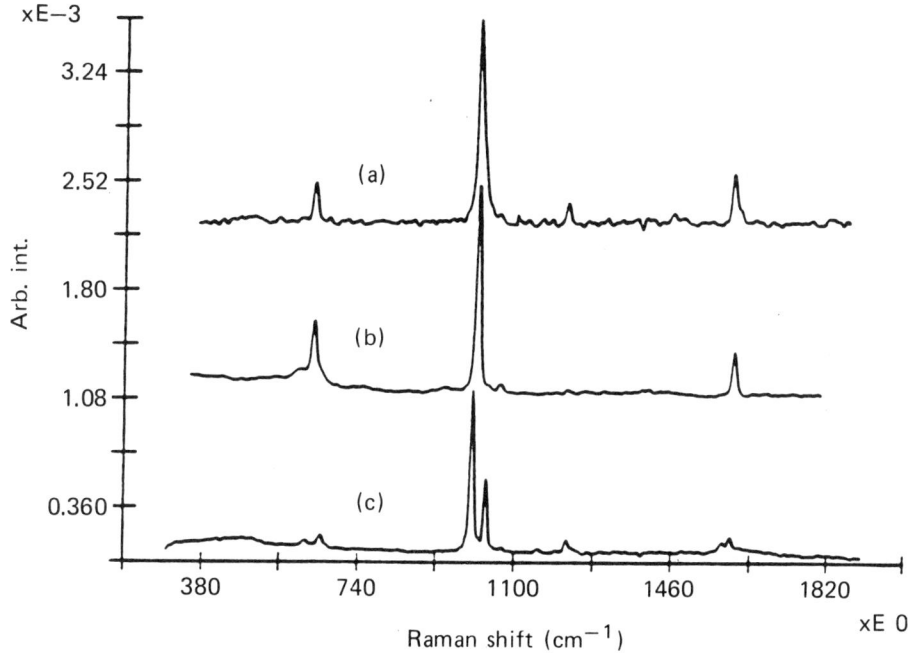

Fig. 10.12 — NIR FT SERS spectra of (a) 0.2-mM pyridine in fresh copper colloid solution, (b) 2-mM pyridine solution adsorbed to a copper electrode at -0.8 V, (c) 10% pyridine solution. Uncorrected data. Reproduced with thanks from S. M. Angel, L. F. Katz, D. D. Archibald and D. E. Honigs, *Applied Spectroscopy*, **43** (1989).

metal colloids (with the exception of chloropyridine on copper) show an intense background.

A very important factor in the SERS enhancement is the size and aspect ratio of the substrate particles. The size of the colloidal particles produced by the redox reaction governs the intensity of the SERS spectra obtained. The optimum diameter seems to be considerably less than the wavelength of the light used to excite the spectra. However, the particle size must not be too small. As the excitation wavelength is increased (i.e. it is moved into the near infrared) the optimum particle size required also grows [12]. This may explain why SERS spectra were obtained from 'coarser' gold and silver colloids, and the precipitation problems encountered with copper colloids.

The SERS effect has been demonstrated to work extremely well using 1064-nm exciting radiation, and excellent quality spectra have been obtained using FT Raman instrumentation. Experiments on both electrode surfaces and colloids have been very successful, and it will presumably be only a matter of time before the technique is applied to other SERS substrates. One very clear advantage which FT Raman SERS experiments have over their conventional visible counterparts is speed. The SERS spectra are so strong that very acceptable spectra can be obtained with just a few scans of the interferometer taking much less than 30 sec, whereas comparable

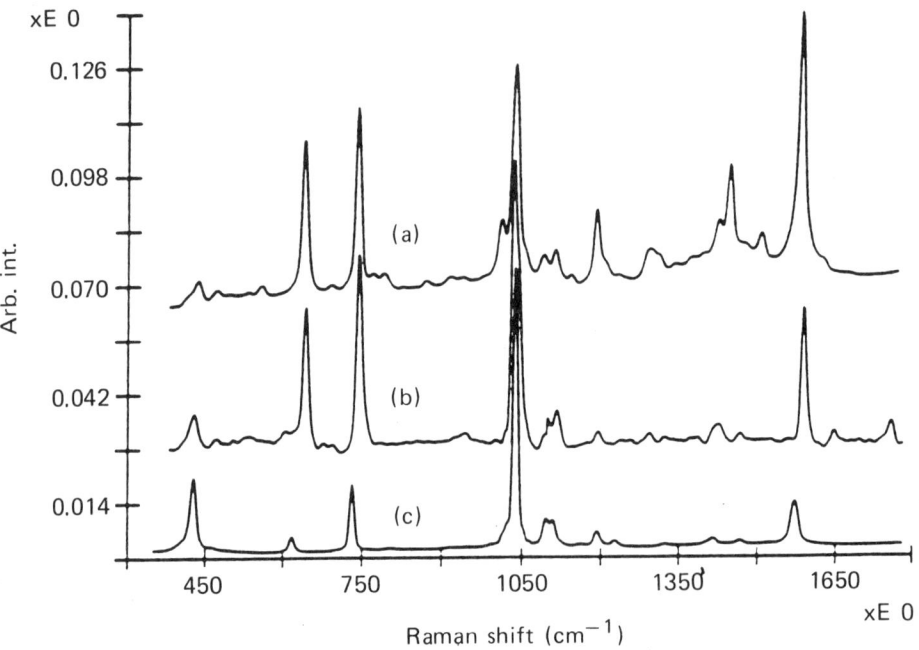

Fig. 10.13 — NIR FT SERS spectra of (a) 0.1-mM 3-chloropyridine in copper colloid solution containing 0.1 mM KCl, (b) 1-mM 3-chloropyridine solution adsorbed to a copper electrode at -0.8 V, (c) neat 3-chloropyridine. Uncorrected data. Reproduced with thanks from S. M. Angel, L. F. Katz, D. D. Archibald and D. E. Honigs, *Applied Spectroscopy*, **43** (1989).

measurements using a scanning monochromator many minutes (less with a spectrograph instrument fitted with a CCD detector). This has a number of important implications. Firstly, it is possible to perform electrode kinetics experiments if the reaction is not too fast. Secondly, it has been observed that SERS spectra obtained in the visible are time-dependent. This is due to photochemical affects of the laser. With FT Raman not only is the photon energy lower but the measurement time is very much less so this problem can be eliminated. Also, as the time needed to record spectra is dramatically reduced it is possible to perform far more experiments and investigate SERS systems with increased rigour, rather than being forced to take short cuts and miss variations. In the case of electrochemical experiments it is possible to fully investigate the effect of changes in potential, pH and electrolyte concentration. This would require exchanging the electrolyte, and possibly moving the working electrode away from the cell window to perform an ORC, necessitating subsequent realignment of the cell. Unfortunately, when using a conventional Raman system aligning an electrochemical cell can be a very tricky procedure. FT Raman not only reduces the spectral acquisition time, but the electrochemical cell can be aligned in seconds, greatly simplifying the whole experiment. To date all the FT Raman SERS spectra obtained have been excited using an Nd^{3+}:YAG laser. As the SERS enhancement is a function of excitation wavelength, an FT Raman

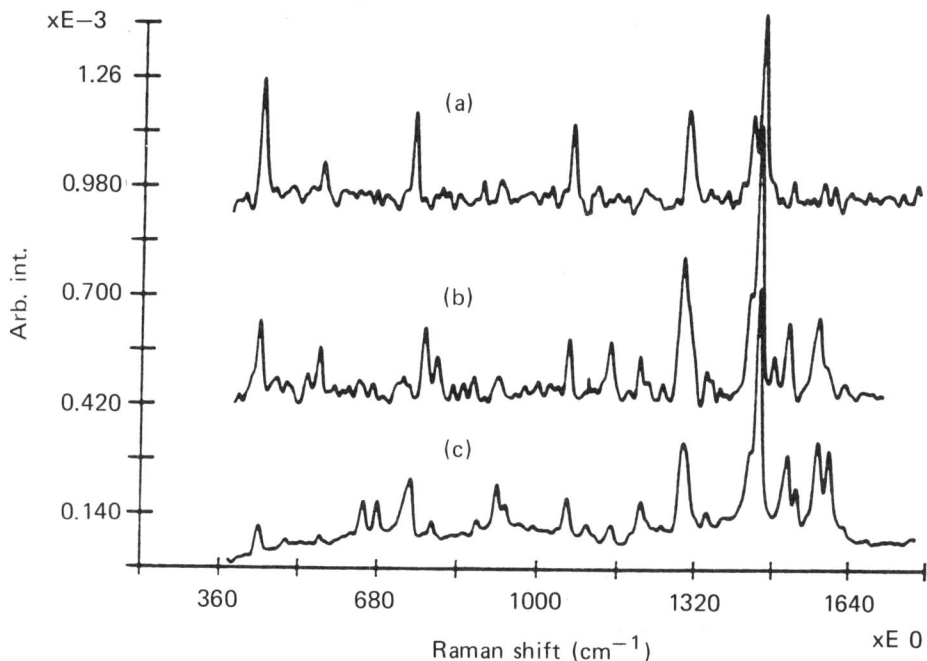

Fig. 10.14 — (a) NIR SERS spectrum of 0.025-mM Ru(o-phen)$_3^{2+}$ in gold colloid solution containing 0.1-mM KCl, (b) NIR SERS spectrum of 0.025-mM Ru(o-phen)$_3^{2+}$ in copper colloid solution containing 0.1-mM KCl, (c) unenhanced spectrum of Ru(o-phen)$_3^{2+}$Cl$_2$ powder. Uncorrected data. Reproduced with thanks from S. M. Angel, L. F. Katz, D. D. Archibald and D. E. Honigs, *Applied Spectroscopy*, **43** (1989).

spectrometer fitted with a tunable near infrared laser would make a very powerful tool in SERS research, and allow spectra to be obtained from the surfaces of metals other than copper, silver, gold and platinum. This prospect is discussed in Chapter 12.

10.5 INFRARED SPECTROSCOPY OF CATALYST SYSTEMS

The use of vibrational spectroscopy to study molecules adsorbed to heterogeneous catalysts is well established [26,27]. The original work in the field was pioneered by Eischens [28,29] in the 1950s, who used transmission infrared spectroscopy. The experiments were performed using the dispersive infrared spectrometers of the day and were hampered by very poor light throughput. Nevertheless, Eischens showed that it was possible to obtain vibrational spectra of molecules adsorbed onto catalyst surfaces and the subject gradually developed.

However, there is one severe problem which hinders the infrared experiments — whether the spectrum is recorded by transmission or reflection it is still predominantly that of the bulk catalyst and not the adsorbed molecules. The majority of experiments performed to date have been carried out on pressed discs of catalyst powder. Heterogeneous catalysts are usually based on oxide materials such as silica or alumina, and these have very strong infrared absorptions below about 1100 cm^{-1} due to skeletal modes. Also, if hydroxyl groups are present on the surface there may be significant absorptions in the OH stretching region. In comparison to the mass of the bulk catalyst, there is very little adsorbate present on the surface. Thus, one is forced to study small adsorbate bands over an intense catalyst background, and the signal-to-noise ratio is not high. The situation is analogous to trying to observe small Raman bands on an intense fluorescent background. In practice the infrared spectrum of the adsorbate can only be studied in narrow 'windows' in the spectral range where the catalyst transmits the infrared. The advent of FTIR spectrometers greatly improved the technique but could not eliminate the fundamental limitation of the catalyst infrared absorption. Despite this limitation much work has been successfully completed using infrared spectroscopy in this field, but this is mainly because it was the only spectroscopic method suitable. In recent years solid-state NMR and desorption mass spectrometry experiments have been introduced.

10.6 RAMAN SPECTROSCOPY OF CATALYST SURFACES

In principle there are several good reasons for trying to use Raman spectroscopy in catalytic studies. Firstly, infrared spectroscopy can only study vibrations where there is a change in the dipole moment as the vibration proceeds. Raman, however, relies on changes in polarizability during the vibration and thus can be used to study different vibrational modes. From the viewpoint of a catalytic chemist reactions involving CO may be best studied with infrared, but for unsaturated hydrocarbons Raman is a more obvious choice. Secondly, many catalysts are extremely poor Raman scatterers† (e.g. silica and alumina). Thus, Raman spectroscopy has the overwhelming advantage that the spectrum obtained is principally that of the adsorbed molecules alone, so the technique is surface-selective.

For these reasons, attention turned to using Raman spectroscopy for the study of catalytic systems in the 1960s and 1970s. Initial publications looked promising [30–42], but the technique never rose to popularity owing to a severe fluorescence problem [38]. When a catalyst is exposed to the atmosphere it invariably absorbs large quantities of water, air and trace organics. For the materials to be catalytically active they must be degassed, often with heating, on a vacuum line for several hours. After this pretreatment, the catalyst became extremely fluorescent under visible laser radiation, and the Raman spectrum of any adsorbate loaded on to it could not be obtained.

Mechanisms were suggested to account for this fluorescence. Under activation the catalyst may getter minute quantities of grease from the vacuum line, and these form highly fluorescent materials on the surface [39]. Grease-free vacuum lines fitted

---

† This point has been made before — many non-crystalline materials give very weak Raman spectra indeed and hence small concentrations of the appropriate materials supported in or on them give clear, usable spectra. Thus, it is possible to observe bands due to additives in the Raman spectra of commercial polymers.

with molecular sieves were used, but often the addition of the organic adsorbate of interest had the same effect. Attempts to quench this fluorescence took three main forms. Firstly, the sample could be left in the laser beam for extended periods of time in an attempt to photo-degrade the fluorescent molecules. Hardin et al. [39] tried 'burning off' any fluorescent organics by heating the catalyst in a flow of oxygen. This met with success in some cases, though the fluorescence tended to reappear with time and in some cases oxygen pretreatment made matters worse [41]. Hendra et al. [34–36] found that activating the catalyst above 900°C reduced the fluorescence, but this and the two previous methods may well damage the catalyst, thus modifying the very system which is under study. For these reasons the study of catalyst systems with Raman spectroscopy never lived up to its early promise. Elegant work has been done in favourable cases, but as a routine method it failed [43].

Recent publications by Warnes et al. have reported the utility of FT Raman in the study of catalysts [44,45]. Excellent quality spectra have been obtained from pyridine adsorbed to samples of silica gel and several different zeolites. These are all white materials of high surface area, and pyridine is a strong Raman scatterer. Pyridine absorption has long been used in infrared spectroscopy as a method of characterizing the acid sites found on the surfaces of catalysts [46]. It is these acid sites which are thought to be responsible for the catalytic action of the material. Both Lewis and Brønsted acid sites are found on catalyst surfaces, and the vibrational frequencies of adsorbed pyridine are dependent upon the nature of the absorption site (see Fig. 10.15). Vibrational frequencies of pyridine in different environments are shown in

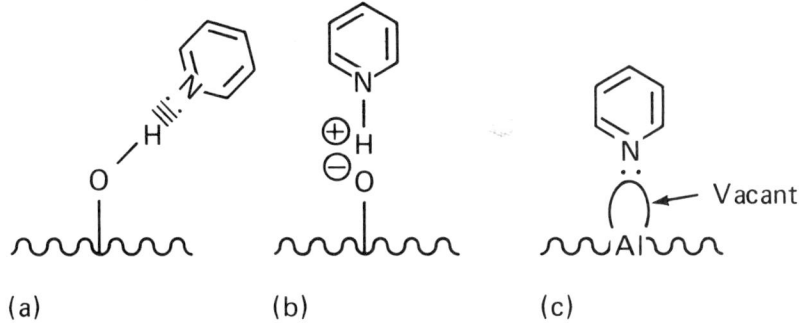

Fig. 10.15 — Pyridine adsorbed to (a) hydrogen-bonding, (b) Brønsted acid and (c) Lewis acid sites.

Table 10.1. Aqueous pyridine, pyridinium chloride and pyridine–zinc chloride complex are used as models for hydrogen bonding, Brønsted and Lewis acid sites respectively. By studying the Raman spectrum of adsorbed pyridine it is possible to determine which types of acid site are present.

The apparatus required for FT Raman catalyst studies is simple and straightforward. Samples of catalyst (approximately 0.5 g) are placed in a pyrex cell fitted with an optically flat window at one end, and a greaseless tap for connection to a vacuum

Table 10.1 — Vibrational frequencies of pyridine in different environments

	v_1 (cm^{-1})	v_{12} (cm^{-1})	$v_{2,13}$ (cm^{-1})
Liquid pyridine	990 (vs)	1029 (vs)	3056 (vs)
Pyridine/H$_2$O	1000 (vs)	1033 (s)	3069 (s)
Pyridine/conc HCl	1009 (vs)	1026 (s)	3099 (s)
Pyridine/ZnCl$_2$ complex	1020 (vs)	—	3072 (s)

line at the other (see Fig. 10.16). The cell is plugged with glass wool to prevent catalyst powder being sucked into the vacuum system. The catalyst is activated by connecting it to the vacuum system and heating in a tube furnace. Liquid adsorbates are degassed by subjecting them to several freeze–pump–thaw cycles. The adsorbate is then vacuum distilled into the cell until the catalyst (the adsorbent) is homogeneously covered. The cell is then weighed and the FT Raman spectrum recorded. The adsorbate is removed by aliquots on the vacuum line, the cell being weighed and a spectrum recorded after each removal. If the surface adsorbate is left to equilibrate with the vapour in the vacuum system at a constant temperature, pressure measurements can be made, allowing the spectra recorded to be related to the absorption isotherm. The cell is held on an aluminium mount which can be easily and reproducibly slotted into the spectrometer. The cell used is simple, and more elaborate apparatus such as a 'flow cell' could easily be devised. All that is required is a glass window through which the laser beam can enter and the Raman scatter exit.

The informative regions of the pyridine spectrum are the ring breathing modes around 1000 cm^{-1} and the CH stretching modes around 3000 cm^{-1} (see Table 10.1). These modes are labelled according to ref [47]. Figure 10.17 shows the spectra obtained of pyridine adsorbed to silica gel (HNO$_3$ washed, 250–425 μm particle size) as the pyridine coverage is gradually reduced. For clarity these spectra have been auto-expanded to the same height. In reality the observed spectral intensity falls as the pyridine is removed, as witnessed by the increased noise level in the spectra at lower coverage. The coverages are expressed as the mass of adsorbed pyridine per gram of adsorbent. At high coverage, the observed spectrum is the same as liquid pyridine as the predominant species on the surface is physisorbed pyridine, the main bands being at 989, 1029 and 3053 cm^{-1}. As the physisorbed pyridine is removed by evacuation, weaker bands arising from chemisorbed pyridine become more significant. These appear at 1004, 1034 and 3070 cm^{-1}. Previous studies in conventional Raman have shown that bands between 1001 and 1009 cm^{-1} are indicative of Brønsted acid sites, the strength of the association between the hydrogen ion and the pyridine being indicated by the closeness of the band to 1009 cm^{-1} [35,36]. Thus, the FT Raman spectra show the presence of Brønsted acid sites on the surface of silica gel. However, the CH stretching frequency is far closer to that of hydrogen bonded pyridine than to that of pyridinium ion.

Zeolites are enormously important industrial catalysts used extensively in cracking reactions in the petroleum industry. They are aluminosilicate materials composed of SiO$_4$ and AlO$_4$ tetrahedra linked together to form a three dimensional

Sec. 10.6] **Raman spectroscopy of catalyst surfaces** 249

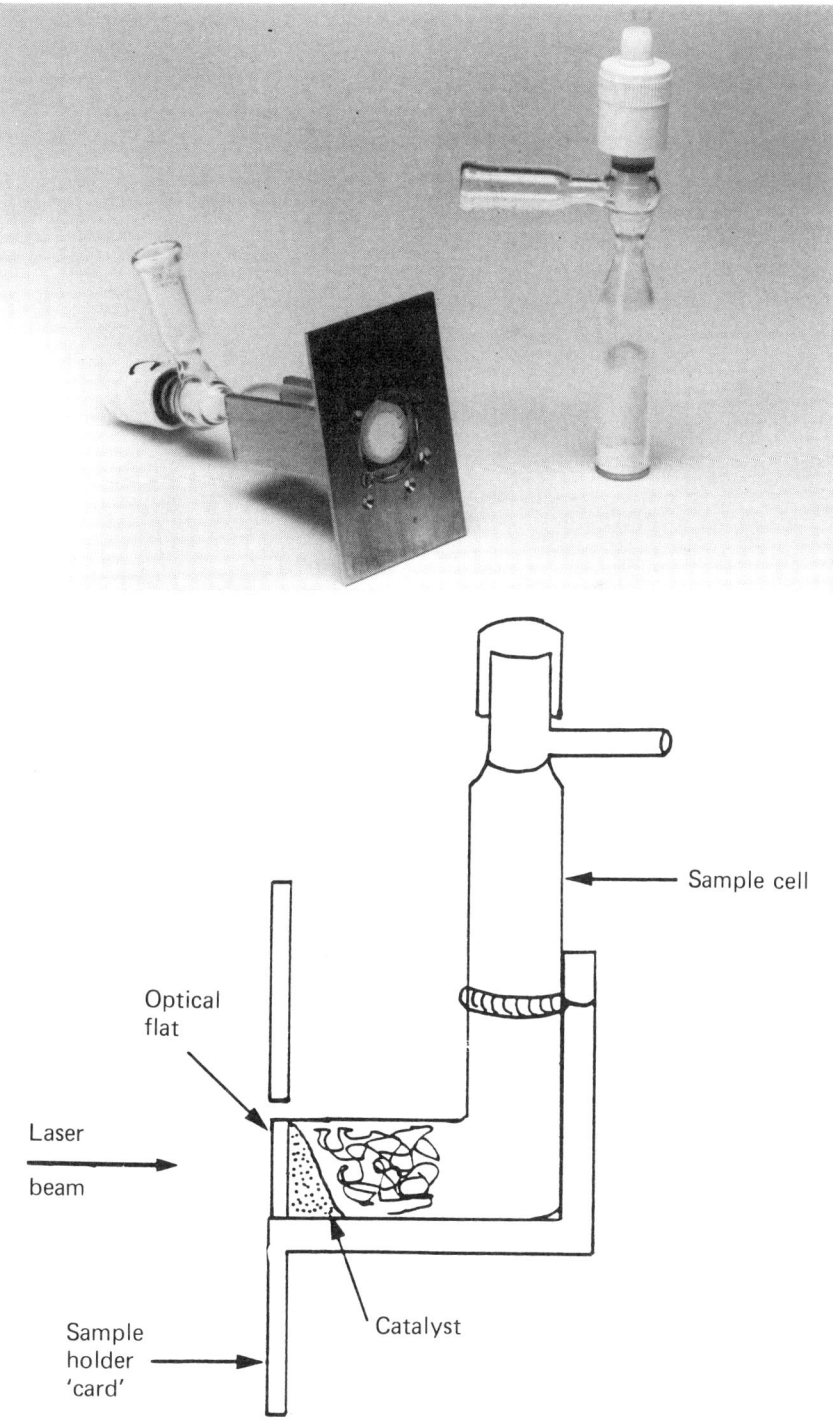

Fig. 10.16 — Adsorbed species cell used for FT Raman experiments.

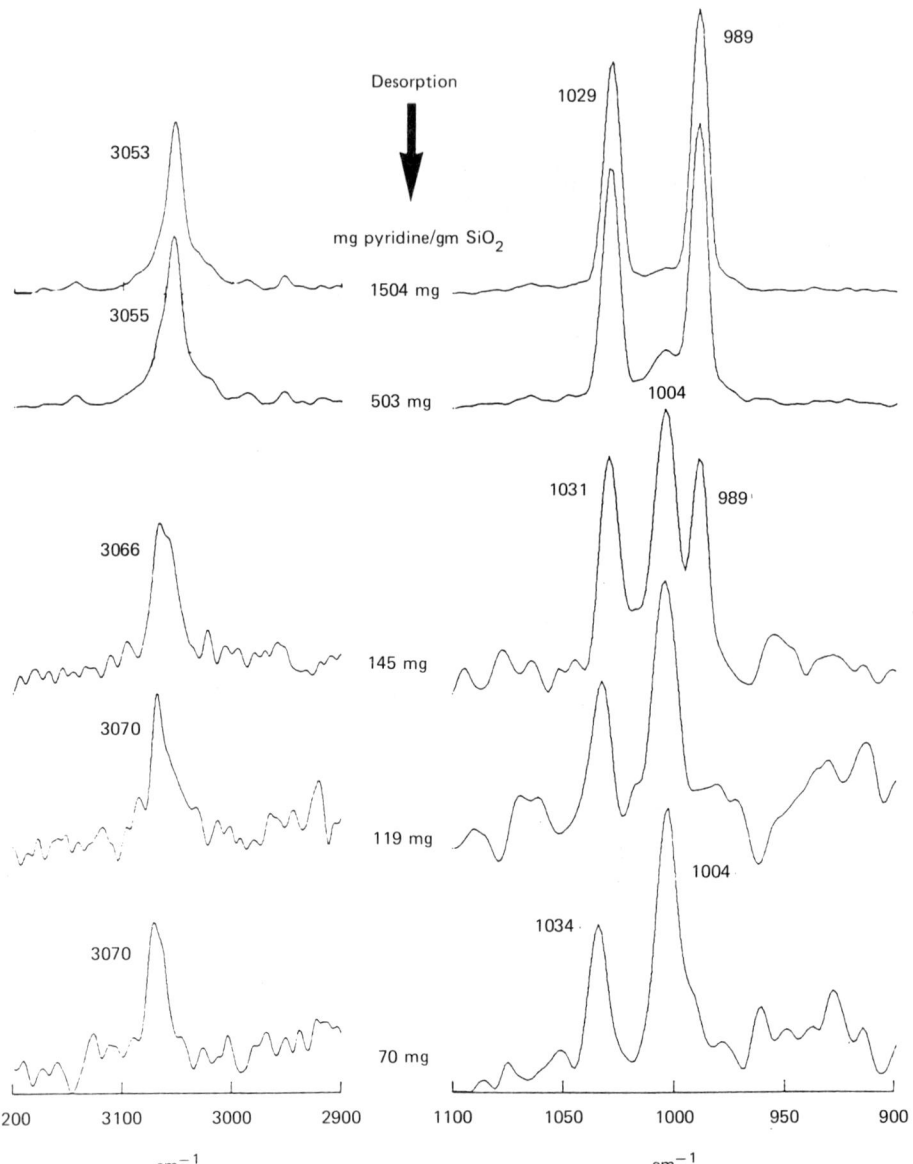

Fig. 10.17 — FT Raman spectra of pyridine adsorbed to silica gel over a range of coverages.

structure with regular pores. The oxygen atoms are shared and the structure can be viewed stoichiometrically as being composed of AlO_2^- and SiO_2 units. Twenty-four of these units link together to form a sodalite cage, a truncated octahedron which has six square faces and eight hexagonal faces (see Fig. 10.18). The whole zeolite structure is composed of sodalite cages linked in different ways. Zeolites formed by

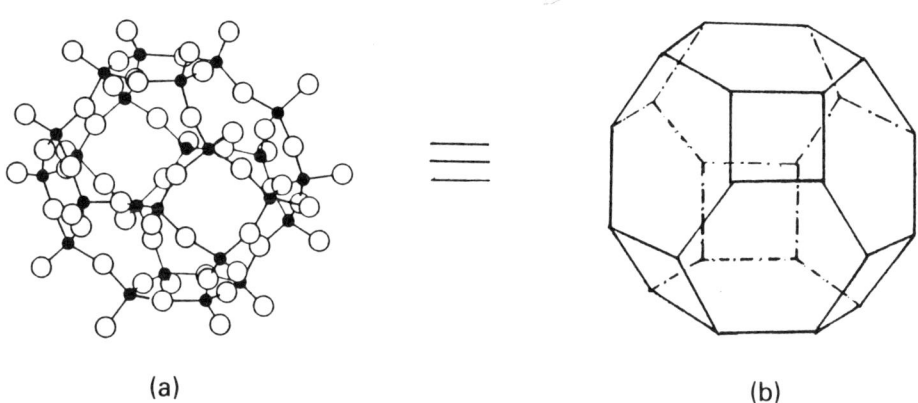

Fig. 10.18 — (a) A sodalite cage composed of interconnected tetrahedra, (b) a line drawing of (a) to emphasize the square and hexagonal faces.

sodalite cages linked by the square faces are labelled A-type, and those linked by the hexagonal faces are termed X- and Y-type zeolites (see Fig. 10.19). In an X- or Y-type zeolite large 'super cages' are formed between ten sodalite units, and these super cages are thought to be the site of catalytic reactions of adsorbed molecules. Since the radius of the super cages can be varied and the size of the entrance to the cages controlled, zeolite catalysts demonstrate very high selectivity and stereospecificity in their reactions. X- and Y-type zeolites differ in composition as measured by their silicon/aluminium ratio. The aluminosilicate network has a net negative charge which is balanced by positive cations. After the first stage of preparation these are sodium ions and the product is known as an NaY zeolite. These sodium ions are then ion-exchanged for ammonium cations and the product heated in a steam atmosphere at 850°C for 30 minutes. This liberates ammonia leaving catalytically active hydrogen ions as the counter ions in the structure.

A series of Y-type zeolites has been investigated using pyridine adsorption and FT Raman spectroscopy [45]. The zeolites were activated by pumping them down on a vacuum line for 2 hours at room temperature. Figure 10.20 shows the series of spectra obtained from an NaY zeolite as the pyridine coverage is decreased. At high coverage (1689 mg/g) the spectrum looks like that of liquid pyridine, except for the presence of a shoulder at 995 cm^{-1}. As the coverage is reduced to 434 mg/g this shoulder becomes a prominent band at 1001 cm^{-1}. The v_{12} band shifts to 1035 cm^{-1} and the $v_{2,13}$ CH stretching band moves to 3067 cm^{-1}. These frequencies are similar to those observed for aqueous pyridine suggesting that only hydrogen-bonded pyridine species are present. Pyridine could be hydrogen-bonded to the residual water found in the zeolite cages and OH groups present on the outer surface of the tiny zeolite particles.

Figure 10.21 shows a related series of spectra of pyridine adsorbed to a so-called 'ultrastable zeolite' (USY). In this material the sodium ions have been exchanged for

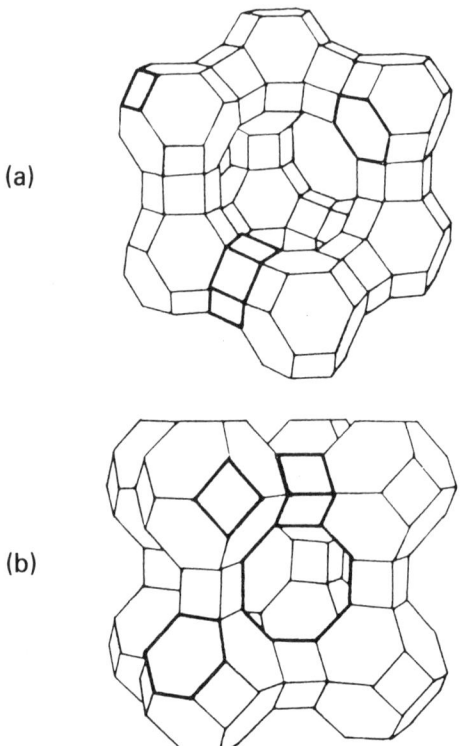

Fig. 10.19—(a) An X- or Y-type zeolite structure formed by the linkage of hexagonal faces; (b) an A type zeolite structure formed by the linkage of square faces. The super cages lie at the centre of the structures shown.

hydrogen ions and then heated at 750°C for an extended period. This treatment is thought to cause a growth of Lewis activity and increase the pK_a of the Brønsted acid sites. The ring breathing mode region of the spectrum is very different from that observed in the NaY spectra. The v_{12} band now appears at 1031 cm^{-1}, and at low coverage a band is present at 1017 cm^{-1} thought to be due to pyridine coordinated to Lewis acid sites. Although the NaY and USY spectra were recorded at the same resolution (nominal 3 cm^{-1}, data points every 1.5 cm^{-1}) the bands are much broader. The apparent broadening of the spectral bands can be attributed to the presence of new chemical species whose Raman bands overlap. The pK_a of the acid sites will influence the vibrational frequencies observed, and in principle, if there were sufficient data points in the spectrum, band shape analysis could provide information on the distribution of different pK_a sites present on the surface. The FT Raman spectra of the two systems show clear differences and these can be attributed to chemical changes in the zeolite structure.

HY zeolites are usually used to crack gas oils at temperatures between 450 and 500°C. At lower temperatures isomerization reactions occur and these too are exploited. Figure 10.22 shows the FT Raman spectra of 1-hexene (a) and 2-hexene

Sec. 10.6] Raman spectroscopy of catalyst surfaces

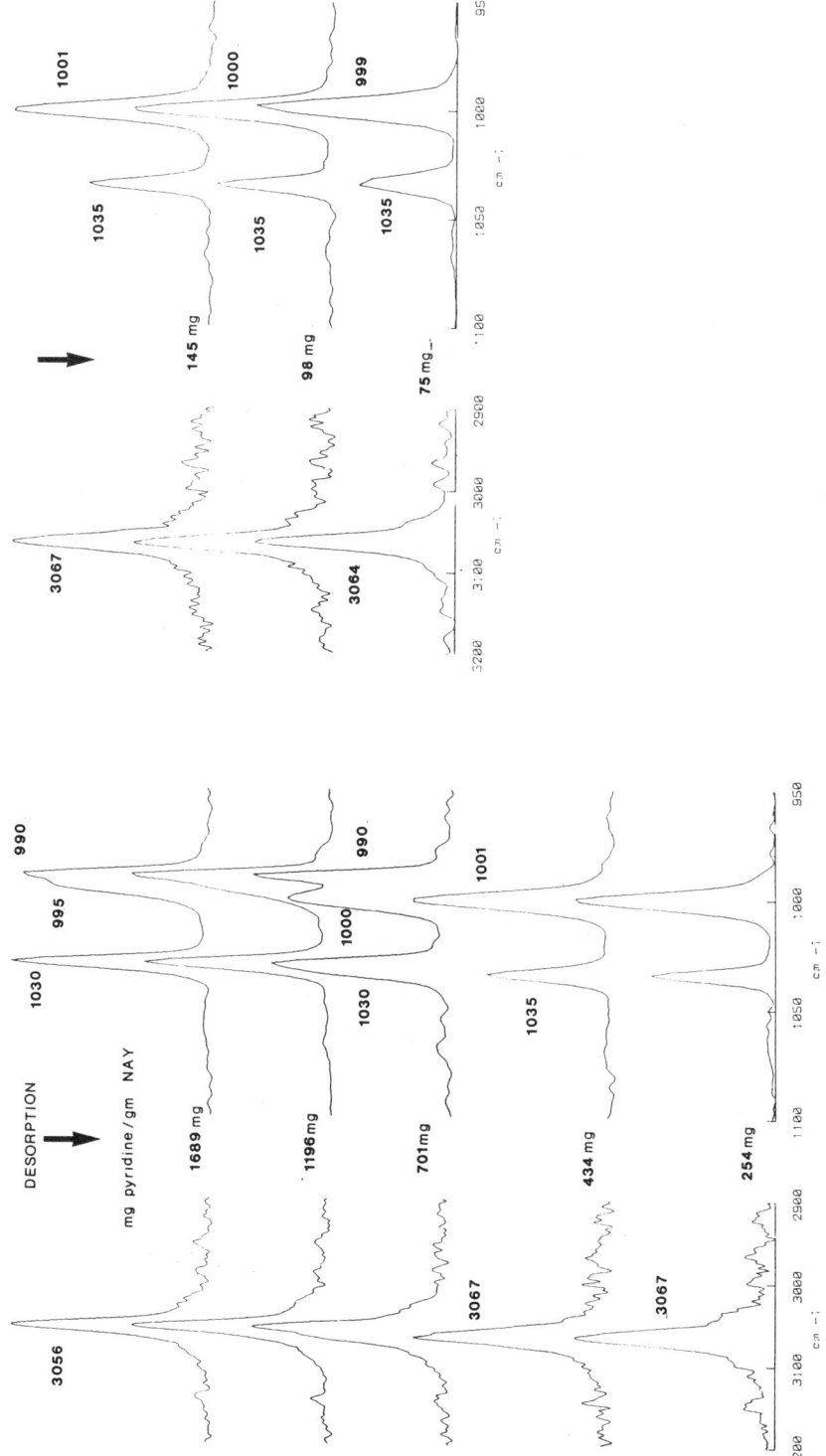

Fig. 10.20 — FT Raman spectra of pyridine adsorbed to NAY zeolite over a range of coverages.

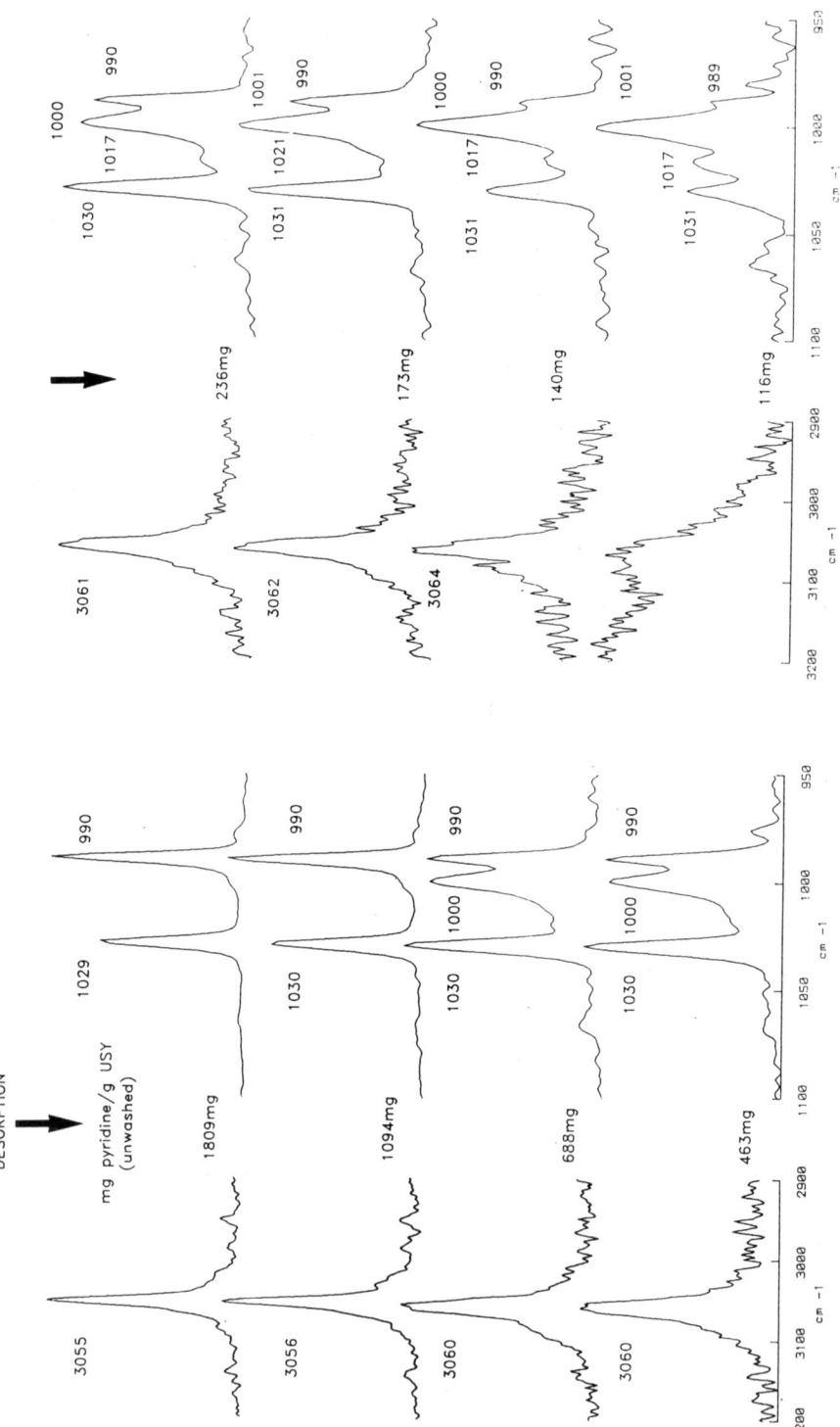

Fig. 10.21 — FT Raman spectra of pyridine adsorbed to USY zeolite over a range of coverages.

Sec. 10.6] Raman spectroscopy of catalyst surfaces

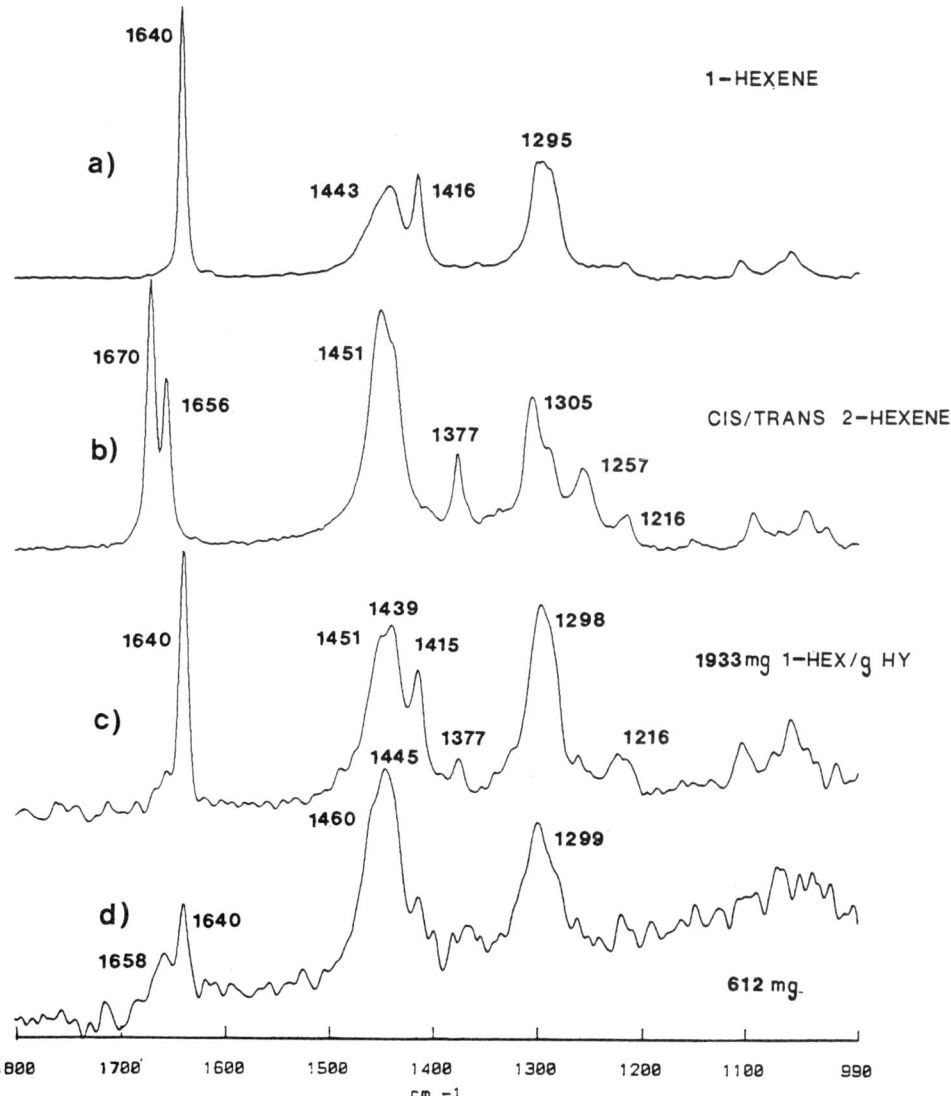

Fig. 10.22 — FT Raman spectra of (a) 1-hexene, (b) a mixture of *cis/trans* 2-hexene and (c) and (d) 1-hexene adsorbed to HY zeolite.

(b), and the spectra obtained when 1-hexene was adsorbed to HY zeolite (c and d). The $\nu_{C=C}$ band of 1-hexene appears at 1640 cm^{-1}, *trans*-2-hexene at 1671 cm^{-1} and *cis*-2-hexene at 1656 cm^{-1}. The spectrum of the adsorbed 1-hexene is different from the pure liquid, indicating that chemical changes have occurred. New bands appear at 1451, 1377 and 1216 cm^{-1}, indicative of isomers of 2-hexene. Unfortunately the spectra of the reactant and possible products overlap, and the single-to-noise ratio is not particularly favourable. However, at a coverage of 612 mg/g there does appear to be a band present at 1658 cm^{-1} showing the presence of *cis*-2-hexene. Attempts to increase the reaction rate by heating the sample caused it to discolour and fluoresce when irradiated with the laser.

In the past Raman spectroscopy has been used not only to study molecules adsorbed to catalysts but also to investigate the catalyst materials themselves. Nielsen *et al.* have studied the Raman spectra of amorphous alumina and compared the results obtained from visible conventional Raman and near infrared FT Raman [48]. In addition to the effects originating from adsorbed organic molecules, the fluorescence observed from catalyst systems can also be due to the presence of any transition metal impurities. These elements have broad *d–d* electronic transitions which can give rise to very intense backgrounds in Raman experiments. These transitions can occur from the ultraviolet to the near infrared. Figure 10.23 (a) shows how iron impurities can cause intense fluorescence, even when the sample is irradiated with an Nd^{3+}:YAG laser at 1.064 μm. The heavy line is the spectrum obtained using an Nd^{3+}:YAG laser from an alumina sample containing 193 ppm iron, whereas the thin line spectrum came from a similar sample but containing less than 20 ppm iron. Figure 10.23 (b) shows the spectrum obtained with Nd^{3+}:YAG excitation from a sample of alumina which had been impregnated with MoO$_3$, and also contained 193 ppm iron and 1000 ppm titanium. Again an intense background is observed.

The Raman spectra of the same samples recorded using a conventional spectrometer and an argon ion laser are of much higher quality than those recorded in the near infrared. This is demonstrated in Fig. 10.24, which shows an equivalent set of spectra to Fig. 10.23 but using conventional Raman instrumentation. When excited in the visible the sample containing no titanium and less than 20 ppm iron shows no fluorescence at all and an excellent Raman spectrum is obtained (Fig. 10.24(a)). The argon laser is still thought to generate some fluorescence, but this is emitted in the near infrared and hence does not hinder visible Raman experiments. The molybdenum-containing alumina sample is not thought to be homogeneously impregnated with molybdenum. The MoO$_3$ is only thought to lie near the surface of the alumina particles. Whereas the near infrared Raman spectrum of this sample shows a very intense background, excitation in the visible again gives a high quality spectrum which is predominantly of the molybdenum species. The near infrared radiation is thought to penetrate deeper into the sample than the visible radiation. Consequently, the visible excited Raman spectrum only shows the outer molybdenum species, whereas the near infrared spectrum is degraded by fluorescence from iron and titanium impurities in the core of the sample.

All the FT Raman spectra of catalyst systems recorded to date have been excited with an Nd^{3+}:YAG laser. Figures 10.23 and 10.24 clearly demonstrate that the 1.064 μm line may not be the ideal source to use in these studies. As well as causing fluorescence, transition metal impurities may be also be responsible for sample heating encountered when studying some zeolite materials [45]. As with SERS studies, an FT Raman spectrometer utilizing a tunable near infrared laser as an excitation source would make a powerful research tool, and permit the experimenter to avoid fluorescence due to transition metal impurities by shifting the excitation frequency. This possibility and others are discussed in Chapter 12.

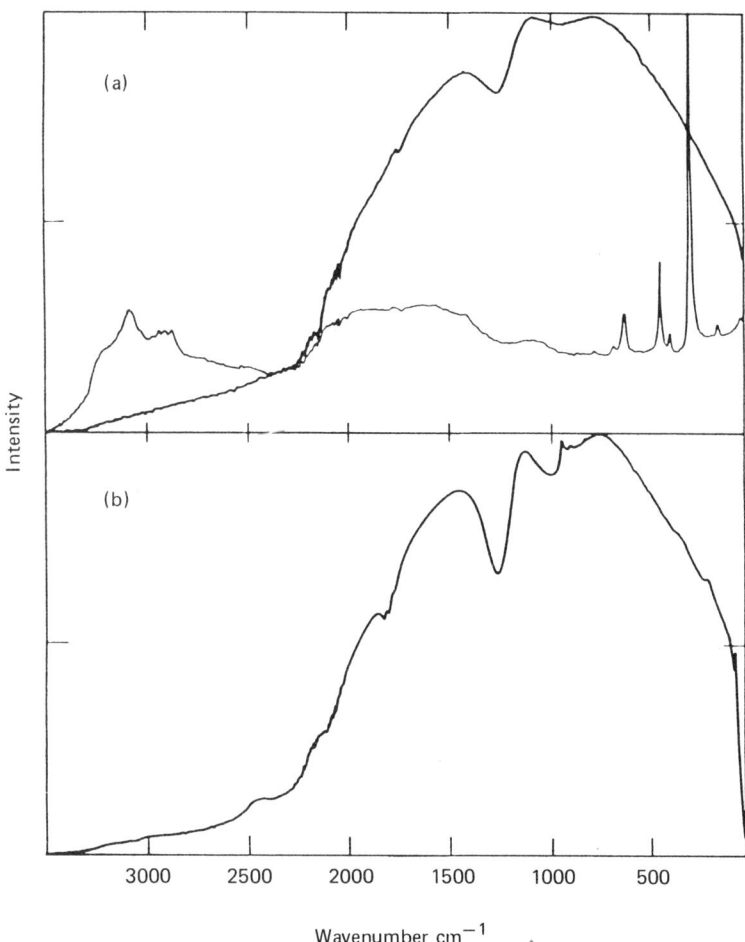

Fig. 10.23 — (a) The effect of transition metal impurities in the near infrared Raman spectra of alumina. The heavy trace is from a sample containing 193 ppm iron and 1000 ppm titanium, the thin trace came from a sample containing less than 20 ppm iron and no titanium. (b) The spectrum obtained with an Nd^{3+}:YAG laser from an Al_2O_3/MoO_3 sample containing 193 ppm iron and 1000 ppm titanium. Uncorrected data. Reproduced by permission from A. Mortenson et al., J. Raman Spectrosc., **22**, 47, copyright John Wiley & Sons Ltd. (1991).

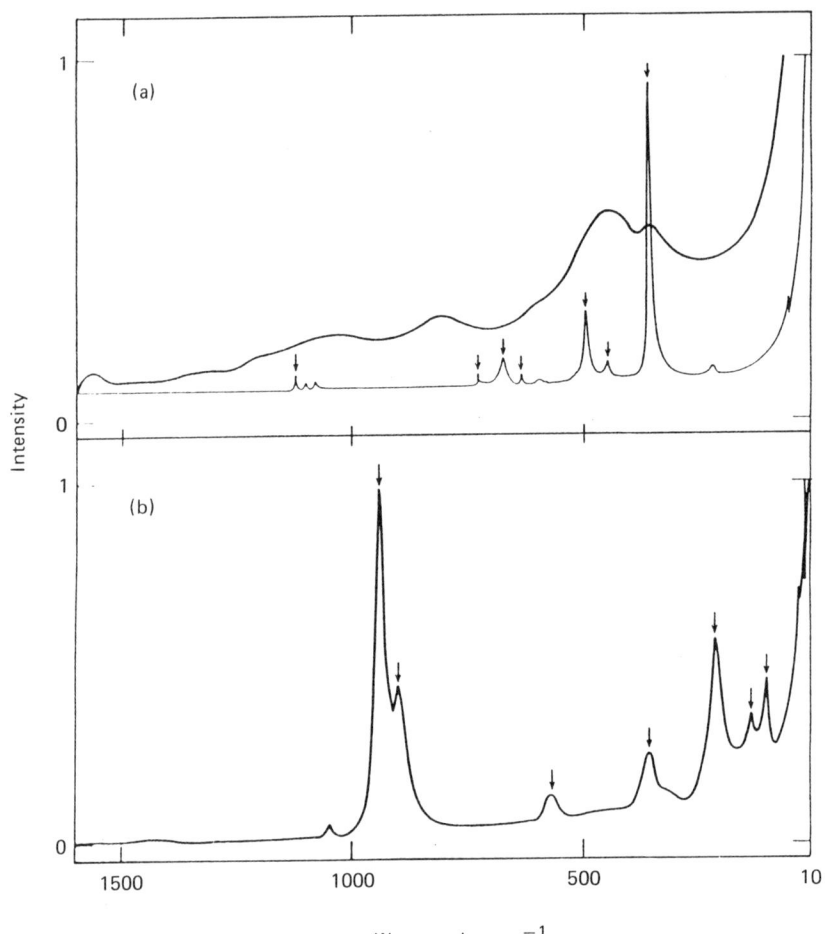

Fig. 10.24 — (a) The effect of transition metal impurities in the visible Raman spectra of alumina. The heavy trace is from a sample containing 193 ppm iron and 1000 ppm titanium; the thin trace came from a sample containing less than 20 ppm iron and no titanium. (b) The spectrum obtained with an Nd^{3+}:YAG laser from an Al_2O_3/MoO_3 sample containing 193 ppm iron and 1000 ppm titanium. Uncorrected data. Reproduced by permission from A. Mortenson et al., J. Raman Spectrosc., **22**, 47, copyright John Wiley & Sons Ltd. (1991).

REFERENCES

[1] *Physical Chemistry of Surfaces*, A. W. Adamson, 4th Ed., John Wiley & Sons, 1982.
[2] A. Bewick and S. Pons in *Advances in Infrared and Raman Spectroscopy*, edited by R. J. H. Clark and R. E. Hester, Vol 12, p. 1, John Wiley & Sons, Chichester, 1985.
[3] K. Ashley and S. Pons, *Chem. Rev.*, 1988, **88**, 673.
[4] M. Fleischmann, P. J. Hendra, and A. J. McQuillan, *J. Chem. Soc. Chem. Commun.*, 1973 80.
[5] M. Fleischmann, P. J. Hendra and A. J. McQuillan, *Chem. Phys. Lett.*, 1974, **26**, 163.
[6] D. L. Jeanmaire and R. P. Van Duyne, *J. Electroanal. Chem.*, 1977, **84**, 1.
[7] M. G. Albrecht and J. A. Creighton, *J. Am. Chem. Soc.*, 1977, **99**, 5215.

References

[8] T. E. Furtak and J. Reyes, *Surf. Sci.*, 1980, **93**, 351.
[9] A. Otto, *Applications of Surface Science*, 1980, **6**, 309.
[10] M. Moskovits, *Rev. Mod. Phys. Part 1*, 1985, **57**, No 3, 783.
[11] D. A. Weitz, M. Moskovits and J. A. Creighton, in *Chemical Structure of Interfaces*, VCH publishers, 1986.
[12] E.J. Zeman and G. C. Schatz, *J. Phys. Chem.*, 1987, **91**, 634.
[13] M. Fleischmann and I. R. Hill *J. Electroanal. Chem.*, 1983, **146**, 353.
[14] D. B. Chase and B. A. Parkinson, *Appl. Spectrosc.*, 1988, **42**, No 7, 1186.
[15] S. M. Angel, L. F. Katz, D. D. Archibald, L. T. Lin and D. E. Honigs, *Appl. Spectrosc.*, 1988, **42**, No 8, 1327.
[16] A. Crookell, M. Fleischmann, M. Hanniet and P. J. Hendra, *Chem. Phys. Lett.*, 1988, **149**, No 2, 123.
[17] S. M. Angel and M. L. Myrick, *Anal. Chem.*, 1989, **61**, No 5, 1648.
[18] D. A. Weitz, S. Garoff, J. I. Gersten and A. Nitzan, *J. Chem. Phys.*, 1973, **79**, 509.
[19] A. Wokaun, H. P. Lutz, A. P. King, U. P. Wild and R. R. Ernst, *J. Chem. Phys.*, 1983, **79**, 509.
[20] M. Fleischmann, D. Sockalingum and M. M. Musiani, *Spectrochimica Acta*, 1990, **46A**, No 2, 285.
[21] S. M. Angel and D. D. Archibald, *Applied Spectrosc.*, 1989, **43**, 1097.
[22] J. A. Creighton, M. S. Alvarez, D. A. Weitz, S. Garoff and M. W. Kim, *J. Phys. Chem.*, 1983, **87**, 4793.
[23] J. A. Creighton, C. G. Blatchford and M. G. Albrecht, *J. Chem. Soc. Faraday Trans.*, 1978, **75**, 2 790.
[24] S. M. Angel, M. L. Myrick and F. P. Milanovich, *Appl. Spectrosc.*, 1990, **44**, No 2, 335.
[25] S. M. Angel, L. F. Katz, D. D. Archibald and D. E. Honigs, *Appl. Spectrosc.*, 1989, **43**, No 3, 367.
[26] *Infrared of Adsorbed Species*, L. H. Little, Academic Press, 1966.
[27] P. C. M. van Woerkom in *Advances in Applied Fourier Transform Infrared Spectroscopy*, edited by M. W. Mackenzie, Chapter 7, p. 323, John Wiley and Sons, Chichester, 1988.
[28] R. P. Eischens, W. A. Pliskin and S. A. Francis, *J. Chem. Phys.*, 1954, **22**, 1786.
[29] R. P. Eischens, S. A. Francis and W. A. Pliskin, *J. Phys. Chem.*, 1956, **60**, 194.
[30] G. Karagounis and R. Issa, *Nature*, 1962, **195**, 1196.
[31] E. V. Pershina and S. S. Raskin, *Optics & Spectroscopy*, Supplement 3, 1964.
[32] P. J. Hendra and E. J. Loader, *Nature*, 1967, **216**, 789.
[33] P. J. Hendra and E. J. Loader, *Nature*, 1968, **217**, 637.
[34] P. J. Hendra and E. J. Loader, *Trans. Faraday Soc.*, 1971, **67**, 827.
[35] P. J. Hendra, J. R. Horder and E. J. Loader, *J. Chem. Soc. A.*, 1971, 1766.
[36] P. J. Hendra, I. D. M. Turner, E. J. Loader and M. Stacey, *J. Phys. Chem.*, 1974, **31**, 300.
[37] R. O. Kagel, *J. Phys. Chem.*, 1970, **74**, 4520.
[38] T. A. Egerton and A. H. Hardin, *Catal. Rev. -Sci. Eng.*, 1975, **11**, 71.
[39] N. Sheppard, A. H. Hardin, T. A. Egerton and Y. Kozirowski, *J.C.S. Chem. Comm.*, 1971, 887.
[40] G. Careri, V. Mazzacurati, G. Signorelli and M. Sampoli, *Phys. Lett.*, 1970, **32A**, 495.
[41] C. L. Angell, *J. Phys. Chem.*, 1973, **77**, 2.
[42] T. A. Egerton, A. H. Hardin and N. Sheppard, *Can. J. Chem.*, 1975, **54**, 586.
[43] J. G. Grasselli, M. Snavely and B. Bulkin, *Chemical Applications of Raman Spectroscopy*, Wiley, New York, 1980.
[44] P. J. Hendra, C. Passingham, G. M. Warnes, R. Burch and D. J. Rawlence, *Chem. Phys. Lett.*, 1989, **164**, 178.
[45] R. Burch, C. Passingham, G. M. Warnes and D. J. Rawlence, *Spectrochim. Acta*, 1990, **46A**, 243.
[46] E. P. Parry, *J. Catalysis*, 1963, **2**, 371.
[47] E. Bright Wilson, *Phys. Rev.*, 1934, **45**, 706.
[48] A. Mortenson, D. H. Christensen, O. Faurskov Nielsen and E. Pederson, *J. Raman Spectrosc.*, 1991, **22**, 47.

11

The analysis and characterization of inorganic compounds

11.1 INTRODUCTION

The message throughout this book has been that the introduction of near infrared sources and FT processing has made Raman spectroscopy a routine, widely available and convenient technique. In this chapter, we must be somewhat more circumspect in our enthusiasm. Although inorganic chemists have found fluorescence and photodecomposition familiar problems when attempting Raman measurements they have also been dogged by problems associated with absorption. The transitions between d states give rise, in almost all transition elements, to intense absorption bands in the visible. These absorptions vary with ligand type and oxidation state, and can frequently preclude measurements from series of related compounds. The traditional method of avoiding the problem is to use a variety of laser lines†. Argon and krypton ion lasers can be employed to produce usefully powerful source lines from the blue to the deep red. An alternative approach is to drive a dye laser (or more recently a titanium sapphire device) with an argon laser and tune the emission from the yellow to the near infrared. Many inorganic chemists have exploited laser absorption by the sample in investigating the resonance Raman spectra of their compounds. This has been particularly prevalent in the study of dilute solutions of bio-inorganic materials such as haemoglobin. Inorganic chemists have always found Raman spectroscopy useful and have habitually switched laser lines to taste. The field is certainly not routine, and is definitely for the dedicated, since the operation of

---

† Strictly this comment is not correct. Long before the laser was introduced as a source, inorganic Raman spectroscopists used 'unusual' sources. Thus, in the days when mercury arcs were the norm people were recording spectra of deeply coloured materials like chromyl chloride using rubidium or helium discharge lamps. The classic work of Stammreich on coloured compounds deserves prominent mention and is reviewed with much else in Nakamoto's invaluable book *Infrared and Raman Spectra of Inorganic and Coordination Compounds* [1].

complex laser sources, their filtering (usually carried out with a wide range pre-monochromator) and the subsequent processing of Raman scatter require skill and practice. Detectors pose problems as they are all much less sensitive in the near infrared. Thus, although modern CCD-based systems work excellently, they perform less adequately as one moves further into the near infrared and this can restrict the freedom of action of the inorganic Raman enthusiast.

During the pioneering phase of FT Raman development it was hoped that the Nd^{3+}:YAG laser would be an ideal source, i.e. it would not be absorbed significantly and as such was unlikely to cause photo- or thermal decomposition. Unfortunately this has not been the case. Studies to date on inorganic materials using FT Raman spectrometers have not always met with the dramatic success encountered in other areas. Either the samples absorb and hence burn in the beam or, in relatively rare cases, massive fluorescence has been encountered. As mentioned in Chapter 6, measurements from aqueous solutions have been precluded by absorption of the Raman light by the solvent. However, excellent spectra can be recorded from accessible samples with ease and speed. Although it is hard to be quantitative, we estimate that success has only been attained in 25% of the samples submitted to us. This, of course, is very different from equivalent experiments in organic analysis where real failure is only encountered in about 15% of the samples submitted. Thus, whole areas of topical inorganic chemistry are inaccessible to our current technique, e.g. the study of clays or compounds of certain well-defined and important oxidation states such as Cu(II).

Early FT Raman spectrometers could measure bands to within some four hundred wavenumbers of the exciting line and hence were immediately acceptable from an organic analytical viewpoint. However, when heavy atoms are involved it is essential to be able to approach much more closely to the exciting line. Normally the range down to $\Delta v = 100 cm^{-1}$ is thought to be useful. Thus, for example, calomel (Hg_2Cl_2) shows an intense v_{Hg-Hg} band at $\Delta v = 153 cm^{-1}$†.

This is not to imply that the method is useless — very fine spectra of *some* inorganic compounds can be recorded with unprecedented ease, but the restriction on sample type and the lack of resonant enhancement using the Nd^{3+}:YAG source means that commercial FT Raman spectrometers are a useful addition to traditional Raman equipment but certainly not a replacement for it. To give readers a feel for the progress made to date and the limitations we will briefly survey the experiments reported at the time of writing.

11.2 SMALL COMPOUNDS

Of the compounds investigated to date, yellow and red chromium compounds are all accessible using Nd^{3+}:YAG excitation, and generally colourless and pale compounds present few problems. These include numerous tetroxometallic anions, sulphur oxy species and many others. The only restriction on the results obtained has been the inability to record bands at low frequencies. Improvements in filter

---

† In a paper to appear shortly Bochmann and Webb have investigated a series of RSeSeR and RTeTeR species. The v_{Se-Se} frequency range is 271–286 cm^{-1} and the v_{Te-Te} range is 167–188 cm^{-1} [2].

technology have almost lifted this restriction since it is now easy to operate down to $\Delta v \approx 100$ cm^{-1} or a little less. However, success is by no means universal. Perhaps the most evocative example comes from manganese chemistry. Although a thoroughly awful spectrum of the permanganate ion was to be found in the literature excited by a He–Ne laser in the early days of laser Raman spectroscopy [3], the near infrared FT Raman spectrum is quite superb (see Fig. 11.1) [4]. On the other hand, the closely

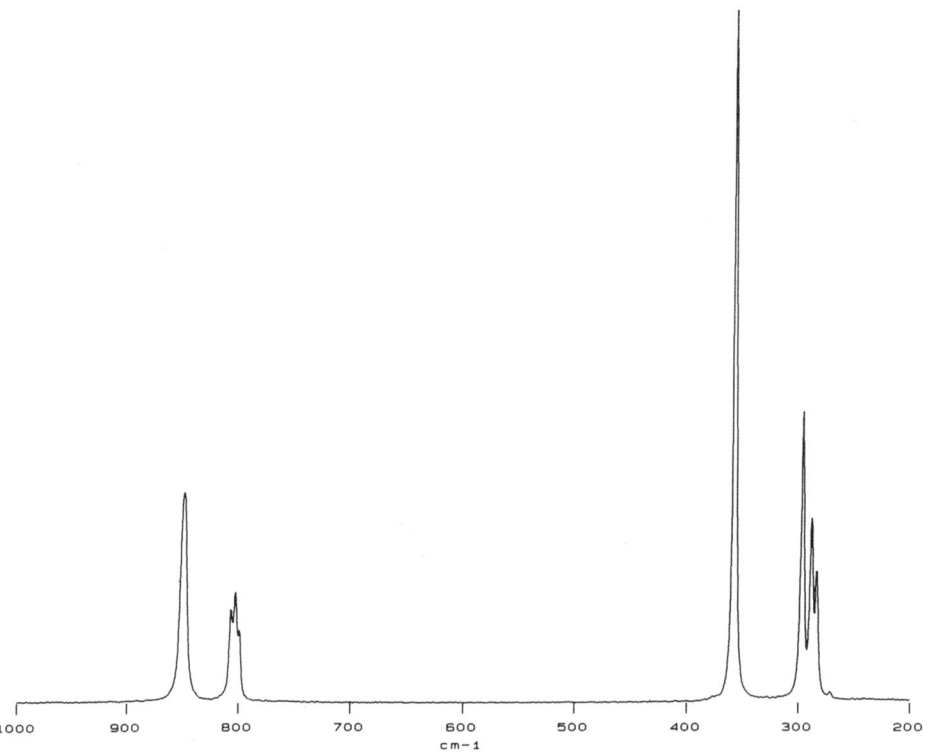

Fig. 11.1 — The FT Raman spectrum of potassium permanganate (KMnO$_4$) excited with an Nd^{3+}:YAG laser. Uncorrected data.

related deep green manganate ion MnO$_4^{2-}$ gives no spectrum at all. Colourless or pale crystalline species such as rhenate, molybdate and tungstate salts gave superb spectra, in general of higher quality than those reported in the literature.† The subject has been described in some detail in a recent paper [4].

Although our comment about spectral quality applies to the study of powders, the aqueous solution spectra are far less promising. As mentioned in Chapter 6 water absorbs in the near infrared, attenuating the Raman bands from dilute solutions.

† The work cited is very early and could be improved upon very considerably by using modern conventional instrumentation.

This is unfortunate, of course, because solution spectra can be examined by polarized Raman experiments which can make our understanding of the origin of the bands, or in favourable cases the structure of the ions, more rigorous.

Many halogen-containing species are coloured, e.g. the platinum and palladium square planar compounds of general formula MX_4^{2-}. Although spectra of very variable quality of the series M=Pd, Pt or Au and X=Cl, Br or I are reported in the literature [5], by no means all can be successfully studied using Nd^{3+}:YAG excitation. The chloroplatinate ion gives an excellent near infrared FT Raman spectrum, but some of the Pd(II) ammino halides burn in the beam. Again, where spectra are produced they are very rapidly recorded, are of high quality [6], and were measured on solids. Two examples are to be found in Fig. 11.2.

Chromyl chloride is an excellent example of a covalent inorganic compound which gives a first class spectrum very easily. One of the most attractive features of Raman spectroscopy has always been that the sample can be held in a sealed glass vessel and studied *in situ*. This advantage applies also to the FT method. The spectrum in Fig. 11.3 was recorded from a sealed ampoule.

Turning now to small organometallics, Almond and his co-workers have published several papers, including a description of their work on several dialkyl tellurides [7,8]. They made the point that photodecomposition due to visible excitation can be a problem in these samples, e.g. (*tert*-butyl)$_2$Te is not normally accessible but the use of a Nd^{3+}:YAG laser solves the problem. They go on to enthuse that in their area of interest, several intriguing possibilities can be defined. Thus, the thermally unstable species diethyl cadmium and its adducts can be studied by near infrared Raman spectroscopy (though only at low laser powers) whereas they had had little success before [7]. More recently Almond *et al.* have extended their studies as part of an investigation into the preparation of the semiconductor gallium nitride from its precursor $(CH_3)_2GaNHR_2$ (R=alkyl) [8]. Many of Almond's samples are toxic and/or extremely unpleasant. The tellurium alkyls smell to a degree and with an odour that is indescribable. As emphasized above, the ease of sampling (or more correctly the lack of sampling) is very valuable and makes our approach to Raman spectroscopy very attractive. Other semiconductors have been studied at Southampton. In these cases the spectrum excited by 1.064−μm radiation consists of an intense luminescence background overlaid by one or more Raman lines. This background may itself be of interest and may well prove eventually to be valuable in monitoring materials by this method.

In collaboration with industry a significant numbers of samples containing relatively simple inorganic ionic species have been studied. These included ranges of borates, concentrated silicate solutions, phosphates and calomel. Preliminary studies on xanthates, related sulphur-containing species and thiosulphates have also been made (see Fig. 11.4).

FT Raman spectra of superconductors have begun to appear. Zamboni *et al.* [9] have reported data on the $YBa_2Cu_3O_{7-x}$ system, varying the oxygen content from $x=0.85$ to 0.15. The species at lowest oxygen content gave several bands in the 500–1500 cm^{-1} region but these were very different from the spectra excited by the blue argon line. The authors therefore conclude that the FT Raman spectra were resonance-enhanced in the near infrared and go on to explain their reasoning.

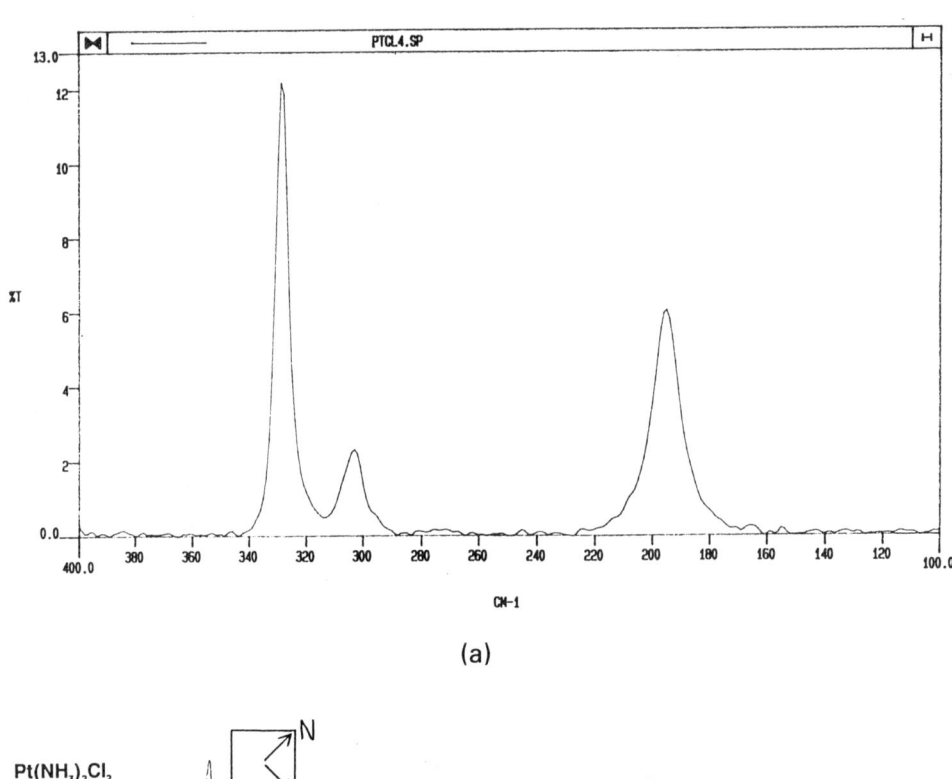

(a)

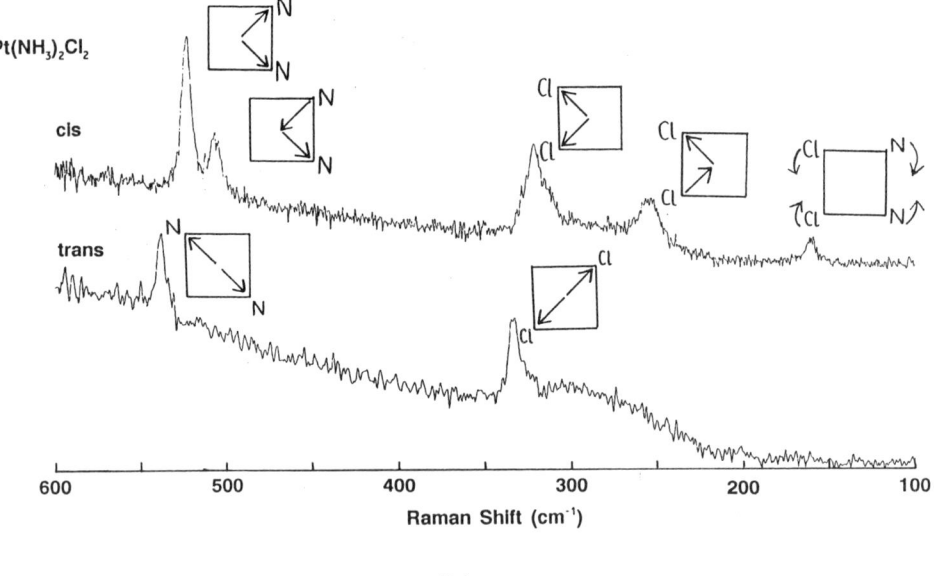

(b)

Fig. 11.2 — The FT Raman spectra of (a) $PtCl_4^{2-}$ and (b) $Pt(NH_3)_2Cl_2$. Courtesy of A. J. Rowlands.

Sec. 11.3] **Co-ordination compounds** 265

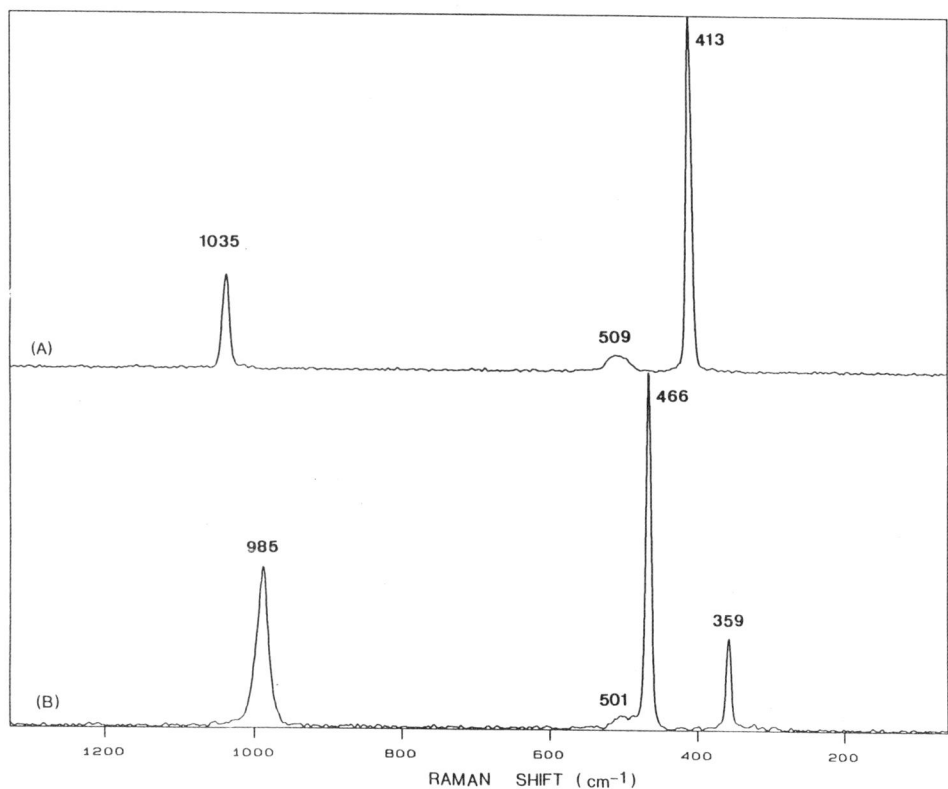

Fig. 11.3 — The FT Raman spectrum of (a) chromyl chloride (CrO_2Cl_2), and (b) vanadyl chloride ($VOCl_3$). Courtesy of A. J. Rowlands.

Several of the elements give good spectra, e.g. sulphur, selenium and diamond (see Figs 11.5 to 11.7), whilst the gases nitrogen and oxygen can be excited weakly in the near infrared. However, colour can again be a significant problem and thus it is not possible to obtain a spectrum from red phosphorus. Very recently the spectra of C_{60} and C_{70} have been recorded. Here, the molecule is hollow, spherical or football-shaped, and highly regular and is formed at discharges between carbon electrodes struck in an inert gas environment. The materials are extremely good scatterers and produce excellent spectra at very low laser powers (see Figs 11.8 and 11.9)

11.3 CO-ORDINATION COMPOUNDS

Spectra of co-ordination compounds of various type have begun to be published. Thus, for example some of the simple platinum (II) and palladium (II) ammines give spectra [6,10]. On the other hand, several are unexpectedly inaccessible, e.g. Magnus's green salt $[Pt(NH_3)_4][PtCl_4]$ gives no spectrum due to absorption but its

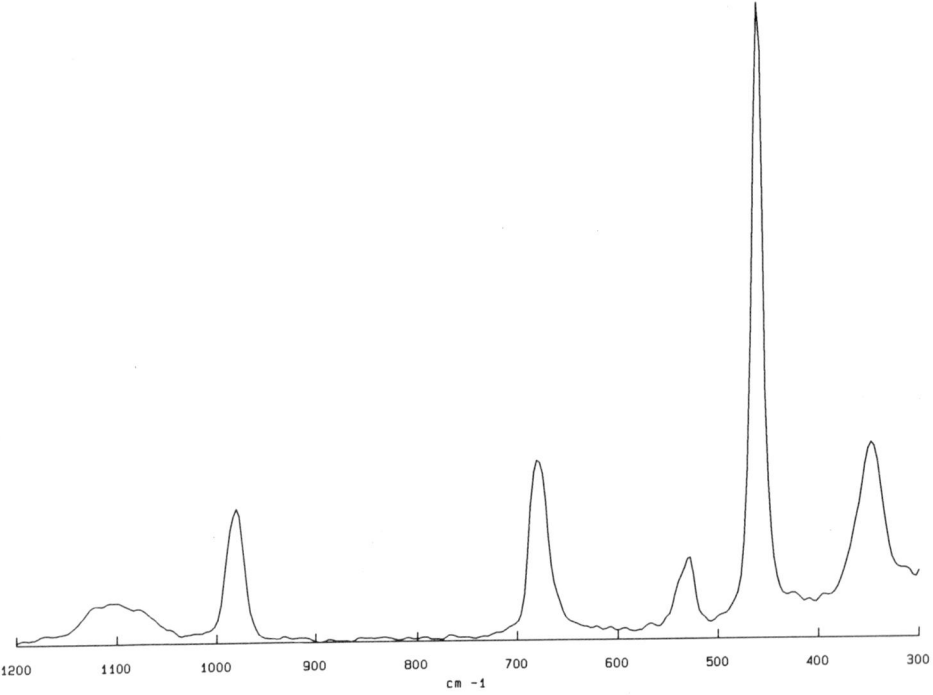

Fig. 11.4 — The FT Raman spectrum of ammonium thiosulphate. Courtesy of A. J. Rowlands.

isomer *cis*-diamminodichloroplatinum (II) gives an adequate spectrum (see Fig. 11.2(b)). The spectrum of the familiar red solid nickel dimethylglyoximate also gives a fine spectrum (Fig. 11.10) [11]. This material is of interest because the hydrogen bond is thought to be anomalously strong and hence its vibrational frequency is of importance [12,13]. Infrared studies in the literature assign the OH stretching vibration to a band at the incredibly low frequency of 1775 cm^{-1}. However, the FT Raman spectrum contains a band near 3100 cm^{-1} which could not be due to anything else other than the hydrogen-bonded v_{OH}, suggesting that the hydrogen-bond vibrates relatively traditionally. Numerous attempts have been made in the past to record the spectrum of this deeply coloured material, and resonance enhancement has been invoked to explain the assignments of the bands to fundamental modes. The near infrared FT Raman spectrum is certainly not resonance-enhanced and is very nearly identical to its visibly excited counterpart, hence resonance enhancement is unlikely to play a role when using visible sources. As a consequence, some of the explanations for the bands need urgent rethinking. Recently, Donohue, Dyer and Swanson [14] have reported the results of some very elegant experiments on the complex salt $[Pt^{II}(en)_2][Pt^{IV}(en)_2Cl_2][ClO_4]_4$. They photolysed the sample in the red and then recorded resonance-enhanced Raman spectra. The latter were analysed with a conventional Raman spectrometer but could just as readily have been studied using an appropriate interferometer.

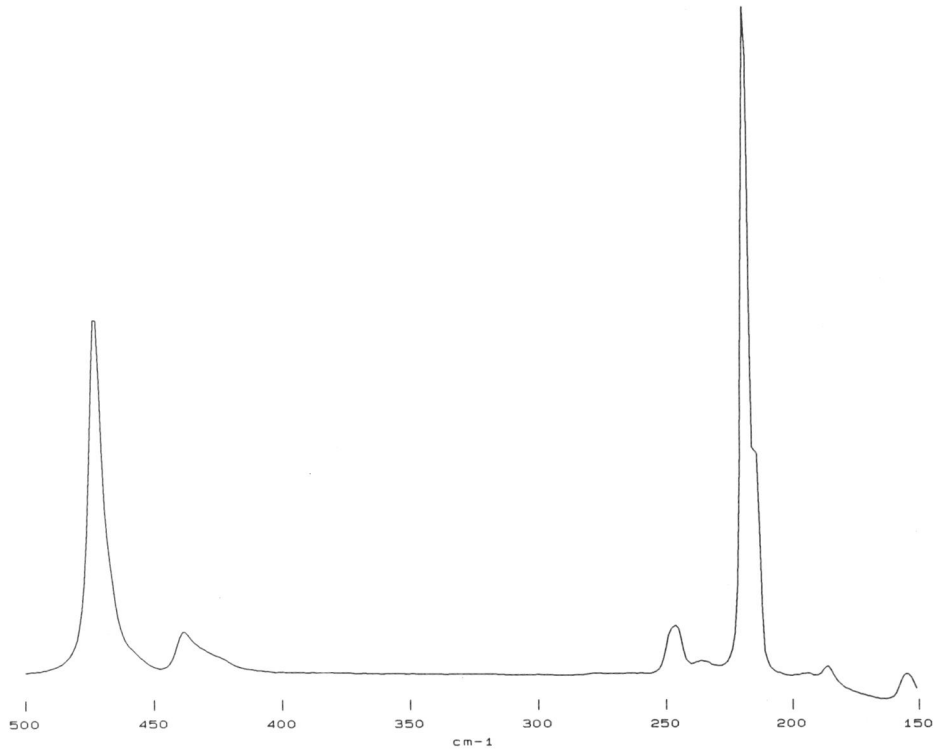

Fig. 11.5 — The FT Raman spectrum of sulphur. Uncorrected data.

11.4 CARBONYL AND RELATED COMPLEXES

Several papers have appeared recently reporting a wide range of spectra of carbonyl compounds. Most have been of a survey nature reporting exploratory work in the near infrared. Thus, Barnett [15] has described several spectra of compounds ArCr(CO)$_3$ where Ar=(phenyl)COONHS, (phenyl)OCH$_3$, (phenyl)CH$_3$, (phenyl)(CH$_3$)$_3$ and cyclopentadienyl manganese tricarbonyl. In another report, Rowlands et al. [16] attempted to survey a vast range of carbonyls and arenes to see which materials gave spectra and which did not. Spectra were successfully recorded and analysed from Cr(CO)$_5$NH$_3$, W(CO)$_5$NH$_3$, W(CO)$_5$P(phenyl)$_3$ and Mo(CO)$_5$P-(phenyl)$_3$, the bipyridyl complex Mo(CO)$_4$(bipyr) and the biphenyl derivatives [Cr(CO)$_3$]$_2$C$_{12}$H$_{10}$ and [Cr(CO)$_3$]$_2$C$_{12}$H$_9$CH$_3$. The paper also includes several arenes, including fluorene, benzene and cyclopentadiene, all involving metal carbonyl fragments. Unfortunately, high quality spectra were only recorded in a relatively small number of cases, the majority of samples giving poor quality spectra or no spectrum at all. Rowlands et al. concluded that the Nd^{3+}:YAG source is useful but certainly not universally applicable. The paper includes quite a detailed analysis on the origin of the bands in several of the carbonyl species and hence can be recommended as a source in this field.

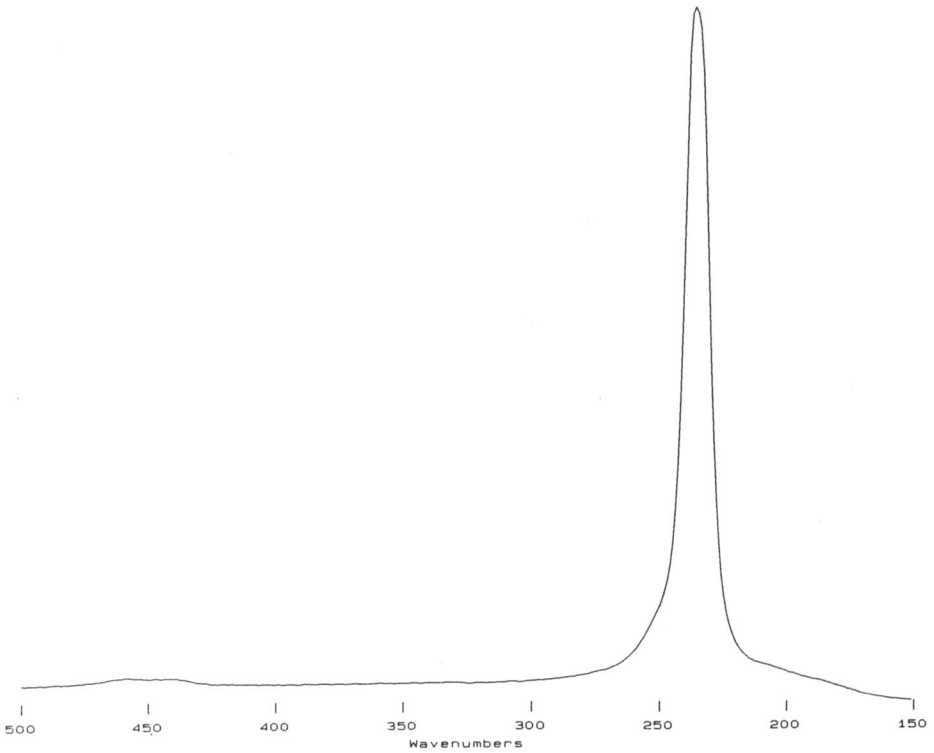

Fig. 11.6 — The FT Raman spectrum of selenium. Uncorrected data.

Nie et al. [17] report Raman spectra of the biosystem organocobalt B12 with particular attention to the cobalt–carbon bond. The authors propose that a sharp, intense band near 500 cm^{-1} results from the Co–CH$_3$ stretching mode. The authors touch on a current deficiency of the FT Raman technique, notably the great difficulty in studying aqueous solutions. The recording of FT Raman spectra of solid samples of biological origin is of limited use as the interesting chemical reactions in which they are involved occur in dilute solutions. Clearly, as sensitivity improves, the study of solutions will inevitably become more important. This is very attractive because depolarization measurements can then be made to assist spectral assignment. However, there is considerable evidence in a wide range of compounds that the structure of the molecules is not the same in solid and solution phases and it is this point that Nie discusses. Biological applications of FT Raman are covered in Chapter 9.

Explaining the spectra of small molecules can be assisted by the judicious use of group theory. It must, however, be remembered that the point group of a molecule or an ion in a crystal is almost always very different from the species in an unperturbed state. As a result, more infrared or Raman bands are seen than expected. The subject is complex but was reviewed many years ago [18].

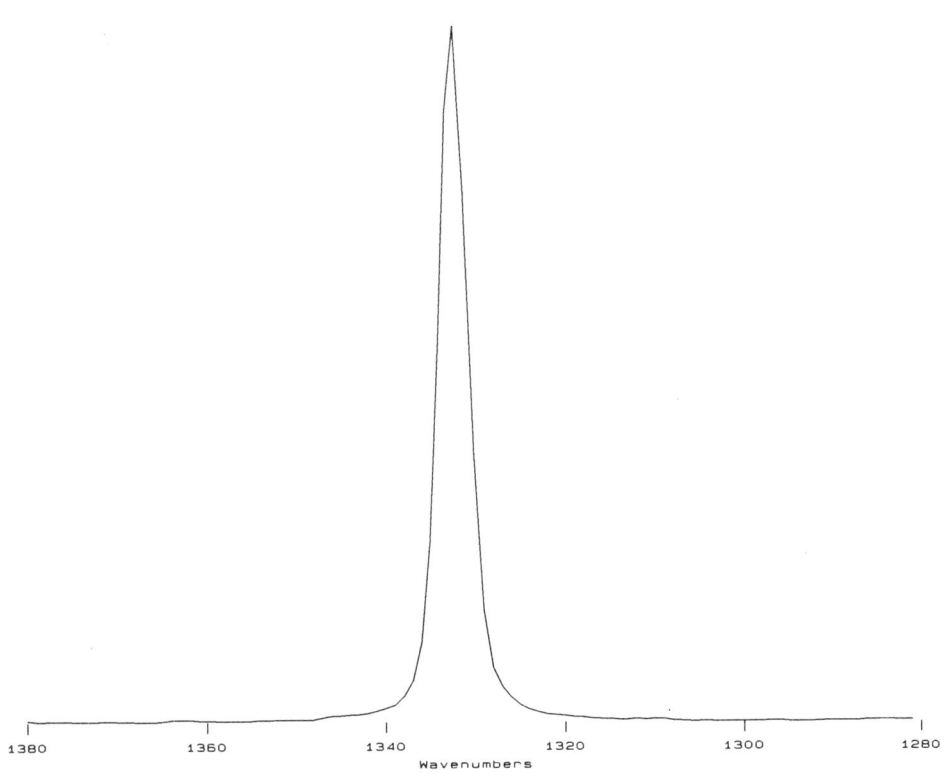

Fig. 11.7 — The FT Raman spectrum of diamond. Uncorrected data.

To conclude, FT Raman spectrometers using Nd^{3+}:YAG laser sources are capable of giving excellent spectra over useful frequency ranges for *some* inorganic compounds. There really is no reason why just the neodymium laser should be used. As new lasers develop and are applied it is certain that the versatility of the technique will dramatically improve.

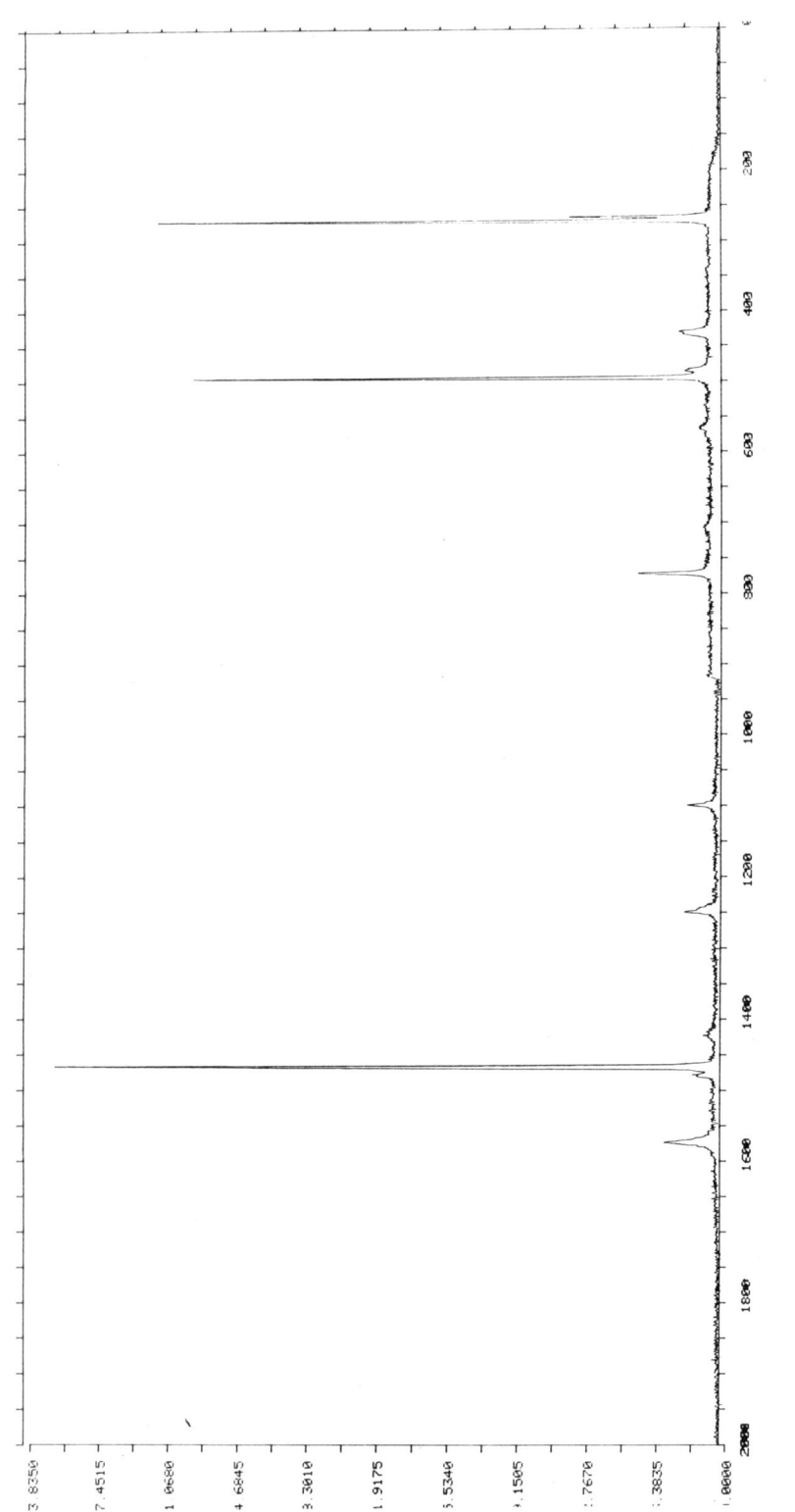

Fig. 11.8 — The FT Raman spectrum of C_{60}. Uncorrected data. Courtesy of Prof. H. Kroto et al.

Sec. 11.4] Carbonyl and related complexes 271

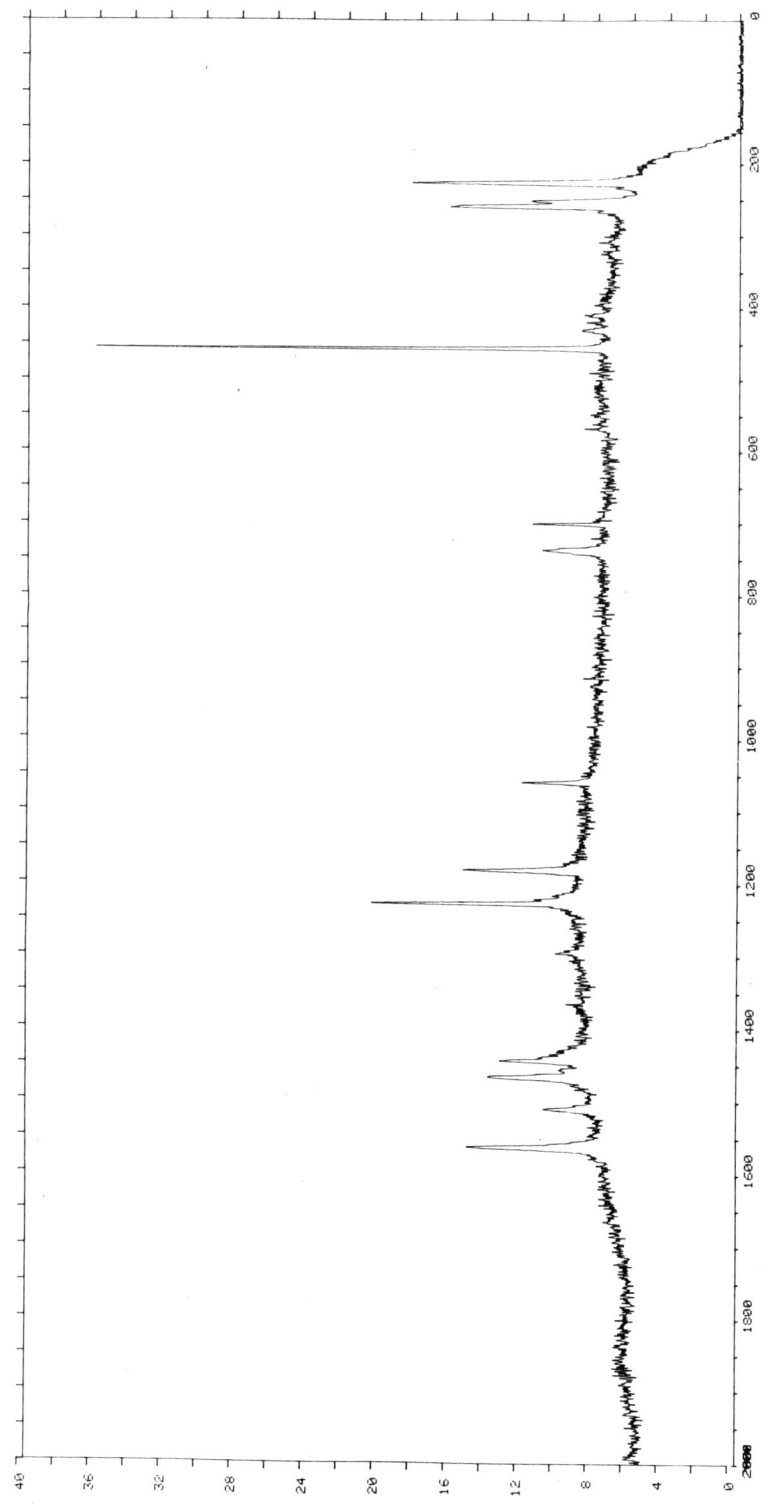

Fig. 11.9 — The FT Raman spectrum of C_{70}. Uncorrected data. Courtesy of Prof. H. Kroto *et al.*

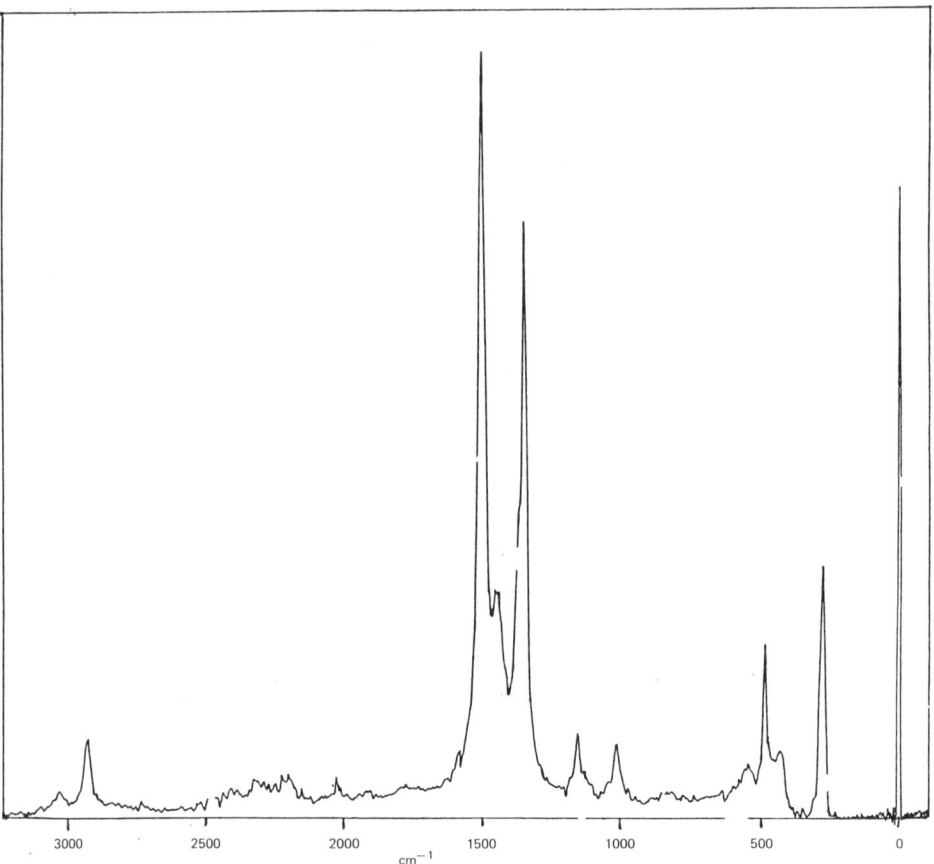

Fig. 11.10 — The FT Raman spectrum of nickel dimethylglyoximate. Courtesy of P. le. Barazer.

REFERENCES

[1] K. Nakamoto, *Infrared and Raman Spectra of Inorganic and Coordination Compounds*, Wiley, New York, 1986.
[2] M. Bochmann and K. J. Webb, *Spectrochim. Acta*, 1991, **47A**, FT Raman special edition, in press.
[3] P. J. Hendra, *Spectrochim. Acta*, 1968, **24A**, 125.
[4] P. J. Hendra, P. Le Barazer and A. Crookell, *J. Raman Spectrosc.*, 1989, **20**, 35.
[5] P. J. Hendra, *Nature*, 1966, **212**, 5058.
[6] A. J. Rowlands and I. A. Degan, *Spectrochim. Acta*, 1991, **47A**, FT Raman special edition, in press.
[7] M. J. Almond, C. A. Yates, R. H. Orrin and D. A. Rice, *Spectrochim. Acta*, 1990, **46A**, 177.
[8] M. J. Almond, D. A. Rice, C. A. Yates, P. J. Hendra and P. T. Brain, *J. Mol. Struct.*, 1991, **239**, 69.
[9] R. Zamboni, A. J. Pal, G. Ruani and C. Taliani, *Physica C*, 1989, 162–164, 1103.
[10] A. J. Rowlands, *Proceedings of the 12th International Conference on Raman Spectroscopy, Columbia, South Carolina, USA*, Ed. J. R. Durig, J. F. Sullivan, Wiley, Chichester, 1990, p398.
[11] P. Le Barazer, *PhD Thesis*, University of Southampton, 1991.
[12] R. E. Rundle and M. Parasol, *J. Phys. Chem.*, 1952, **20**, 1487.
[13] Y. Ohashi, I. Hanazaki and S. Nagakura, *Inorg. Chem.*, 1970, **9**, 2551.
[14] R. J. Donohue, R. B. Dyer and B. I. Swanson, *Solid State Comm.*, 1990, **73**, 521.
[15] S. M. Barnett, F. Dicaire and A. A. Ismail, *Can. J. Chem.*, 1990, **68**, 1196.
[16] P. J. Hendra, T. N. Day, A. J. Rowlands and A. J. Rest, *Spectrochim. Acta*, 1991, **47A**, FT Raman special edition, in press.
[17] S. Nie, P.A. Marzilli, L.G. Marzilli and N. Yu, *J. Am. Chem. Soc.*, 1990, **112**, 6084.
[18] P. J. Hendra and T. R. Gilson, *Laser Raman Spectroscopy*, Wiley, Chichester, 1970.

12

Future developments

The purpose of this chapter is to peer into the crystal ball and suggest to readers what they may reasonably expect in the future. The difficulty, of course, is that the subject is expanding rapidly and many of these predictions will be tested long before this book is out of print. As a result it has been decided to describe those developments that, it is believed, are more likely to occur imminently and then separately to comment on more speculative developments.

12.1 IMMINENT DEVELOPMENTS

All the instruments offered at the moment are in fact accessories usually interfaced to the emission ports of existing FTIR spectrometers. To enable manufacturers to scan the near infrared, appropriate multiple interchangeable beam-splitters are incorporated or in one case a long-range beam-splitter of reasonable performance has been devised. Although all the right decisions have been made in their design, no manufacturer is offering an instrument designed from the start as an infrared/Raman FT machine. FT Raman equipment has been a marketable commodity for about two years, hence it could reasonably be anticipated that the next generation of medium or high quality FT infrared instruments may be designed as combined instruments. However, instrument manufacturers expect a limited sales life from their products, probably between five and eight years, and the gestation period of new designs in many companies can be so long that the replacement of this design may take a number of years to achieve. There are a number of features which should be incorporated in a combined infrared/Raman instrument which are absent in one intended originally for infrared alone.

The principal problem in Raman spectrometry is sensitivity. Assuming all manufacturers have access to similar suppliers of filters and detectors, the only real difference between the performance of their instruments will be the optical throughput. There is no doubt that, although it may be attractive commercially, a long-range-beam splitter is unlikely to perform as well as one optimized for the near infrared. The reasons are complex and involve the index of refraction of the beam-splitter material, its flatness, the perfection of the optical surfaces and the quality of the coatings. The stability, optical perfection and, of course, longevity of quartz beam-splitters make these devices very attractive. As time goes by, the surfaces of uncoated potassium bromide surfaces inevitably deteriorate even if the humidity is carefully controlled, but the deterioration in quartz and calcium fluoride components should be negligible. Hence, it is predicted that all new designs, unless they are aimed at the low cost sector, will incorporate rapidly interchangeable beam-splitters, with a commensurate improvement in FT Raman sensitivity. However, if the beam-splitters have to be moved by the user as part of their replacement, and particularly if they must be lifted out of the machine, deterioration is almost inevitable. Motor-driven interchange within the sealed optical bench is essential.

Another critical point is the étendue of the interferometer. For Raman experiments it must be as large as is feasible, possibly larger than is really necessary for the vast majority of FTIR applications. Hence it is to be hoped that the f number of the interferometer and the Jacquinot stop size used in the Raman experiment will be subject to some improvement in newer designs.

Enhancements in detector sensitivity are also anticipated. Once written off by many, the indium gallium arsenide (InGaAs) device is being continuously improved and it is now a very attractive detector. It is very cheap and can be operated at room temperature, enabling users to scan to $3500\,cm^{-1}$ Stokes shift when using the Nd^{3+}:YAG source. Although, the extended germanium device is excellent and is used by at least one current manufacturer, unless it is reduced in cost and physical size it is unlikely to completely supersede the InGaAs detector. Indeed the performance of the state-of-the art InGaAs devices is rapidly approaching that of the extended-range germanium detector for only 20% of the cost. Several near infrared photomultipliers have been announced, but they have failed, as yet, to match the performance of the semiconductor devices. Near infrared sensitive image intensifiers of high performance have been demonstrated in military applications and these could provide a route to new and better detectors.

New laser systems are constantly being proposed and developed, but which are likely in the short term to find applicability in FT Raman instruments? A change from 1.064-μm excitation is not expected in the near future. The reasons for this are commercial rather than fundamental. Laser manufacturers are keen to develop new, better, smaller, cheaper Nd^{3+}:YAG lasers because these improvements can be made easily with existing technology. It is clear that high power stable solid-state powered lasers will very soon become the normal excitation source. Once their cost is reduced, the freedom they offer from instability, elaborate liquid cooling systems and electrical inefficiency make them overwhelmingly superior to the xenon arc or tungsten/quartz-iodine lamp-powered, liquid-cooled devices used currently. Sufficient investment has already been made in solid-state developments that laser manufacturers must be obliged to recover their research costs before considering the possible benefits of new laser systems. This, in turn, will continue to lead filter manufacturers to have confidence that their investment in developing better and better filters for use with Nd^{3+}:YAG sources will bring them sales.

Filters, as reported in Chapter 5, are of two types, transmission and reflection. At the moment the majority of spectrometers incorporate the much simpler and hence cheaper transmission filters as standard. Currently, transmission filters offer cut-in frequency shifts of around $\Delta v = 100 \, cm^{-1}$ and can be made relatively free from periodic fluctuations in transmission across the spectrum. Reflection filters allow spectra to be recorded at lower Raman shifts (around $50 \, cm^{-1}$) but the extra cost incurred for this improvement will probably mean that they will remain an optional extra for specialist applications. Both types of filter are likely to improve. It is, however, most unlikely that any filters will appear to seriously challenge the frequency shift performance of multiple-scanning monochromator-based spectrometers. Our 12-year-old Coderg triple monochromator machine operates in the green down to $\Delta v = 5 \, cm^{-1}$ when recording longitudinal acoustic modes from polyethylene samples which lie between 5 and $40 \, cm^{-1}$ shift.

As will be explained later, it is thought that future instrument designs are likely to incorporate lasers capable of operating at wavelengths shorter than $1.064 \, \mu m$. The implication for designers is that good performance down to $0.5 \, \mu m$ may well be required. The vast majority of analytical-grade FTIR machines are quite incapable of scanning to these wavelengths for a variety of reasons, including imprecision in the scanning system and inadequate data point density resulting from the wavelength of the laser that monitors the scan. These subjects are covered in detail in Chapter 4 and they too will influence designs of new combined infrared/Raman machines.

Perhaps the area where the instrument makers will make the maximum impact on intending purchasers and users is in the design of the sample area. At present, the commercial instruments are far easier to use than their diffraction precursors, but in our experience, none have yet approached the prototype machines at Southampton in this respect. Manufacturers insist in providing three-axis adjustment of the sample within the sample area, when it has been demonstrated that two of these are redundant if the sample holding system is properly designed. Since knobs are there for twiddling, the skill element required to recover from misalignment becomes a problem. It is anticipated that future designs will adopt one of two possible solutions. Either sample mounting systems should be preset by the service engineer or, far more satisfactorily, adjustments driven by stepper motors, and controlled through the software to optimize the sample alignment position. The sample area system at Southampton allows the correct position of the sample to be found by noting the superimposition of three He–Ne laser 'spots'. These are exactly analogous to the He–Ne beams passing through the infrared sample area. It is found that these are ideal for rapidly aligning unusual samples such as elastomers under strain, Dewars or high temperature cells. The elegant simplicity of this system accounts for the fantastically high throughput of samples attainable with the Southampton prototype. Sadly, this feature is not offered on most of the commercial instruments.

Accessory manufacturers or the instrument makers themselves will certainly produce cells for high and low temperature analysis, for experiments at high pressures or in electrochemical environments. One area of concern is the size and in particular the height of sample area lids. Contemporary designs do not allow the use of Dewars; it is hoped that future designs will be more practical. In particular, space above and below the optical axis is essential if *in situ* measurements need to made on a bulky sample. Often the latter is ignored.

Fibre-optic coupling between the FT Raman spectrometer and the sample has been demonstrated, but as yet it is far from routine. However, since losses down the fibre can be kept very low, particularly if they are carefully selected and laser wavelengths shorter than 0.9 µm are used, on-site *in situ* monitoring of composition or process control in the bulk chemical or pharmaceutical industries should develop rapidly. One could even hope to see multiple analytical sites coupled by fibre to a remote spectrometer itself controlled and monitored remotely by networked computers. Applications in quality control or even in screening of suspect samples at customs posts or airports have been proposed. The incorporation of FT Raman fibrescopes into dental probes and endoscopes has also been suggested.

In favourable cases, Raman methods can be used for high speed sorting and one or two proposals along these lines have appeared in the patent literature. The idea is to excite a really strong characteristic Raman emission, pass the collected radiation through narrow bandpass filters capable of transmitting a single wavelength. A sensor is then used to monitor emission at the particular wavelength chosen. The principle advantage of the interferometer/FT processor approach is that the whole spectrum is accumulated each pass of the interferometer.

A very attractive sample system in Raman spectroscopy is to use a modified microscope as the light collection system and to illuminate coaxially with the laser. It is normal to provide a television system to view the specimen for safety reasons and to provide a crosswire indication of the exact position of illumination of the specimen in the microscope's field of view. Apart from the obvious attraction of being able to microscopically assess specimens, Raman spectra can be recorded with a spatial radiation of a few micrometres.

Microscope attachments, optimized for use in the infrared, are available from several manufacturers. However, it is believed that with the introduction of combined infrared/Raman spectrometers, a combined infrared/Raman microscope should also be designed. A microscope can cost as much as an interferometer, and thus it is unfair to expect customers to purchase two separate systems. Combining the two microscopes into one is, in principle, straightforward as far as the microscope itself is concerned since the reflection optics required of the infrared measurement also operate in the near infrared. The problem centres on the optical layout. The currently available arrangements are of the form given in Fig. 12.1. With some ingenuity, an optical design could be produced that would allow the use of a single microscope attachment for use in both infrared and Raman experiments.

12.2 LONG-TERM DEVELOPMENTS

The longer-term developments described are of a more speculative nature.

12.2.1 New lasers and filters

To avoid the problem of water absorption discussed in Chapters 6 and 9, it is essential to find suitable new laser sources with wavelengths around 800–900 nm. Inevitably, the use of shorter wavelengths will be accompanied by an increased level of fluorescence. Laser diodes which operate in this area are already being developed. GaAs injection devices which operate at 850 nm are becoming available. These and

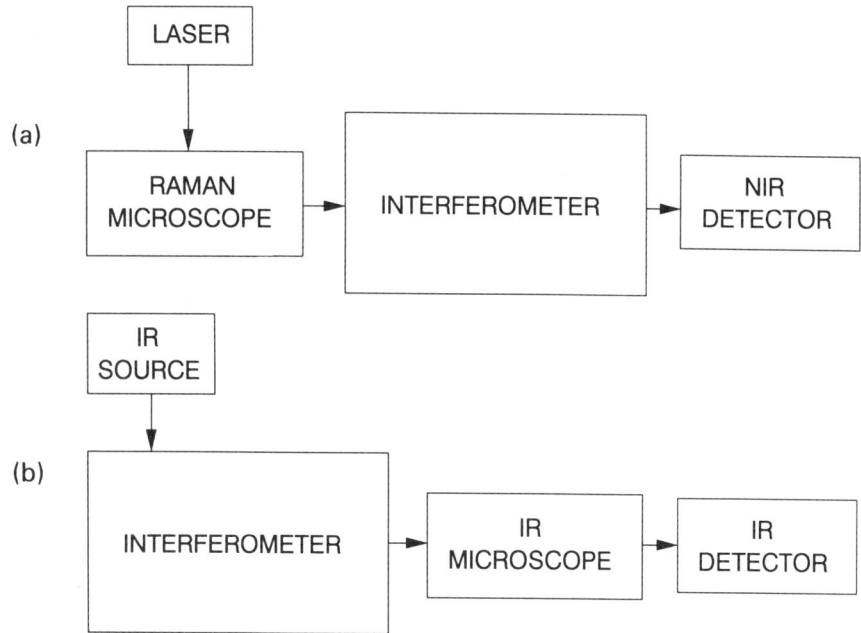

Fig. 12.1 — Existing optical arrangements for (a) Raman microscopes and (b) infrared microscopes.

other systems being developed are likely to be powerful and cheap. Currently, they have unattractive, non-circular beam profiles, but this problem will be overcome. It is impossible to predict the extent of the increase in the level of fluorescence which will be encountered at these wavelengths and the research remains to be done. Clearly, recording spectra from a random selection of samples using a shorter wavelength, near infrared source will result in more failures than using 1.064-μm excitation, but we do not know how many more failures are to be expected and whether this will impair the utility of the technique.

A number of samples fluoresce badly even when using a 1.064-μm source. Thus, inevitably for some types of analytical problem the FT Raman approach has failed to satisfy and for these samples longer wavelength sources would be desirable. Unfortunately, the problem is suitable detectors. The Nd^{3+}:YAG laser emits at 1.3 μm and in principle this source could be used, but the extended germanium or InGaAs detectors fail to respond beyond 1.7 μm and the maximum shift that they would be able to record would be 1800 cm^{-1}. It is therefore not clear if the benefits of using longer wavelength lasers would be worthwhile. Detector manufacturers would have to be very convinced before investing capital in developing high performance devices operating out to 2.5 μm. InAs detectors would be sensitive over the full Raman shift range, but these are very noisy and coupled with the reduction in Raman scatter due to the v^4 relationship it is likely that dreadful spectra would result. Certainly, the self-absorption of Raman light due to vibrational overtones will be considerably more troublesome.

Inorganic chemists have special problems since the transition elements absorb in the visible and near infrared. The approach normally adopted to record Raman spectra of ranges of inorganic materials is to switch from laser line to laser line. Clearly, if a significantly large market is thought to exist, new FTIR/Raman instruments capable of operating off argon and krypton ion laser lines as well as He–Ne or devices emitting in the near infrared will appear. One particularly suitable source would be a titanium sapphire laser which is tunable over the range 1000–650 µm. This would also be beneficial in SERS research where the ability to change the excitation wavelength is very appealing. This has recently been demonstrated by Rabolt [1].

The problem with all of these proposals is that a high performance filter set has to be developed for each laser line. Since it is impossible to couple a premonochromator to an interferometer without ruining the latter's favourable étendue, the problem of designing and manufacturing the filters is inescapable. Tunable filters would be nice, of course, but tunability would have to be combined with a consistently sharp cut in, a most daunting requirement. However, new filter systems appear regularly, and new multilayer geometries, Bragg filters and holographic devices have all been announced recently, so new and ingenious approaches to these problems are always in prospect.

One debilitating problem when using an Nd^{3+}:YAG laser to excite FT Raman spectra is the inability to study samples at temperatures much greater than 150°C. At this temperature thermal blackbody emission from the sample becomes much more intense than the Raman signal and manifests itself as a broad background. At best, this drastically reduces the signal-to-noise ratio; at worst, it saturates the detector completely. In current instrumentation the laser irradiates the sample continuously and the detector signal is integrated slightly before taking each data point. Recently, the possibility of using a pulsed laser as an excitation source has been demonstrated [2,3]. By using the reference laser interferogram (usually He–Ne) it is possible to trigger a Q-switched laser to produce a large pulse synchronized with the collection of a data point. The detector is 'gated' such that its signal is only available when the Q-switched laser is emitting a pulse. Thus, the Raman scatter is only produced when the Nd^{3+}:YAG laser is on, whereas the thermal emission is emitted continuously. The detector signal is only integrated for the duration of the Nd^{3+}:YAG laser pulse, but the energy of the pulse is adjusted such that the average power is the same as that emitted by a continuous wave laser, and consequently the Raman signal detected remains constant. However, because the signal measured for each data point is integrated over a much shorter time period, the contribution of the thermal background to the overall signal is very much less. The effect is to reduce the thermal background intensity by a factor of eighty and greatly improve the signal-to-noise ratio. Although this method allows much better quality spectra to be recorded from heated samples, it is as yet uncertain whether it will significantly increase the temperature at which samples can be studied.

12.2.2 Specialist instruments

Interferometers can be compressed into very small volumes and a hand-held interferometer may well become a reality. Lasers of adequate output, small weight

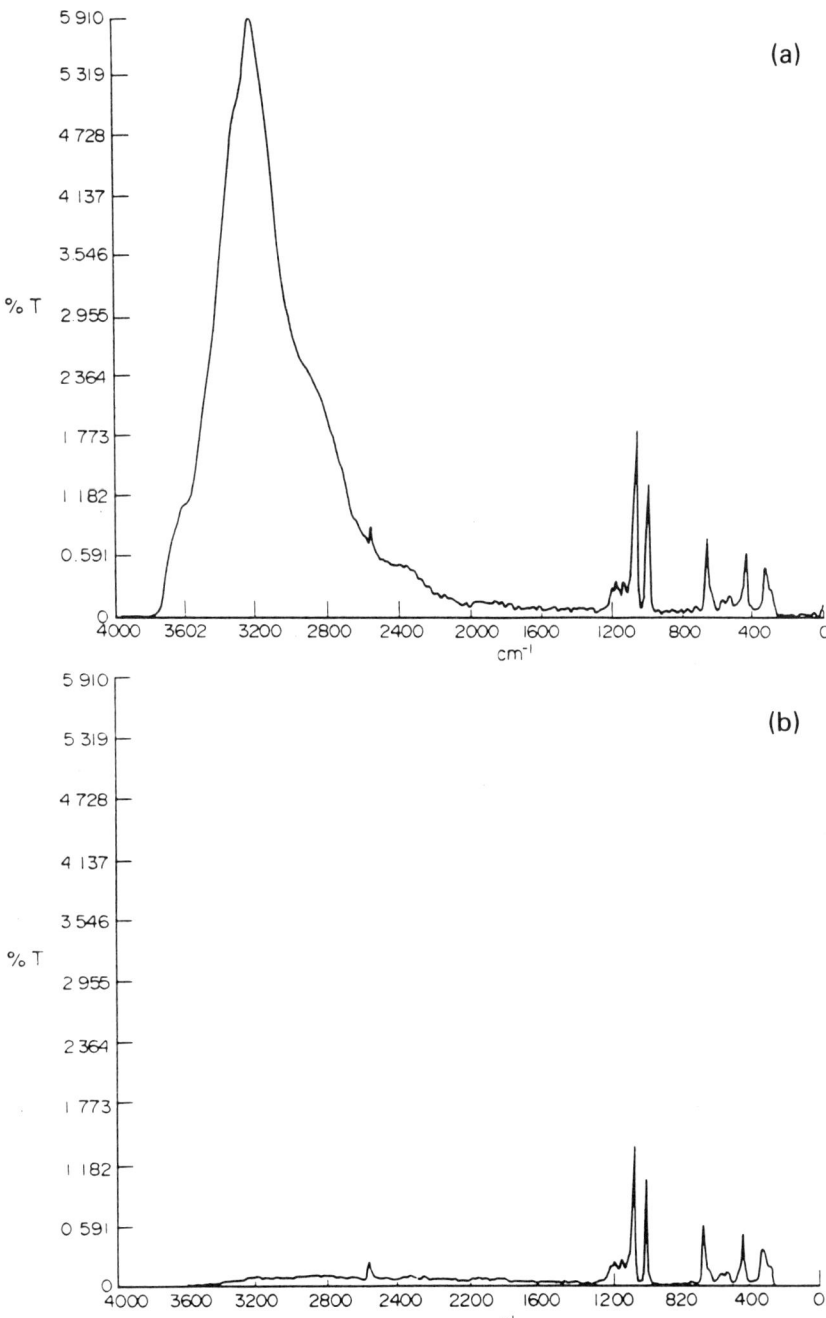

Fig. 12.2 — (a) The FT Raman spectrum of calcium sulphate at 140° recorded with a continuous wave Nd^{3+}:YAG laser producing 22 mW, and (b) the FT Raman spectrum of the same sample obtained using a Q-switched Nd^{3+}:YAG laser pulsing at 3.16 kHz giving an average power output of 22 mW. Uncorrected data. Reproduced with permission from *Spectrochimica Acta*, **46A**, C. J. Petty and R. Bennett, copyright 1990 Pergamon Press PLC.

and size and with very modest power requirement are already available and could be incorporated into a miniature Raman instrument. Data processing systems are also capable of miniaturization if this market developed. Thus, small instruments for specific analysis in defined analytical areas such as medical analysis, water quality control or food production can be envisaged. Trace analysis using enhancement based on SERS (Chapter 10) or possibly resonance techniques is also known to be practicable.

12.3 ALTERNATIVES TO FT METHODS FOR NEAR INFRARED RAMAN SPECTROSCOPY

Protagonists of the charged-coupled device (CCD) point out that as soon as it is sensitive to near infrared radiation, these devices mounted in spectrographs will sweep all before them. However, the CCD approach seems unattractive because the devices are small in physical size. To obtain the multiplex advantage a reasonably large part of the spectrum must be surveyed. However, to gain acceptable resolution the entrance slit of the spectrograph must be narrow. In other words, the dispersion of spectrographs containing these devices is low and thus the examined patch at the sample is small and sample alignment is critical. This poses no problem for the specialist, but will deter the analyst. Alternatively, the whole spectrum can be obtained in one go, rather than record it in sectors, but the penalty is poor resolution spectra containing currently 1024 data points at most. Further, if the CCD/spectrograph approach is adopted, a dedicated Raman machine and an infrared absorption spectrometer would be required. This would be a rather expensive prospect. However, in an academic environment where time is not the most pressing problem the sensitivity of CCD spectrometers makes them superlative research tools.

The large instrument makers seem to have all decided that the FT Raman method is here to stay. Subject to the impressive expertise and resources of these manufacturers and stimulated by intense competition it is clear that the field of FT Raman spectroscopy will expand very rapidly, as FTIR did during the 1980s.

12.4 SOURCES OF INFORMATION

Publications on FT Raman spectroscopy appear throughout a wide range of journals because their prime interest lies in the applications rather than in pure spectroscopy. In an effort to provide some sort of centre of interest, the important spectrochemical journal *Spectrochimica Acta* has started an annual special edition exclusively dedicated to FT Raman applications and developments.

In 1990, this journal produced the first of these highly successful special editions, describing the principles and applications then demonstrated in FT Raman spectroscopy. The title pages are given in Table 12.1. This edition contained 20 papers, most of which surveyed applicability and described hopes for the future right across science, including polymers, elastomers and paints, surfaces and a range of analytical applications from forensic science to Mars bars. In 1991, the second edition reflects the increasing maturity of the subject in that it contains 43 papers nearly all describing specific detailed applications. The titles of the papers are listed in Table 12.2. It is hoped these volumes will become regarded as the source of information on the latest applications and developments in FT Raman spectroscopy.

Table 12.1 — Papers in the first special edition of *Spectrochimica Acta* (**46A** no. 2)

The development of Fourier Transform Raman spectroscopy.
Fourier Transform Raman instrumentation.
A microscope for Fourier Transform Raman spectroscopy.
Quantitative analysis using Raman methods.
Fourier Transform Raman spectroscopy with polarization modulation.
Fourier Transform Raman spectroscopy — a tool for inorganic, organometallic and solid state chemists?
Future directions for Fourier Transform Raman spectroscopy in industrial analysis.
Applications of Fourier Transform Raman spectroscopy in the synthetic polymer field.
Fourier Transform Raman spectroscopy of elastomers: an overview.
The application of Fourier Transform Raman spectroscopy to the study of paint systems.
The use of Fourier Transform Raman spectroscopy in the characterization of catalysts.
Fourier Transform Raman studies of materials and compounds of biological importance.
Fourier Transform Raman spectroscopy of polymeric biomaterials and drug delivery systems.
The use of near infrared Fourier Transform techniques in the study of surface enhanced Raman spectra.
Studies of dyestuffs in fibres by Fourier Transform Raman spectroscopy.
The use of Fourier Transform Raman spectroscopy in the forensic identification of illicit drugs and explosives.
The characterization of polycyclic aromatic hydrocarbons by Raman spectroscopy.
Fourier Transform Raman and infrared spectroscopy of N-phenylmaleimide and methylene dianiline bismaleimide.
The Fourier Transform Raman and infrared spectra of naphthazarin.
Future advances in near infrared Fourier Transform Raman spectroscopy.

Several books are also in preparation including one edited by Dr Bruce Chase and another by Dr Ken Williams.

12.5 DATA COLLECTIONS

For analytical use and especially for routine analysis reliable comprehensive and easily accessible collections of spectra are invaluable. Unfortunately, these are simply not available in Raman spectroscopy. Several very old books exist containing spectra, but few new ones. Perhaps the best collection available at present is that due to Schrader [4]. It contains spectra of more than 800 compounds. Infrared and Raman spectra in each case are superimposed on one diagram and the quality of the data is very high. For polymer spectra the book by Bower and Maddams is recommended [5].

Attempts to generate complete series of Raman spectra are in hand at the time of writing, but they are unlikely to appear in a published form for several years. This delay is anticipated because at present no decision has been made on standard presentation and data format. Quite obviously the need for such data libraries is urgent and it is hoped that significant progress will be made very quickly.

Table 12.2 — Papers in the second special edition of *Spectrochimica Acta* (**47A**)

The use of polarised FT Raman spectroscopy in morphological studies of uniaxially oriented PEEK fibres — some preliminary results.
Simultaneous application of FT Raman spectroscopy and differential scanning calorimetry for the *in situ* investigation of phase transitions in condensed matter.
Comprehensive analysis of oils and fats by FT-Raman spectroscopy.
The application of Q-switched lasers in NIR FTR spectroscopy II.
A comparison of FTIR and NIR Raman spectroscopy for quantitative measurements of polymorphine.
Infrared spectroscopic–rheological correlation for thin polystyrene films.
NIR FT Raman spectroscopy: facing absorption and background.
FT Raman spectroscopic study of main-chain thermotropic liquid crystalline polyesters.
Tunable near infrared Raman spectroscopy.
Analysis of high explosive samples by FT Raman spectroscopy.
Intensity calibration of single beam FT Raman spectrometers.
A standard intensity scale for FT Raman spectra of liquids.
FT Raman spectroscopy of sulphur-containing heterocycles.
A Raman investigation of wood.
The use of Fourier Transform Raman spectroscopy to study semi-crystalline polychloroprene.
The Raman spectra of some aromatic nitro compounds.
The FT Raman spectra of a series of Pt(II), Pd(II) and Au(III) square planar complexes.
An overview of the application of FT Raman spectroscopy to the analysis of inorganic materials.
FT Raman studies of materials and compounds of biological importance. Part II: Identification of anomeric changes in glucose.
Near-infrared FT-SERS of azole copper corrosion inhibitors in aqueous chloride media.
A study of the crystallization processes in natural rubber using FT Raman spectroscopy.
A study of the autoxidation of some unsaturated fatty acid methyl esters using FT Raman spectroscopy.
1064 nm excited Fourier Transform Raman spectroscopy of conducting polymers.
1064 nm excited FT Raman studies of bacteriochlorophyll-a in solid films and in a blue–green mutant of rhodobactersphaeroides.
FT Raman and infrared and surface enhanced Raman spectra of rhodamine 6G.
Studies of adducts of main group organometallic compounds by FT Raman spectroscopy.
Bulk radical homo-polymerisation studies of commercial acrylate monomers using FT Raman spectroscopy.
The application of FT Raman spectroscopy to the identification and characterization of polyamides II: double number nylons.
The application of near-infrared FT Raman spectroscopy to the analysis of polymorphic forms of cimetidine.
Application of FT Raman spectroscopy to the analysis of poly(anhydride) homo and co-polymers.
Surface enhanced Raman scattering at a flake-silver surface.
Structural studies of gel phases (4). An infrared reflectance and FT-Raman study of silica and silica/titania gel glasses.
SERS and FT-SERS of the ferri/ferrocyanide couple.
Near-infrared excitation of Raman scattering by chromophoric proteins.
Quantitative analysis with FT-NIR Raman spectroscopy.
Low frequency NIR-FT-Raman studies of ellipticines and DNA.
FT-Raman spectroscopic studies on sterically hindered aryl dichalcoginides.
Combined Fourier Transform infrared and Raman spectroscopic studies of some carbonyl-diphosphine-indenyl iron cations.
Vibrational Raman spectroscopy of C_{60} and C_{70}.

REFERENCES

[1] A. Schulte, T. J. Lenk, V. M. Hallmark and J. F. Rabolt, *Appl. Spectrosc.*, 1990, **45**, 325.
[2] C. J. Petty and R. Bennett, *Spectrochim. Acta*, 1990, **46A**, 331.
[3] D. J. Cutler, *Spectrochim. Acta*, 1991, **47A**, FT Raman special edition, in press.
[4] B. Schrader, *Raman/Infrared Atlas of Organic Compounds*, VCH Verlagsgesellschaft, Weinheim, Germany, 1989.
[5] D. I. Bower and W. F. Maddams, *The Vibrational Spectroscopy of Polymers*, Cambridge University Press, Cambridge, UK., 1989.

Appendix 1: The processing history of a spectrum — from ADC to VDU

A.1 INTRODUCTION

In this appendix, we will follow the sampling and digital processing of a spectrum from its collection at the analogue-to-digital converter (ADC) to its presentation to the spectroscopist. Words and phrases which are not defined in the appendix will be printed in bold, and will be briefly defined in a glossary at the end. At each stage, the processing will be examined and the associated sources of error highlighted. The possible solutions to these errors will be identified. The processing chain is shown in Fig. A.1.

A.2 THE COLLECTION OF DATA POINTS

Data is continuously available as an electrical output from the detector. The magnitude of this output is to a first approximation proportional to the radiation level impinging on the detector element. In order to sample an interferogram, data points must be collected from the detector at regular intervals. In **FTNMR** experiments [4,14,15] these are intervals of time. In an **FTIR** [29–31] or **FT Raman** experiment, the interferogram is a plot of intensity *vs.* optical path difference through the interferometer, so each data point collected must be at an exactly known **optical path difference** (opd). Whilst some interferometer mechanisms are designed to move with a linear motion (e.g. Nicolet 6000, 7000 and 8000 series), others are designed to move in an arc (e.g. Perkin–Elmer 1700 series). If the speed with which such an interferometer moves were to be kept constant, the rate of change of opd would vary throughout the movement, and with it the rate of data point collection. For this reason, and lack of perfectly stable driving mechanisms, collection of data at regular time intervals in FTIR and FT Raman experiments is not possible and a means of measuring the opd is required. Monochromatic laser radiation is passed through the interferometer, giving an interference pattern which has maxima and minima at precise intervals of opd. A second, separate detector analyses this sinusoidal interference pattern and triggers data collection at regular intervals, e.g. each time the sinusoidal signal becomes zero, which occurs twice per laser wavelength.

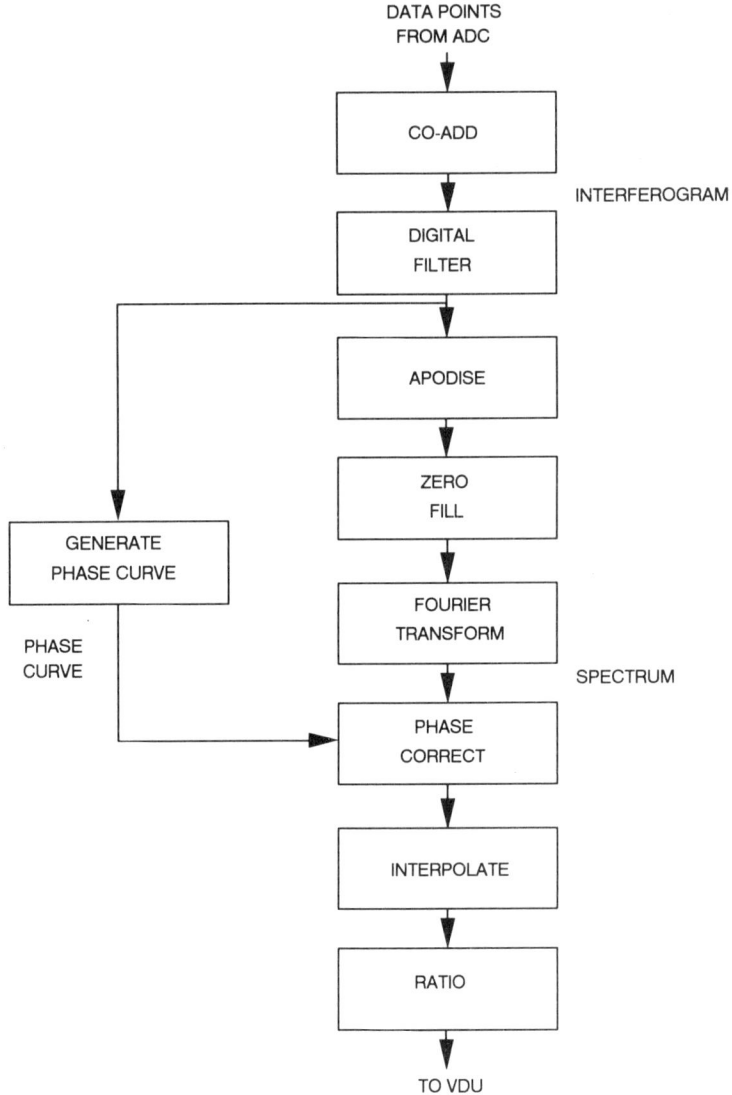

Fig. A.1 — Schematic diagram of data processing in FTIR and FT Raman.

In this way an interferogram is collected as the interferometer mechanism is moved between the required limits of opd. This single interferogram can be Fourier-transformed to yield a spectrum, but the **signal-to-noise** characteristics of the spectrum can be improved by co-adding many interferograms. This process of signal averaging will be discussed in Section A.4.

Most interferometer mechanisms move so as to measure data at opds either side of the point of zero path difference (ZPD), and thus double-sided interferograms can be collected. In practice, interferometer mechanisms have to move over a range of opds greater than that required to collect the interferogram, to allow for deceleration and acceleration in the opposite direction. Two terms are often used to describe interferogram collection, and are easy to confuse: the sidedness and the directionality. If an interferogram is collected from the point of ZPD to one limit of maximum opd it is said to be single-sided. If it is collected between the positive and negative limits of maximum opd, with the point of ZPD somewhere in the middle, it is said to be double-sided. This concept of sidedness is completely separate to that of directionality. If an interferogram has been collected unidirectionally, then during data point acquisition, the interferometer mechanism has only been moved in one direction. A bidirectional interferogram has been collected with the interferometer mechanism moving in both directions. These collection modes are illustrated in Fig. A.2.

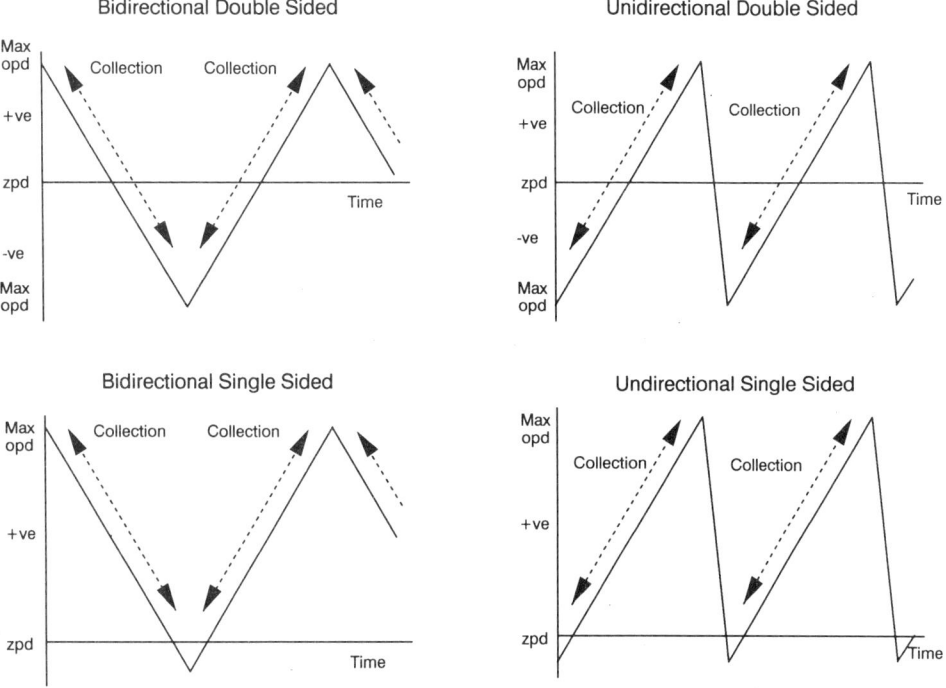

Fig. A.2 — Modes of data collection in an interferometer.

A.3 DIGITIZATION

The **analogue** signals from the detector are converted to a number system which can be understood by computers by a piece of hardware known as an analogue-to-digital converter (ADC). The number system used is **binary**. The conversion of real signal values into binary numbers is referred to as digitization. A binary number is split into bits, e.g. the binary number 11010 is a 5-bit number. It is not usual to represent fractions in binary, so binary numbers normally represent integers. This is a potential source of errors because analogue values are not integers — they are **real numbers**, and in converting them to binary form, we lose the fractional part of each number. The accuracy with which an analogue value can be represented as a binary number with B bits is given by Eq. (A.1).

$$\text{in } 2^B - 1 \tag{A.1}$$

The error produced by truncating the data points to their nearest binary equivalent is known as digitization noise. If a given ADC has an input range of for example $+1$ to -1 V and it converts its input to 6-bit numbers, it may convert $+1$ V to the binary number 111111 and -1 V to 000000. This representation is referred to as unipolar. A more useful representation is bipolar, where 0 V is converted to 000000, which requires that a convention is adopted so that negative numbers can be represented in binary. Usually the most significant bit (the left-most one) is used as a sign bit. It adopts the value 0 if the number is positive and 1 if it is negative, thus $+1$ V would be represented as 011111 and -1 V as 100001. It might first appear that 111111 should represent the -1 V signal, but the convention adopted for creating negative binary numbers is to write the magnitude in binary (e.g. for -1 V the magnitude is 1 V, represented as 011111) then subtract in binary that number from zero ($000000 - 011111 = 100001$). This is equivalent to inverting all the bits in the magnitude of the number, then adding 1, and is known as 2's complement. The reason for adopting bipolar digitization is that zero volts is represented as zero and will thus not add a bias to a running total during the addition of many bipolar numbers.

Typically the ADC has a somewhat smaller number of bits (or ADC resolution) than the number of bits that the computer can use for each data point, e.g. the Perkin–Elmer 1700 series has a 16-bit ADC but stores each data point using 32 bits. This may at first appear surprising, but we have not yet considered the effects of co-adding many interferograms.

A.4 SIGNAL AVERAGING

Signal averaging [1,2] is a process in which successive interferograms are summed in memory. Using this technique, weak signals are enhanced, since the signal is assumed to be coherent and the noise random. Hence the signal grows linearly with the number of scans, N, according to Eq. (A.2).

$$\text{signal} = k'N \tag{A.2}$$

whilst the noise grows at a rate proportional to the square root of the number of scans. (refs. [7,8] explain the gross simplifications in this assumption).

$$\text{noise} = k''\sqrt{N}$$

$$\frac{\text{signal}}{\text{noise}} = \frac{k'N}{k''\sqrt{N}} = k\sqrt{N} \tag{A.3}$$

$$\text{where } k = \frac{k'}{k''}$$

i.e. the signal-to-noise ratio increases with the square root of the number of scans. For a more detailed consideration of the signal-to-noise considerations see ref. [5]. Signal averaging depends on the fact that each scan starts at the same opd and thus the addition of successive points at the same opd is possible. In fact there is a small difference in the exact opd at which successive points of nominally the same opd are collected, and this is another source of error, which is also referred to as digitization noise [3]. The true signal averaging process requires that the final total must be divided by the number of scans. This is seldom done, however, since in practice it is easier to alter the ordinate scale of the spectrum after Fourier transformation. Since no division by the number of scans occurs during the signal averaging process, the data points increase in magnitude and the computer's memory may overflow under certain conditions. This is why the output from the ADC comprises fewer bits than the computer can store for each data point. If the number of bits that can be stored for each data point is W and the number of bits in the output from the ADC is D, then it would appear that N scans can be collected before overflow occurs where

$$N = 2^{(W-D)} \tag{A.4}$$

In other words a 32-bit computer having an ADC with 16 bits of resolution would allow 65 536 scans to be collected. In practice the maximum number of scans only tends to that given by Eq. (A.4) if the signal-to-noise ratio tends to infinity. In the cases where most signal averaging is needed, the initial signal-to-noise ratio is very low, and a great number of scans can be collected.

Suppose for example we have an initial signal-to-noise ratio of 1:1, a 16-bit storage capacity, and a 12-bit ADC. The largest number we can store is $(2^{16} - 1) = 65\,535$. On the basis of Eq. (A.4) we would expect a limit of 16 scans, but consider what the ADC sees in a single scan — in the most intense region of the interferogram the information will be half signal and half noise. A 12-bit ADC would thus produce 2048 counts of signal and 2048 counts of noise. Consider the situation after 16 scans. Using Eq. (A.3):

Counts due to signal $= 16 \times 2048 = 32\,768$

Counts due to noise $= 4 \times 2048 = 8192$

Total counts $= 40\,960$

Comparing this with the largest number which can be stored, it is clear that there is room for more averaging before overflow occurs. We can calculate the total number of scans possible for any initial signal to noise ratio by generalizing Eq. (A.3). The ADC produces 2^D counts divided between signal, s, and noise, n. Thus

$$\text{signal counts} = \frac{2^D s}{s+n} \qquad (A.5)$$

$$\text{noise counts} = \frac{2^D n}{s+n} \qquad (A.6)$$

After N scans, the total number of counts due to signal will be given by

$$\text{total signal counts} = \frac{2^D s N}{s+n} \qquad (A.7)$$

$$\text{total noise counts} = \frac{2^D \sqrt{N}}{s+n} \qquad (A.8)$$

The memory storage will overflow when the sum of these two first exceeds the largest number which can be stored, $2^W - 1$. Thus the overflow condition is

$$2^W - 1 = \frac{2^D(Ns + n\sqrt{N})}{s+n} \qquad (A.9)$$

We generally refer to signal-to-noise ratios as being 'something to 1', so we can replace n with 1.

$$\sqrt{N} = \frac{-1 + \sqrt{(1 + 4s(s+1))2^{W-D}}}{2s} \qquad (A.10)$$

Note that this reduces to Eq. (A.4) in the case where $s \gg 1$. A tabulation of the total number of scans possible for various values of signal-to-noise ratio and $(W - D)$ has been reported [6]. Normally we do not know the initial signal-to-noise ratio and so acquisition software must be designed to keep a track of potential overflow and prevent it before it occurs. This may be done by dividing the running average by 2 as memory overflow becomes imminent, then dividing all further ADC readings by 2 as well. This process can be continued until there is only one bit of resolution left in the ADC. In fact, if the initial $S:N$ is less than 1, a 1-bit ADC is quite sufficient for signal averaging. Whilst no commercial data systems boast of 1-bit ADCs, most of them have software which divides the ADC reading in this way as needed to prevent overflow, eventually resulting in just that.

So what of the shapes of interferograms? Practitioners of FTIR will be familiar with the interferogram shown in Fig. A.3, which is produced by co-adding a vast number of frequencies together, their intensities being those characteristic of the emission properties of the infrared source modified by the absorption characteristic of the sample. It has a very clearly defined centreburst and a very low intensity in the

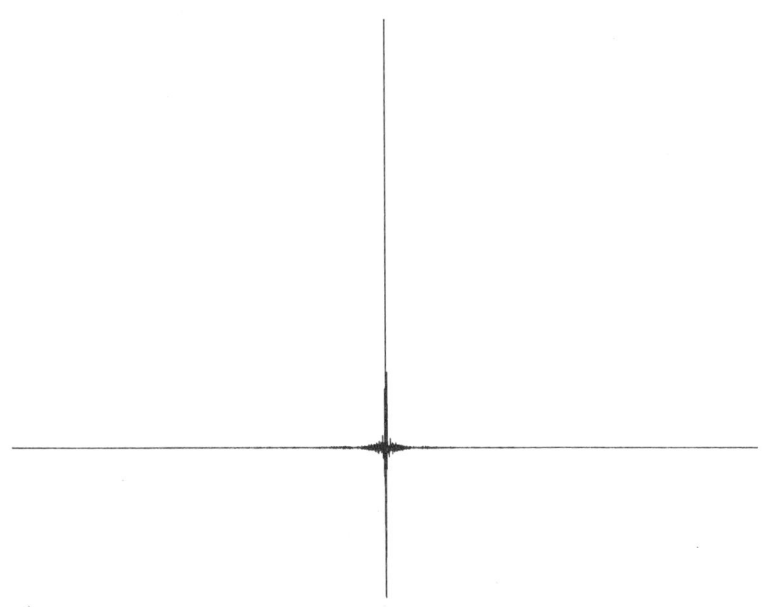

Fig. A.3 — The interferogram derived from the spectrum of the emission source in an FTIR spectrometer.

wings of the interferogram. The reason for this will be discussed later. If the ADC is set up so that the centreburst fills the ADC, the wings of the interferogram will only vary over a few bits, and the digitization noise here will be much more significant than across the centreburst. This problem may be overcome by gain switching. The **ADC gain** is controlled over the range of the interferogram so that the gain is higher at large opd and smaller at the centreburst. This allows more accurate digitization of the wings. Gain switching gives less of an advantage in digitizing interferograms produced in FT Raman experiments. To understand why this is, we must consider the composition of the data in an FT Raman experiment. Raman emission occurs at a few well-defined frequencies (neglecting **fluorescence** and **thermal background** effects) and thus the interferogram results from the co-addition of the frequencies. Normally the emission at the Rayleigh frequency is several orders of magnitude greater than the Raman frequencies, but after passing the collected frequencies through the Rayleigh rejection filters, the Rayleigh line is of a similar or lower intensity than the Raman emissions.

Thus the interferogram is composed of several discrete frequencies of reasonably similar intensities. Such an interferogram is shown in Fig. A.4. Note that the centreburst is much less clearly defined and extends over a larger range of opd than shown in Fig. A.3. Figure A.3 represents the interferogram generated by the source in an FTIR spectrometer, and those of FTIR spectra have the same general shape. Since the centrebursts are made to fill the ADC, the wings of the FT Raman

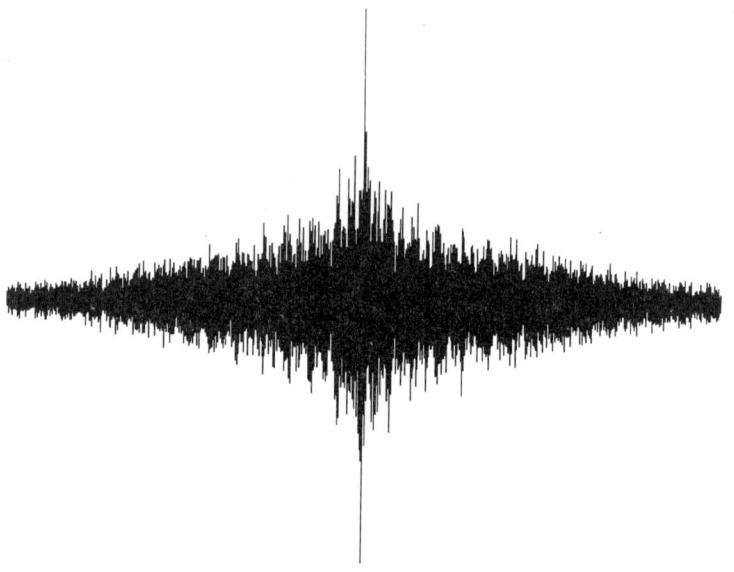

Fig. A.4 — The inteferogram derived from the FT Raman spectrum of anthracene.

interferogram will be digitized by more bits of the ADC than those of the FTIR interferogram, and the digitization noise will be less drastic, thus the effects of gain switching will be less beneficial. However the shape of the FT Raman interferogram presents us with a problem which is more significant than in the case where the intensity in the wings is very low. This is the effect of the finite length of the collected interferogram. The functions from which the interferogram is composed extend to infinitely large opd, but the interferogram can only be collected over finite opd and therefore it 'stops' suddenly. Put more mathematically, the interferogram is convoluted with a rectangular or boxcar function. The effect of this is the spectrum series of oscillations known as 'ringing' either side of each peak. This is demonstrated in Fig. A.5. A single frequency in (c) is represented by the interferogram (a). Since this is sampled over finite opd, the collected interferogram is (b), and the Fourier transform is shown in (d).

A.5 APODIZATION

The spectral characteristics of a sample are generally better represented by Gaussian [22] or Lorentzian [23] functions than the **delta function** shown in Fig. A.5, and this manifests itself in the interferogram as a diminishing of intensity with increasing opd. If the peaks in the spectrum are broad the intensity will decrease slowly with opd. As the peaks narrow, tending towards a delta function, so the intensity dies off more slowly. Measurement and prediction of spectral lineshape parameters from Fourier transform instruments has been investigated by Dana and Valentin [24]. The

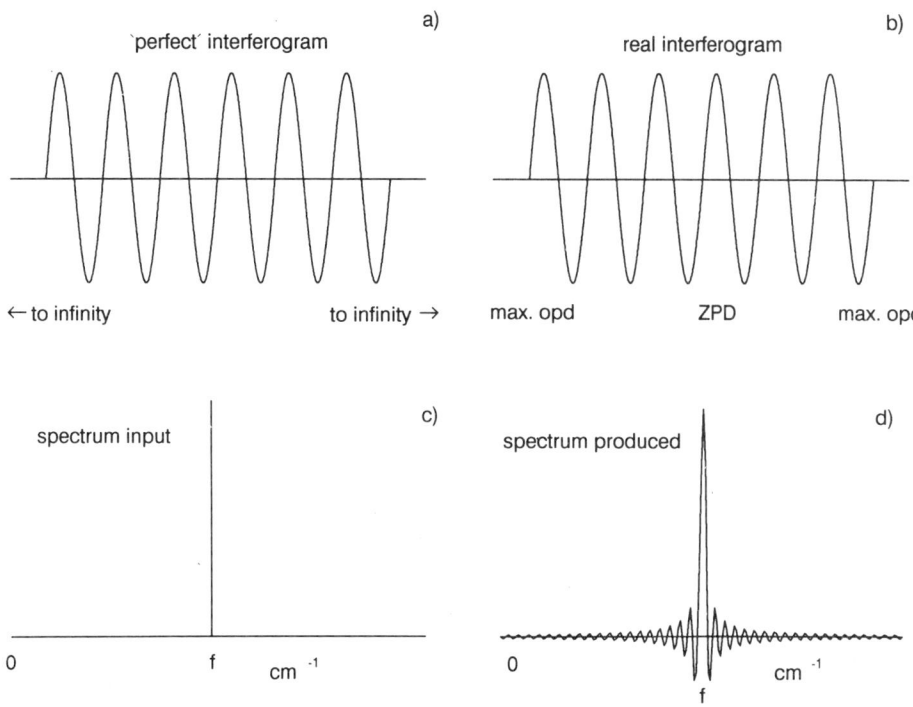

Fig. A.5 — (a) An ideal interferogram of infinite length, (b) a real truncated interferogram, (c) the Fourier transform of (a), (d) the Fourier Transform of (b).

intensity of the interferogram at maximum opd is usually relatively insignificant and the ringing either side of peaks is not generally a problem unless the spectral peaks are very narrow. As we have seen, the interferogram produced in an FT Raman experiment has a broader centreburst and in general diminishes in intensity more slowly. The intensity at the limits of opd thus tends to be higher and the ringing more severe.

The universally adopted solution to this problem is apodization. The interferogram is multiplied by a weighting function which decreases artificially the intensity of the wings, so that there is no observable 'step' at the ends. This idea is certainly not new. Lord Rayleigh suggested the use of triangular and bell-shaped apodization functions in 1912 [9]. More complex functions such as the Norton–Beer functions [10] reduce the ringing more efficiently. Gaussian and trapezoidal functions are also commonly used, as are cosine functions [11] such as

$$1 + \cos\left(\frac{\pi x}{x_1}\right) \tag{A.11}$$

$$\cos\left(\frac{\pi x}{2x_1}\right) \qquad (A.12)$$

Apodization reduces the oscillations either side of spectral peaks, but at the expense of an increased peak width, leading to decreased resolution. Thus one is faced with a compromise between loss of resolution and of reduction of spurious oscillations in the spectrum.

We have seen that narrow spectral peaks give rise to interferograms which diminish slowly in intensity with increasing opd. This leads us to a more general relationship. Spectral resolution is inversely proportional to the maximum opd. The relationship is approximately inverse, a maximum opd of 2 cm leads to resolution of 0.5 wavenumbers and so on. Note that the maximum opd has no effect on the range of spectral frequencies collected. This is affected by the 'closeness' of the points in the interferogram. Here we have to understand some of the consequences of Nyquist's theory [11,12], which states that to sample any frequency accurately, data must be collected at a rate greater than twice that frequency.

It was stated previously that data can be collected at the zero crossing points of the reference laser interference pattern; thus two data points are collected per unit laser wavelength, and frequencies up to the laser wavelength can be sampled. Using a He–Ne reference laser spectra can thus be obtained up to a high wavenumber limit of 15 802 wavenumbers. The high wavenumber limit is known as the Nyquist frequency. If we only require data up to for example 7901 cm^{-1}, then one of three methods of collection can be used:

(1) The interferometer can be made to collect on every other laser fringe zero crossing point. This is known as fringe skipping.
(2) Data can be collected at every fringe zero crossing point, and alternate data points rejected.
(3) Data can be collected at every fringe zero crossing point, and pairs of data points combined together. This is know as digital filtering. A treatment of this subject is given by Savitzky and Golay [25]

A.6 DIGITAL FILTERING

The simplest form of digital filter is the block average. Here, groups of data points are added, and the mean value taken. If we require half the number of points in the interferogram, we could add pairs of points. This will remove frequencies above 7901 cm^{-1}. There is a problem associated with simple block averaging, however. What we are actually doing is multiplying the interferogram repeatedly by a rectangular function. To produce each point in the output interferogram, we add a block of points together. This is equivalent to multiplying all the points in the block by 1 and all the other points in the input interferogram by 0, then adding all the points in the interferogram so produced to obtain one point for the output interferogram. This is repeated for each point in the output interferogram. The effect of multiplying the interferogram by a rectangle has been seen before — it produces ringing (consider the effect of finite limits of opd). This suggests that simple block averaging is inadequate for effective digital filtering. What we really want is a rectangular function convoluted with the *spectrum*, i.e. a sharp high wavenumber cutoff. Because Fourier transforms are reversible (more of this later) the implication is that the filter we use should be the shape of the ringing function rather than a rectangle. The filter we require is shown in Fig. A.6. It is the function $\sin(x)/x$ or $\mathrm{sinc}(x)$. If we

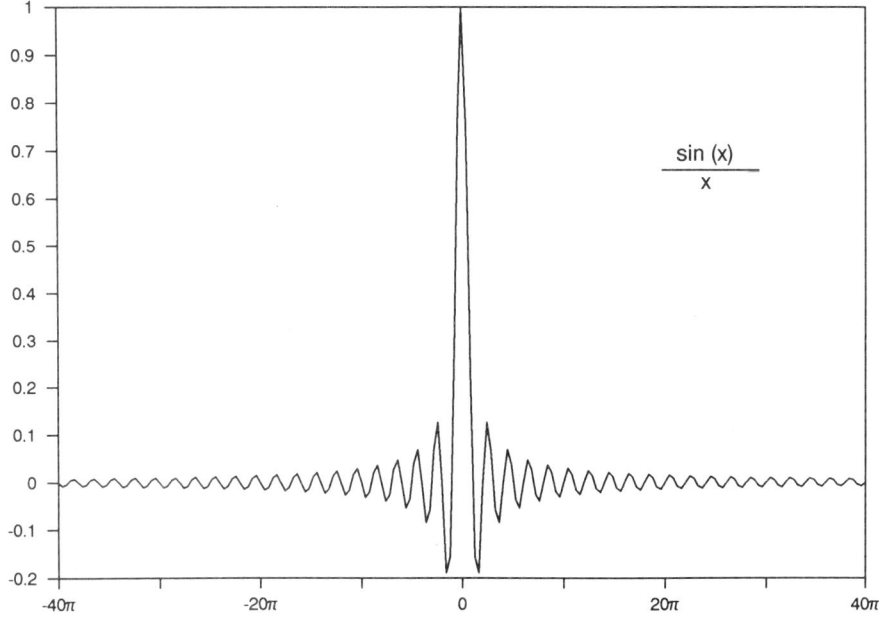

Fig. A.6 — A $\mathrm{sinc}(x)$ function.

require a low wavenumber cutoff to remove spurious information below the region of interest, this may be achieved by using the function $1-\mathrm{sinc}(x)$, and a bandpass filter function giving both high and low wavenumber cutoffs can be produced by convoluting a $\mathrm{sinc}(x)$ function with a $1-\mathrm{sinc}(x)$ function.

What we do is to digitize the filter function, typically using between 20 and 50 points to give the filter envelope, then, starting at one end of the interferogram, apply it to get a filtered interferogram point, then reapply it, e.g. two points further on if we want to get half the number of output points, and continue to reapply it until all the interferogram has been filtered.

If radiation of a higher frequency than the Nyquist frequency is sampled, it will be 'folded back' or 'aliased' into the spectrum at a frequency exactly as much lower than the Nyquist frequency as the radiation frequency is higher than the Nyquist frequency. If the Nyquist frequency is Q, radiation of frequency $Q+f$ will appear in the spectrum at $Q - f$. Clearly we must remove all frequencies above the Nyquist frequency from the interferogram before the spectrum is produced. This, then, is the other reason for digital filtering. Filtering up to the Nyquist frequency is achieved by applying the filter envelope to every point in the interferogram, giving as many output data points as were in the input interferogram. At this point we have collected, filtered and apodized our interferogram. A typical benchtop FTIR will filter and apodize the data in around 1 sec.

The next task to be performed is the conversion of the interferogram into a spectrum. This is the process of Fourier transformation, named after J. B. J. Fourier [13]. The principles of the Fourier transform as applied to spectroscopy are discussed by Wayne [26] and programs to demonstrate Fourier transforms and associated data processing are listed and explained in an article by Williams [27].

A.7 FOURIER TRANSFORMATION

An interferogram is said to be in the distance domain — it is a plot of intensity versus opd, and opd has units of distance. The Fourier transform converts it to a spectrum — a plot of intensity versus wavenumbers, with units of reciprocal distance. In general Fourier transformation converts data from one domain to its reciprocal domain. This property of Fourier transformation means that Fourier transforming an interferogram twice will regenerate the original interferogram.

In Chapter 4, the equation

$$I(x) \propto \sum_{\lambda=0}^{\lambda=\infty} I\cos\left(\frac{2\pi x}{\lambda}\right) \tag{A.13}$$

was used to represent the intensity of the interferogram at opd x. To understand how the spectrum may be generated from this, a basic mathematical property of cosines must be understood, that of orthogonality. Cosine waves are said to be orthogonal because they only yield positive integrated areas when multiplied by themselves. If two cosine waves of different frequencies are multiplied and the resulting curve integrated, the integrated area is zero, because of destructive interference. We can use this property to 'interrogate' the interferogram, which Eq. (A.13) shows is composed of cosine waves. If the entire interferogram is multiplied by a frequency p (the probe frequency) then integrated, the components of frequencies other than p will integrate to zero. Only the component of frequency p in the interferogram will give a positive area when multiplied by the probe frequency. The magnitude of the integrated area is clearly proportional to the amount of p component in the interferogram. If we now repeat this process for a range of values of p, what we get is a set of values which are functions of p, and which are proportional to the magnitude of each frequency p in the interferogram, i.e. a plot of 'amount of frequency present' vs. frequency. This is precisely what a spectrum represents.

Mathematically, the spectrum as a function of wavenumber, $S(f)$, is given by

Fourier transformation

$$S(f) = \int_{-\infty}^{+\infty} I(x)\cos(2\pi px)dx = 2\int_{0}^{\infty} I(x)\cos(2\pi px)dx \qquad (A.14)$$

However, the interferogram is collected over a non-infinite range of opd,

$$S(f) = 2\int_{0}^{x_1} I(x)\cos(2\pi px)dx \qquad (A.15)$$

where x_1 = maximum opd.

From Eq. (A.13),

$$I(x) = k\sum_{f} I\cos(2\pi xf) \qquad (A.16)$$

where $f = \dfrac{1}{\lambda}$, k = constant.

Thus we evaluate

$$2k\int_{0}^{x_1} \cos(2\pi xp)I\cos(2\pi xf)dx \qquad (A.17)$$

for each value of p and f, which gives

$$S(f) = \frac{2kI\sin 2\pi(f-p)x_1}{2\pi(f-p)x_1}$$

$$S(f) = 2kI\,\mathrm{sinc}(2\pi(f-p)x_1) \qquad (A.18)$$

It can thus be seen that the effect of the finite length of interferogram is to replace each single frequency f in the spectrum with a function of the form $\mathrm{sinc}(2\pi(f-p)x_1)$, known as the scanning function [11].

So far we have assumed that the interferogram can be represented using only cosine waves, but as a result of instrumental imperfections and sampling errors, the interferogram may not be symmetrical, and it should be regarded as the sum of two interferograms, one symmetrical, composed of cosine waves, and one antisymmetrical and composed of sine waves. These are known as the real and imaginary interferograms respectively, for reasons which will later be apparent.

The sine transform has the form

$$I(x) \propto \sum_{\lambda=0}^{\lambda=\infty} I\sin\left(\frac{2\pi x}{\lambda}\right) \qquad (A.19)$$

and can be 'interrogated' by multiplying by a sine wave probe frequency.

We can use complex numbers to represent both sine and cosine components :

$$S(f) = \int_0^{x_1} I(x)\exp(-2\pi i x f)\,dx \qquad (A.20)$$

since

$$\cos(\omega) - i\sin(\omega) = \exp(-i\omega) \qquad (A.21)$$

where $i = \sqrt{-1}$, $\omega = 2\pi f x$

It is in this form that computers handle data whilst performing Fourier transforms. Direct integration is not possible using a microprocessor and thus summations are performed instead. The spectrum can be expressed as

$$S(f) = \sum_{n=0}^{n=2N-1} I(n\Delta x)\exp(-2\pi i f n \Delta x) \qquad (A.22)$$

for one side of a symmetrical interferogram with $2N$ data points, having Δx as the data point interval, since

$$x_1 = 2N\Delta x, \quad \Delta x = \frac{x_1}{2N} \qquad (A.23)$$

The method of Fourier transformation most commonly used is that of Cooley and Tukey [17] which gives a $N\log_2 N$ advantage in speed for processing N data points over evaluating each value of $S(f)$ separately. It is generally referred to as the fast Fourier transform (FFT).

The smallest resolvable interval in the spectrum δf is related inversely to the maximum opd — the larger the maximum opd, the better the resolution. The resolving power, r, is usually defined as

$$r = \frac{f}{\delta f}, \quad \therefore r = f x_1$$

$$f = \frac{r}{x_1} \qquad (A.24)$$

thus by substituting Eq. (A.23) and (A.24)

$$\exp(-2\pi i f n \Delta x) = \exp\left(-2\pi i \left(\frac{r}{x_1}\right) n \left(\frac{x_1}{2N}\right)\right) = \exp\left(\frac{-\pi i r n}{N}\right) = W^{nr}$$

$$(A.25)$$

where $W = \exp\left(\dfrac{-\pi i}{N}\right)$

hence
$$S(f) = \sum_{n=0}^{n=2N-1} I(n\Delta x) W^{nr} \qquad (A.26)$$

we can write
$$S(f) = I_0 + I_1 W^r + I_2 W^{2r} + \ldots \qquad (A.27)$$

to $2N$ terms. This can in turn be split into even and odd parts:

$$S(f)_{\text{even}} = I_0 + I_2 W^{2r} + I_4 W^{4r} + \ldots \qquad (A.28)$$

to N terms and

$$\begin{aligned}S(f)_{\text{odd}} &= I_1 W^r + I_3 W^{3r} + \ldots \\ &= W^r(I_1 + I_3 W^{2r} + I_5 W^{4r} + \ldots)\end{aligned} \qquad (A.29)$$

to N terms with

$$S(f) = S(f)_{\text{even}} + S(f)_{\text{odd}} \qquad (A.30)$$

This is the key to the fast Fourier transform algorithm, since $2N$ complex samples of the FT of the interferogram can be computed from N complex samples of the FT of each of the two subsets (even + odd). Similarly each of these subsets can be computed from two further subsets of $N/2$ complex samples and the process can be repeated until each subset contains one value of $I(x)$, i.e. there are $\log N$ levels of subsets. This leads us to the conclusion that N must be a power of 2 for the FFT algorithm to work. For further details of the FFT see refs [6 or 17]. There are a variety of other Fourier transform algorithms [16,18] but these are rarely used for FTIR and FT Raman work. Using the Cooley and Tukey algorithm, we have a mechanism of obtaining a spectrum efficiently assuming we have a number of data points equal to a power of 2. This is rarely the case however, since the number of points collected is a function of the required resolution (related to the maximum opd) and the reference laser wavelength.

A.8 ZERO FILLING

What is done in practice is to pad out the ends of the interferogram with zeros until an appropriate number of data points is reached. This process (zero filling) has been shown by Ernst [1] to be entirely mathematically justifiable so long as the block of zeros does not exceed the length of the original sampled data block. Despite this, zero filling to an extent greater than the original data is often done. Zero filling to any extent is equivalent to interpolating the spectral data (see Section A.10).

A.9 PHASE CORRECTION

The output from the Fourier transform routine is not, in fact, one spectrum but two. Just as the interferogram contains both real and imaginary data, the Fourier transform produces a real and an imaginary spectrum. One is the cosine transform of the interferogram, and one the sine transform. If there were no phase and sampling errors introduced by the optics of the spectrometer and sampling arrangements, the sine transform would be zero and the cosine transform would represent exactly the frequency characteristics of the sample. The phase errors which are introduced can be removed by phase correction. In principle what is done is to rotate each frequency in the sine and cosine transforms by a phase angle until the sine transform tends to zero. At this angle the cosine transform is corrected for phase errors. In practice phase correction is done using a phase curve which is a low-resolution (typically 32 cm^{-1}) spectrum extracted from the interferogram and stored as a set of phase angles.

$$\text{phase angle } A = \tan^{-1}(\text{real/imaginary}) \tag{A.31}$$

The real and imaginary spectra are then rotated using these phase angles to give the phase-corrected spectra. The residual phase errors can be seen by examining the phase-corrected imaginary (sine) spectrum, which should tend to zero.

An alternative method of removing the phase errors is to calculate the magnitude spectrum. If the cosine and sine transforms respectively have intensities at frequency f of $S(f)\cos(\phi)$ and $S(f)\sin(\phi)$, where ϕ is the angle of phase error introduced, then the magnitude spectrum is given by

$$[(S(f)\cos\phi)^2 + (S(f)\sin\phi)^2]^{1/2} = S(f)[\cos^2\phi + \sin^2\phi]^{1/2} = S(f) \tag{A.32}$$

and is independent of ϕ. The magnitude spectrum is thus the square root of the sum of the real spectrum squared and the imaginary spectrum squared. The drawback of magnitude phase correction is that the noise level in the magnitude spectrum is greater than that in the real or imaginary spectrum. This is because noise in the real and imaginary spectra fluctuating around an intensity of zero is made all positive by the process of squaring the spectra.

A.10 INTERPOLATION

The next stage in the processing of the spectrum is to obtain data points at a 'useful' interval. It is normal to require an integer number of points per wavenumber or wavenumbers per point. If the number of points provided as input to the FFT is N, then the resulting spectrum has a data point interval before interpolation equal to

L/Nr where L = reference laser wavelength (usually a He–Ne), and r = interferogram sampling interval (i.e. the number of reference laser fringe zero crossing points per interferogram data point). Clearly this expression does not lead to integer spectrum data point intervals and a method of calculating points in-between the existing points must be used.

The process which achieves this is known as interpolation. Interpolation increases the number of data points in the spectrum, or decreases the interval between data points. It is easy to think that the resolution of the spectrum has been improved by the interpolation, so the spectroscopist must beware. The actual resolution has *not* increased. We have already noted the inverse relationship between the maximum opd of the interferogram and the resolution of the spectrum, and this relationship leads us to the conclusion that the process of interpolation is equivalent to extending the interferogram to a greater maximum opd. We have also noted that the interpolation process does not provide us with more information, merely more points. It is thus easy to see that interpolation of the spectrum is identical to the process of zero filling the interferogram [28] mentioned earlier. Why, then, do we perform a complicated interpolation process to extract data points at the required interval rather than zero filling the interferogram until the transformed spectrum has the required number of points? This question has been answered earlier. The Cooley and Tukey algorithm requires a number of input points equal to an integer power of two, so we cannot zero fill to any arbitrary extent. Thus we are faced with the problem of extracting points from the spectral data. Mechanisms of interpolation vary, but in principle what is done is to take a block of spectrum points and to calculate the coefficients of an equation such as a **polynomial** or sinc function to fit the data points. When the coefficients of the equation have been calculated, any other points in the block can be derived from the equation. This process is repeated using blocks of points throughout the spectrum to obtain all the necessary interpolated points.

An important point to consider when assessing interpolation algorithms is that a peak should appear the same shape if it is shifted by a fraction of a data point in the spectrum. If this criterion is not met, then the size of a peak will vary when its position shifts slightly, purely because of the interpolation routine being used. Imagine the effect this would have on the interpretation of spectra from experiments where small shifts in peak position and size are crucial, such as studies in surface adsorption.

A.11 RATIOING

In order to obtain an infrared spectrum of the sample free from the emission characteristics of the source and the absorption of the beam-splitter and atmospheric water vapour and CO_2, spectra are usually divided by a background spectrum taken with no sample in the beam. This is ratioing.

Source characteristics are not a problem in FT Raman spectroscopy: however, the detector does not respond linearly with wavenumbers, and the transmission of the beam-splitter is not linear with wavenumbers, so the intensities of peaks in the Raman spectrum are a function of the detector response characteristics and beam-splitter transmission curve. A process similar to ratioing can be used to remove these instrument response characteristics. If the instrument response curve is available, and is used as a background spectrum, ratioing will scale the peaks so as to make the spectrum detector and beam-splitter independent. Obtaining the instrument response curve is possible if a spectrum is recorded using a pseudo-blackbody emitter for a source (i.e. the source should have the same intensity over the spectral range of interest).

It is important to note that after this ratioing process, the noise in the spectrum will be much greater in the regions where the instrument response is smallest, because the spectrum has been divided by a small number. The signal-to-noise ratio is not affected by the ratioing, however, since both signal and noise are scaled by the same amount.

A.12 ADDITIONAL PROCESSING

The basic processing chain from data points to finished spectrum has been outlined in the preceding pages, but there are a variety of optional processing stages which are sometimes used to further enhance or correct the collected data. Two such routines which will be discussed are detector non-linearity correction and Fourier self-deconvolution.

A.12.1 Non-linearity correction

At a given frequency, the signal from the detector does not increase linearly with the intensity of incident radiation on the detector surface. In general the detector response approximates to linear behaviour over the range of intensities used for FTIR, but this approximation is not always sufficiently good, and non-linearity correction must be applied. Rather than assuming the detector response to be linear with intensity, it is fitted to a curve such as $y = x + ax^2$ (Perkin–Elmer 1800) or $y = a^2/(a - x)$ where a is a constant, very much greater than x, and related to the total energy in the spectrum. The problem thus becomes one of determining what value of a to use for a given interferogram. One way to do this is to generate two low resolution spectra from copies of the uncorrected interferogram. One spectrum is obtained after correcting with a value of a which is thought to be larger than that required, and the other after correcting with a value of a smaller than that required. The region of the spectra between zero wavenumbers and the low wavenumber limit of the instrument response (where the intensity would be expected to be zero) is examined in both cases, and by a process of extrapolation, a value of a is determined which would be expected to result in the average energy in this region being zero. This value of a is then used to correct the intensities in the original interferogram before Fourier transformation. Low resolution spectra are used for the estimation of the correction factor, a, simply to increase the speed with which a can be determined.

An alternative method of obtaining the correction factor is to have stored in the computer a table of values which have been calibrated to correspond to different centreburst heights. This may be done because the average energy in the spectrum is related to the centreburst height since the intensities of all the cosine components add constructively at zero opd. If such a table is stored, a value of a can be selected from the table for each interferogram processed.

A.12.2 Fourier self-deconvolution

It has already been shown that the width of a peak in a spectrum is related to the rate at which its frequency components in the interferogram diminish in intensity away from the centreburst. Thus, to represent sharp peaks, a large maximum opd, leading to a high resolution, is required. If this argument is inverted, the width of the peaks in a spectrum may be decreased, leading to an artificial enhancement of the apparent resolution, by decreasing the rate at which the interferogram diminishes in intensity away from the centreburst. Such an apparent increase in resolution would potentially lead to the observation of peaks which were previously unrecognizable because of the widths of adjacent peaks.

The first stage in the process of Fourier self-deconvolution is to multiply the interferogram by some function, such as an exponential, which increases the relative intensity of the wings of the interferogram, leading to the situation shown in Fig. A.7(b). This stage is known as resolution enhancement. Clearly the interferogram now has a large step at the maximum opd, which will lead to very bad ringing effects in the spectrum which could obscure weak peaks and distort large ones. The solution to this problem is to apply an apodization function to the interferogram after resolution enhancement, reducing the intensity of the interferogram to zero towards maximum opd. The interferogram can then be Fourier transformed to give the deconvoluted spectrum. Papers discussing the deconvolution of Raman spectra [19] and infrared spectra [21,22] have been published, and these clearly show the advantages (resolution of previously unresolved features) and disadvantages (appearance of spurious peaks and increase in noise level) associated with this technique.

A.13 DATA PROCESSING FOR FT RAMAN SPECTRA

New fields such as FT Raman develop rapidly in their formative years. In this section some of the specialized data processing necessary to handle Fourier-transformed Raman spectra will be discussed. In future years, the data handling will probably be done differently. Possibly FT Raman may not develop at all, but consider the words of Twyman [32] in 1930

> I do not think a spectroscope for the infra-red will be a much purchased instrument. Everyone who makes these most difficult measurements will construct his own apparatus.

The Raman spectrum is excited using laser emission at 9398 cm^{-1} (Nd^{3+}:YAG 1.064-μm line) and vibrational features appear to Raman shifts of around 4000 cm^{-1}. The Stokes lines appear at wavenumbers lower than those of the exciting line, so the Raman spectrum is entirely contained in the region 9398 cm^{-1} to 5398 cm^{-1}. This is not the most convenient range to work with, since all the points below 5398 cm^{-1} are not needed. One way of getting rid of this data is to make use of Nyquist's theory. If a digital filter function is applied to the interferogram so that in the spectral domain

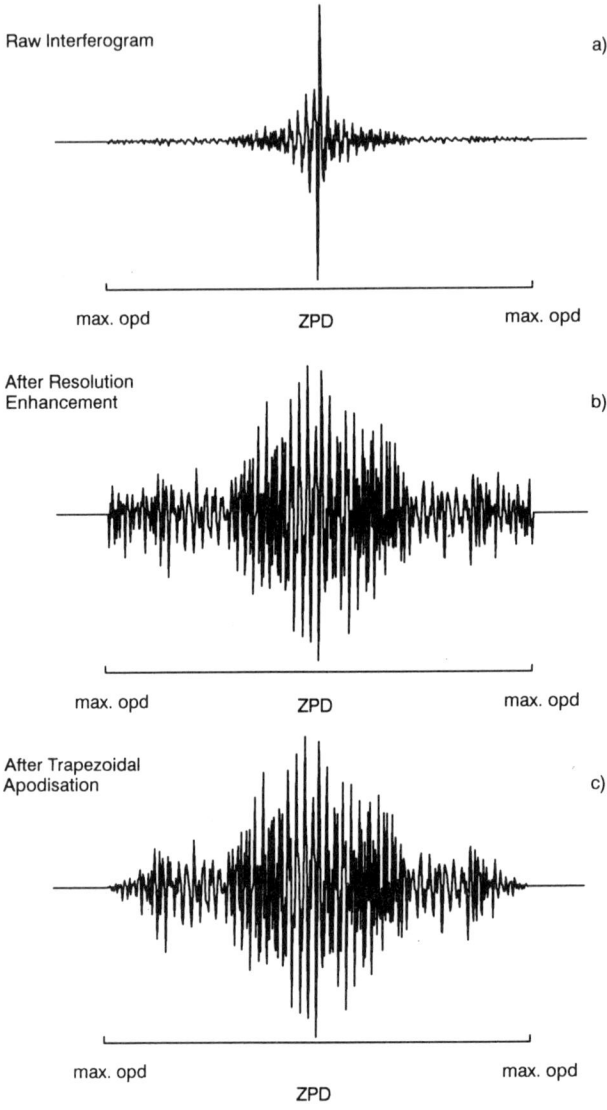

Fig. A.7 — The process of Fourier self-deconvolution: (a) raw interferogram, (b) raw interferogram after resolution enhancement increasing the intensity of the wings, and (c) resolution-enhanced interferogram after multiplication with a trapezoidal apodization function.

frequencies in the range 5400 to 10 300 cm^{-1} were passed, but all others rejected (corresponding to a filter bandwidth of -1000 to $+4000$ cm^{-1} shift), and the filter is applied so as to produce one output point for every three input points, then the Nyquist frequency is reduced to 5268 cm^{-1} and the filtered data appears (reversed) as a result of aliasing in the range 5100 cm^{-1} to 250 cm^{-1}. The 1.064 μm laser line now appears at 1136 cm^{-1}. A shift can then be applied to the data so that the Rayleigh line appears at 0 cm^{-1}, giving a high wavenumber limit of 3964 cm^{-1}.

Phase correction for FT Raman spectra is usually of magnitude. The reason for this is fundamental. The spectrum does not contain enough information (only a few peaks and a lot of low level noise) to generate a useful phase curve. We have seen that the noise level in the spectrum is increased by magnitude phase correction, and thus a method of phase curve phase correction would be beneficial. This can be achieved if a white-light source is used to collect a background spectrum and the phase curve is generated from this. Since the phase curve depends on the instrument optics and is not sample-dependent the phase curve can be used to phase correct Raman spectra generated subsequently. There is one drawback, which is that the phase curve is a function of the instrument alignment. As the alignment of the optics changes, the phase curve changes very rapidly. Small changes in alignment produce very large changes in the instrument phase curve. Instrument alignment rarely stays constant over a long period of time, since thermal and atmospheric effects are involved. The phase curve generated from the white light source, and accurate for the instrument when produced, becomes inaccurate as time progresses. Hence the phase curve must be regenerated at regular intervals if phase curve phase correction is to be used for the Raman experiment. In some cases this is not practical — it may involve removing the Raman experiment from the sampling optics to replace it with a lamp, or some other disturbance to the experiment — and in these cases magnitude phase correction is probably the best option.

A.14 GLOSSARY OF TERMS

ADC gain
The analogue-to-digital converter takes analogue inputs and creates digital outputs. Clearly there is some scaling factor which relates the magnitudes of the analogue values input to the magnitudes of the digital values output. This factor is not necessarily unity. The largest voltage input may be 5 V, whilst the digital output for this voltage may well be 32767 (decimal). When the signals input are weak the scaling factor, which is referred to as the ADC static gain, can be increased.

Analogue
Numbers which can vary infinitely are referred to as analogue; thus the speed of an accelerating object is analogue — it does not change in steps. Note that we are not considering quantum effects here; analogue values exist where there is a continuum of real values.

Binary
The number system in which all numbers are represented by a series of 1s and 0s. Just as decimal numbers are split in to hundreds, tens, and units, etc., binary numbers are split into eights, fours, twos and units, etc. Thus the number 13 would be represented as 1101 ($1 \times 8 + 1 \times 4 + 0 \times 2 + 1 \times 1$).

Delta function
A delta function has the property of having an infinitely narrow bandwidth. Thus it represents a single frequency in the spectrum (frequency) domain.

Fluorescence
In Raman experiments, the term fluorescence is often used to describe a variety of phenomena which cause the baseline of the spectrum to be non-zero. Fluorescence in a sample may be caused by degradation products and impurities/additives as well as the bulk materials, and are almost always much more intense than Raman emission.

FTIR (abbr.)
Fourier transform infrared spectroscopy. The examination of spectral absorption or reflection characteristics of a sample, giving information about the vibrational behaviour of the sample and leading indirectly to information about its molecular structure.

FT-Raman (abbr. FTR)
Fourier transform Raman spectroscopy. The examination of the polarizability behaviour of a sample under near infrared laser excitation, leading to information about its molecular structure. The data is complementary rather than supplementary to that produced in the FTIR experiment.

FT NMR (abbr.)
Fourier transform nuclear magnetic resonance. Examination of the environments of nuclei in the sample by examining the free-induction decay following the application of a wide band radio-frequency energy pulse to the sample. Gives direct information about molecular structure.

Optical path difference (opd)
If two photons reach the beam-splitter, and one is reflected while the other passes through, they will both be reflected by the mirrors in the interferometer back to the beam-splitter. They travel different paths through the interferometer, and the difference in the distances they travel is known as the optical path difference.

Polynomial
An expression in powers of x. Usually has the form

$$a + bx + cx^2 + dx^3 + ex^4 + \ldots$$

Any curve can be fitted to a polynomial. The quality of the fit depends on how many powers of x are used.

Rayleigh emission
The radiation elastically scattered by the sample, which thus is detected at the same frequency as the excitation laser in Raman experiments.

Real numbers
Similar to analogue values. Real numbers can have any value: fractional, integer, or irrational. Thus 6, π, and 0.75 are real numbers. This is in contrast to integers which are whole numbers; the above numbers would be 6, 3 and 1 when represented as nearest integers.

Signal-to-noise ratio (S:N)
Any spectrum contains both noise (from the detector, spurious radiation, etc.), which is seen as 'jitter' in the baseline regions and, normally, real spectral features such as peaks. The signal-to-noise ratio as understood by spectroscopists is the ratio of the peak height — the signal — to the amplitude of the jitter. Physicists and engineers have different concepts of signal and noise, which involve measuring RMS values over the interferogram. These are not intended here.

Thermal backgrounds
When a sample is heated (in the case of a Raman experiment this can often be caused by the excitation laser) it emits a wide spectrum of radiation which causes a lift in the baseline of spectra. This effect is not often seen in FTIR experiments because the sample is usually placed after the interferometer, so the thermal emission is in principle not modulated and therefore not detected. In practice a small amount of thermal emission is detected because of multiple reflections within the interferometer. The other potential problem in FTIR is 'blinding' the detector — exciting all the available charge carriers from the valence to the conduction bands — making the detection of spectral features impossible. In the case of FT Raman experiments, the sample is placed before the interferometer and thus thermal effects will appear in the spectrum. For this reason the sample must be kept cool, which may require liquid nitrogen cells, sample spinners or a chopped laser source.

Wavenumbers (cm^{-1})
The abscissa scale used most commonly for FTIR and FT Raman spectra. The units are reciprocal centimetres, so a wavelength of 0.1 cm corresponds to 10 wavenumbers, or 10 waves per centimetre.

REFERENCES
[1] E. Bartholdi and R. R. Ernst, *J. Magnetic Resonance*, 1973, **11**(9).
[2] J. W. Cooper, *Topics in carbon-13 NMR*, vol. 2, p.392. Wiley, NY, 1976.
[3] E. V. Loewenstein, *The ASPEN conference on Fourier Spectroscopy*, p.3. 1970.
[4] R. R. Ernst and W. A. Anderson, *Rev. Sci. Instrum.*, 1966, **37**(1), p.93.
[5] H. Sakai, *The ASPEN Conference on Fourier Spectroscopy*, p.19. 1970.
[6] J. W. Cooper, *Computers in Chemistry*, 1976, **1**, p.55.
[7] R. Clark Jones, *Advances in Electronics*, 1953, **V**, 6.
[8] R. D. Hudson, *Infrared System Engineering*, p. 264, Wiley, 1968.
[9] Lord Rayleigh, *Phil. Magn.*, 1912, **24**, 864.
[10] R. H. Norton and R. Beer, *J. Opt. Soc. America*, 1976, **66**(3), 259.
[11] A. E. Martin, *Vibrational Spectra and Structure*, Elsevier, 1980, **8**, p.37.
[12] B. P. Lathi, *Communic. Systems*, p.89. Wiley, NY, 1968.
[13] J. B. J. Fourier, *Theorie Analytique de la Chaleur*, Paris: Fremin Didot 1822.
[14] M. B. Comisarow and A. G. Marshall, *Chem. Phys. Lett.*, 1974, **25**(2), p.282.

[15] E. D. Becker and T. C. Farrar, *Science*, 1972, **178**, 361.
[16] M. O'Neill, *BYTE*, 1988, April 293.
[17] J. W. Cooley and J. W. Tukey, *Mathematics of Computation*, 1965, **19**, 297.
[18] R. N. Bracewell, *Proceedings of the IEEE*, 1984, **72**(8), 1010.
[19] H. J. Bowley, S. M. H. Colin, D. L. Gerrard, D. I. James, W. F. Maddams, P. B. Tooks and I. D. Wyatt, *Applied Spectroscopy*, 1985, **39**(6), 1004.
[20] D. Compton, FTIR application notes No.s 8311 and 8312, Nicolet Corp.
[21] D. Compton, *Spectral Lines*, 1983, **5**(1), 4.
[22] R. A. Caruana, R. B. Searle, T. Heller and S. I. Shupack, *Anal. Chem.*, 1986, **58**, 1162.
[23] R. A. Caruana, R. B. Searle and S. I. Shupack, *Anal. Chem.*, 1988, **60**, 1896.
[24] V. Dana and A. Valentin, *Appl. Optics*, 1988, **27**, 4450.
[25] A. Savitzky and M. Golay, *Anal. Chem.*, 1964, **36**(8), 1627.
[26] R. P. Wayne, *Chem. in Britain*, 1987, May, 440.
[27] R. Williams, *Intelligent Instruments and Computers*, 1987, Jan/Feb, p.49.
[28] M. L. Forman, *Appl. Optics*, 1977, **16**(11), 2801.
[29] P. Jacquinot, *Rep. Prog. Phys.*, 1960, **23**, 267.
[30] P. Connes, *The ASPEN Conference on Fourier Spectroscopy*, 1970, 121.
[31] P. Fellgett, *Le Journal de Phys. et le radium*, 1958, **19**, 187.
[32] F. Twyman, *Chem. and Industry*, 1930, **49**, 578.

Index

A–D converter, 83, 99, 285–290
adsorbed species cell, 247, 249
alumina, sorption to, 246, 256–258
anharmonic oscillator, 18, 19, 20
animal tissue, 208
    eyes (FT Raman study), 208
    human artery, 209
    bone, 209
    conventional Raman spectrum, 210
    FT Raman spectrum, 210
    teeth, FT Raman spectra, 211
anthracene FT Raman with symmetry labels, 160
anti-Stokes bonds
    definition, 25
    intensities, 28
    intensity ratio, 28
    Stokes/anti-Stokes comparisons, 28
apodization, 290–292
assignments
    X–H, 157
    saturated rings, 157
    unsaturated groups, 158
    aromatics, 159
    oxygen containing species, 159, 161
    S. P, Si and halogenic species, 160, 162

bandpass, 62, 63
    definition, 60
beamsplitters, future developments
    (interchangeable), 273
Beer–Lambert law, 129
benzene
    symmetry properties, 34–36
    character table, 38
    modes, 37
    infrared/R, 39
BIO-ROD commercial instrument, 116, 117
blackbody emissions, 135–138
Bomem commercial instrument, 124–126
Brönsted acids, 235–236, 247, 248
Bruker commercial instrument, 122–123

C_{60}, C_{70}, 265
    C_{60} FT Raman spectrum, 270
    C_{70} FT Raman spectrum, 271

calibration of FT Raman intensities, 134
carbon dioxide modes, 33
carbon tetrachloride, meaning excited early
    spectrum, 12
carbonyl complexes, 267
catalysts, infrared spectra of, 245
cells
    spinning, 77
    liqid, 106
    solids, 107
    adsorbed species, 108
    microscopes, 108
    kinetic (thermostatted), 109
    explosives, 165
chemometrics, 151
chlorine Raman spectrum, 26, 27
chloropyridine, 239, 240, 241, 243, 244
chromium
    chromyl chloride CrO_2Cl_2, 263
    FT Raman spectrum, 265
CIRCOM, 151, 172, 223
Connes advantage, 89
conventional Raman intensity measurements, 129
corn oil, conventional low Raman spectrum, 15
correction of FT Raman spectra, 134–143
corrosion inhibitors, Raman spectra of, 238–240
crystallinity determination, 183
    LA modes, 184
    polychloroprene, 184
    natural rubber, 184
    (spectrum)←FT Raman, 185
crystals, Raman spectra of, 47

$\partial\alpha/\partial q$ and Raman activity, 24, 25, 33
$\partial\mu/\partial q$ and infrared activity, 33
dark current in PM tubes, 63, 66
data collections, 281
data point density, 86
depolarization ratio, 45, 46
    polarized and depolarized sources, *N*-phenyl
    maleimide FT Raman spectra, 173
detectors (*see also* germanium and indium
    gallium arsenide detectors), 63–66
    future developments, 274
diamond, FT Raman spectrum, 269

2:5 dichloro acetone, infrared and Raman
    spectra, 41
dichloric ratio, 43
digital filtering, 293–294
o-dinitro benzene, Raman spectrum, 104
DNA, 204, 220
dyes
    polymer fibres, 200
    FT Raman spectrum of dyed Courtelle fibre,
        168–169

electrically conducting polymers, 197–198
    near-infrared FT Raman studies, 198
    resonance Raman studies, 198
electrode potential, effect on SERS spectrum,
    237
electron micrographs, 232, 242
environmental monitoring, 239
errors in frequency, 84
étendue of optical instruments, 87, 274
explosives, 163–167
    FT Raman spectrum of diaminotrinitrobenzene
        (DATB), 165
    FT Raman spectra of RDX and PETN, 167
    FT Raman spectra of different Semtex
        samples, 167

fact Fourier transform, 83, 296, 297
Felgett advantage, 88
Felgett disadvantage, 89
ferricyanide ion $[Fe(CN)_6]^{3-}$, 238–239
ferrocyanide ion $[Fe(CN)_6]^{4-}$, 238–239
fibre optic coupling, 74
    future developments, 275, 276
    in biological applications, 205
fluorescence in catalysts, 246
foreshortening in sample area, 131
Fourier self-deconvolution, 301
Fourier transform, 82, 83, 294–297
Fourier transform Raman spectrometers,
    specialist instruments, 280
FTR spectra of foodstuffs, 221 *et seq.*
    brazil nuts, 223
    butter, 222
    carbohydrates, 224
    cashew nuts, 223
    ficoll, 226
    glucose
        α and β forms, 224
        structures, 225
    margarine, 222
    olive oil, 222
    walnuts, 223
fundamental modes of vibration
    definition, 17
    energy, 20
    H_2O, 22
    CCl_4, 32
    C_6H_6, 37
fuel mixtures, 151
future developments, 273–282
    a combined microscope, 276

beamsplitters, 273
detectors, 274
lasers, 274, 276–278
fibre-optic coupling, 275–276
filters, 274–278

germanium detectors, including extended Ge
    detectors, 90–92, 98
gels, FT Raman study, 187
glossary of terms, 303–305
gratings, 55
group frequency approach
    applications, 156
    explanation, 38, 40

harmonic oscillator, 18
hexanal infrared spectrum, 23
histogram in FT spectra, 84

InGaAs detectors, 90–92, 97, 98
    effect of cooling, 98, 99
intensity formulae, 127–129, 140–144
Interferogram, shape,
    FT infrared, 289
    FT Raman, 290
interferometer
    block diagram, 81
    Chase's early instrument, 91
    how it works in detail, 283–285
    interference formulae, 81
    Michelson, 81
    Southampton prototype, 95
intermolecular interactions, 150, 151
interpolation, 298–299
infrared dichroism and orientation, 42, 43
infrared spectroscopy, 11, 12
    polymers, 176
    problems with paints, 190
    typical infrared interferogram, 289

Jacquinot advantage, 87
Jacquinot stop, 61, 274

lasers
    Ar^+ and Kr^+ wavelengths, 50
    external resonator mode, 73
    future developments, 274, 276–278
    He/Ne, 83
    near-infrared — NdYAG, 101
    Q-switched pulsed, 278
    sources in spectrometers, 49
    tuneable, 51
low stability, 101
Lewis acids, 235–236, 247, 248
light-sensitive systems, 213–215
    cobalamins, 213, 268
    rhodopsins, 214
    chlorophylls, 214
liquid crystalline polymers
    poly(bis-4-methoxy-phenoxy phosphazine),
        FT Raman study, 184
    structure, 186

polyheptamethylene
  terephthaloyl-bis-4-oxybenzoate
    FT Raman study, 184–186
    structure, 186
    smectic phase structure, 186
longitudinal acoustic modes (LAM), in polyethylene, 57, 184, 275

manganese
  potassium permanganate ($KMnO_4$), FTR spectrum, 262
  MnO_4^{2-}, 262
margarines, FTR of, 148–150
membranes, 205–208
metals, SERS
  colloids, 229, 231, 240–245
  metal island films, 229, 231
  SERS over electrodes
    copper, 232, 234, 236
    gold, 232, 234, 236, 241, 242
    platinum, 234, 236
    silver, 229, 231, 232, 234
    backgrounds over metals, 233, 234, 235
microscope as Raman illuminator and viewer, 52
  future developments, 276
  Kevlar fibre analysis, 195
  study of cell walls, 215
monochromators, 55
  multiple, 56–60
Morse
  curve, 19
  function, 19
multiplex advantage, 66, 89, 280
mutual exclusion rule, 32
MX_4^{2-}
  Magnus's green salt $[Pt(NH_3)_4][PtCl_4]$, 265
  $PtCl_4^{2-}$ FT Raman spectrum, 264
  cis-$Pt(NH_3)_2Cl_2$ FT Raman spectrum, 264
  trans-$Pt(NH_3)_2Cl_2$ FT Raman spectrum, 264

natural rubber, spectra of
  different grades, 189
  filled with oil, 190
  filled with silica, 177
  vulcanized, 191
nickel
  nickel dimethyl glyoxime, 266
    FT Raman spectrum, 272
Nicolet commercial instrument, 118–119
noise, 66, 68
  and resulting spurious lines, 103, 293
noise equivalent power, 65
non-linearity correction, 300
Nyquist limit, 105, 293, 294, 301–303
nylon
  identification, 179
  infrared and Raman, 181
nylon number, 154

oils, FTR of, 148–150
on-axis mirror collection, 94

on-axis lens collection, 96
optical films in Raman spectroscopy, 94, 109
organometallic compounds, 263
  $(CH_3)_2GaNHR_2$, 263
  TeR_2, 263

paint systems
  alkyl resins, 192
  copolymer latex formation, FT Raman study, 191–192
  FTR Raman spectrum as a function of time, 194
  weathering, 193
particle size
  effect on Raman activity, 152
  effect on SERS activity, 243
Peak picking, errors therein, 84
Perkin Elmer, commercial instrument, 120–121
  in Southampton prototype, 93–97
photodiodes, 64, 65
photomultipliers, 63, 64
Planck distribution, 135–138
pharmaceuticals
  alkaloids, 162–163, 220
  amphetamine sulphate, 164
  1,4-benzodiazepine, 220
  codeine, 163
  diazepam, 170
  diclofenac, 170, 200
  elliptocines, 217, 220
    structure, 221
  heroine, 163
  indomethacin, 169
  morphine, 163
  polymer drug supports, 200, 220
  promethazine, 170
phase correction, 298
phase transitions
  poly(bis-4-methoxy-phenoxy phosphazine), 184
  poly(di-n-hexylgermane), 182
  poly(heptamethylene terephthaloyl-bis-4-oxybenzoate), 184
  polyterrafluoroethylene, 184
platinum
  Magnus's green salt $[Pt(NH_3)_4][PtCl_4]$, 265
  $[Pt^{II}(en)_2][Pt^{IV}(en)_2Cl_2][ClO_4]_4$, 266
  $PtCl_4^{2-}$, 264
  cis-$Pt(NH_3)_2Cl_2$, 264
  trans-$Pt(NH_3)_2Cl_2$, 264
polarizability, 24, 25
polyethylene
  LAM spectrum, 57
  Raman spectrum of polyethylene duck, 178
  v_{CH} modes, 156
polyethylelne terephthalate
  polarized infrared spectrum, 43
  polyimides, 195, 196
polymorphism, 172
  cimetidine, 172
polypeptides, 209–212
  FTR spectrum
    human nail, 213

polytryptophan, 212
L-tryptophan, 212
proteins, 211
polystyrene
  first spectrum of polymer, 175
  FT Raman spectrum,
    atactic, 183
    isotactic, 183
    tacticity, 182
polysucrose (ficoll), FT Raman spectrum, 226
polytetrafluoroethylene (PTFE)
  density of vibrational states, 40, 113, 114
  modes of vibration, 37
  phase transitions, 184
polyurethane, reacted-injection-moulded, study of, 188–190
Porto nomenclature, 47
position sensitive detectors, 58
  CCDs, 66, 280
  diode arays, 66, 280
  TV systems, 66, 280
pyridine symmetry elements, 34
  Rama spectrum, various detectors, 92

quantitative analysis
  copolymer analysis, 178
  ES:EES, 193
  in BRS, 178
  margarines, 221
  nylons, 179
  polyurea/polyurethane, 178
  vinyl in SBR, 178
quantization of energy, 17, 20
quartz beamsplitters (in interferometers), 94

Raman effect
  classical description, 24, 25
  quantum description, 28
Raman spectroscopy
  history, 11, 12, 13
  laser-, 13–15
ratioing, 299–350
refraction, 130
rejection filters for Raman work (*see also* reflection filters), 93
  future developments, 275
  tilting of filters, 100, 101
reproducibility in FT Raman intensities, 133
resolution, 61, 62
resonance, Raman, 198, 199
  bacteriochlorophyll-a, 214
  $[Pt^{II}(en)_2][Pt^{IV}(en)_2Cl_2][ClO_4]_4$, 266
rubidium complexes, 242, 245

safety considerations,
  interlocked lids, 95, 111
  precautions, good practice, 110, 111
sample alignment, 53, 105
  commercial instruments, 115–126
sample fluorescence
  biological samples, problems, 205
  eyes, 208
  methods to reduce, 175
  new laser lines to overcome problem, 276–278
  polymers (general), 175
selenium, FT Raman spectrum, 268
self-absorption, 147–149
semiconductors, 263
sensitivity
  comparison of instruments, 102
  correction for variations with wavelength, 100
  example of *o*-dinitrobenzene, 104
  SERS (surface-enhanced Raman spectroscopy), (*see also* metals), 228–245
    4-dicynomethylene-2-methylene-6(*p*)dimethylaminostyryl-4H-pyran (DCM), 169
    electrochemical cell for, 233
    electrochemistry, studies by, 228–240
    electromagnetic model, 229
    excitation frequency, 232
    instrument response, 235
    orientation of molecules, 230
    polymer surfaces, 200
    pyridine, 231, 233, 234, 235, 236, 238, 239, 241, 247, 248
    rhodamine 6G, 169
    self-absorption, 236
    water, 236
sesquiterpene lactone, 217
  FT Raman spectra
    linofolin-A, 219
    helenalin, 219
    general structure, 218
signal:noise ratio, 66, 70, 113, 284, 287, 305
silica, sorption to, 246, 248
slitwidths in monochromators and spectrographs, 60, 61
small samples,
  polymer inclusion, 180
  spectrum, 182
smoothing, 139, 143
Snell's law, 130
solids, quantitative studies, 152
Spectrochimica Acta (special editions devoted to FT Raman), 280–282
spectrographs, 58, 280
spurious lines in spectra, 103
standards, internal and external, 132, 145
standardization of FT Raman spectra, 143–147
Stokes bonds
  definition, 25
  intensity, 28
  Stokes/anit-Stokes, ratios, 28
stray light, 57
sulphur
  FT Raman spectrum, 267
  use in vulcanization, 183
superconductors, 263
synthetic rubbers
  polybutadiene
    microstructure, 178
    FTR spectra, 180
  nitrile rubber, FTR spectrum, 188

styrene-butadiene rubber,
  FTR spectrum, 188
  FTR/FTIR spectrum, 31
styrene butadiene rubber (SBR)
  FT Raman spectrum, analysis, 188
  infrared and Raman spectra of, 31
sucrose, Raman spectrum of, 113, 114, 226

thermoplastics/thermosets
  maleimides, 172
  N-phenyl maleimide, FT Raman spectrum, 173
  poly (aryl ether ketone), 195
  poly (aryl ether ether ketone), 195
  poly (aryl ether sulphone), 193, 195
  poly (aryl ether ether sulphone), 193, 195
thiosulphates, 263
  FT Raman spectrum of ammonium thiosulphate, 266
transmission filters, 90–97
  future progress in, 276–278
transitional metal impurities, 256
tuneable laws, rhodamine dye devices, 51

unsaturation, determination of, 147–150

vanadium
  vanadyl chloride ($VOCl_3$), FT Raman spectrum, 265
vibration, effect on FT instrument performance, 104

water
  fundamental modes, 22
  new laser lines to avoid self-absorption, 276–278
  self-absorption, 205, 211, 262
  symmetry class, 34, 35
waveguide, FT Raman, 200
wood, 215–216
  FT Raman spectrum
    apple wood, 218
    Douglas fir, 218

zeolites, 248–256
  HY, 252–255
  isomerization reaction, 252, 255, 256
  NaY, 251, 253
  USY, 251, 252, 254
zero filling, 86, 297